THE METABOLIC SYNDROME

Dietary Supplements and Food Ingredients

THE METABOLIC SYNDROME

Dietary Supplements and Food Ingredients

Edited by

Raj K. Keservani, MPharm

Durgavati Yadav, PhD

Rajesh K. Kesharwani, PhD

Sippy Singh, DPhil

Kumar Sandeep, PhD

AAP APPLE ACADEMIC PRESS

First edition published 2023

Apple Academic Press Inc.
1265 Goldenrod Circle, NE,
Palm Bay, FL 32905 USA

760 Laurentian Drive, Unit 19,
Burlington, ON L7N 0A4, CANADA

CRC Press
6000 Broken Sound Parkway NW,
Suite 300, Boca Raton, FL 33487-2742 USA

4 Park Square, Milton Park,
Abingdon, Oxon, OX14 4RN UK

Library and Archives Canada Cataloguing in Publication

Title: The metabolic syndrome : dietary supplements and food ingredients / edited by Kumar Sandeep, PhD [and four others].
Other titles: Metabolic syndrome (Palm Bay, Fla.)
Names: Kumar, Sandeep (Medical scientist), editor.
Description: First edition. | Includes bibliographical references and index.
Identifiers: Canadiana (print) 20230156207 | Canadiana (ebook) 20230156282 | ISBN 9781774911112 (hardcover) | ISBN 9781774912096 (softcover) | ISBN 9781003329732 (ebook)
Subjects: LCSH: Metabolic syndrome—Diet therapy. | LCSH: Dietary supplements. | LCSH: Food—Vitamin content.
Classification: LCC RC662.4 .M48 2023 | DDC 616.3/90654—dc23

Library of Congress Cataloging-in-Publication Data

CIP data on file with US Library of Congress

ISBN: 978-1-77491-111-2 (hbk)
ISBN: 978-1-77491-209-6 (pbk)
ISBN: 978-1-00332-973-2 (ebk)

About the Editors

Raj K. Keservani, MPharm

Faculty of Bachelor of Pharmacy, CSM Group of Institutions, Allahabad, India

Raj K. Keservani, MPharm, is a Faculty of Bachelor of Pharmacy, CSM Group of Institutions, Allahabad, India. He has more than 12 years of academic (teaching) experience from various institutes in India in pharmaceutical education. He has published 31 peer-reviewed papers in the field of pharmaceutical sciences in national and international journals, 43 book chapters, three co-authored books, and 18 edited books. He is also active as a reviewer for several international scientific journals. Mr. Keservani graduated with a pharmacy degree from the Department of Pharmacy, Kumaun University, Nainital (Uttarakhand), India. He received his Master of Pharmacy (MPharm) (specialization in pharmaceutics) from the School of Pharmaceutical Sciences, Rajiv Gandhi Proudyogiki Vishwavidyalaya, Bhopal, India. Mr. Keservani is a life member of the Society of Pharmaceutical Education and Research (SPER). His research interests include nutraceutical and functional foods, novel drug delivery systems (NDDS), transdermal drug delivery, health science, cancer biology, and neurobiology.

Durgavati Yadav, PhD

Scientist, All India Institute of Medical Sciences, New Delhi, India

Durgavati Yadav, PhD, has more than 10 years of research experience at various institutes in India. She obtained her doctoral degree in Biochemistry from Banaras Hindu University, Varanasi, India. Currently, she is working at the All India Institute of Medical Sciences, New Delhi as a Scientist. She was awarded a CSIR fellowship. She has worked and published extensively on phytomedicine and diabetes. Her current research interests are prostate cancer and liquid biopsies in cancer prognosis and diagnosis.

Rajesh K. Kesharwani, PhD
Faculty, National Institute of Technology, Warangal, Telangana, India

Rajesh K. Kesharwani, PhD, MTech, has more than 10 years of research and eight years of teaching experience at various institutes in India, imparting bioinformatics and biotechnology education. He has received several awards, including the NASI-Swarna Jayanti Puruskar by The National Academy of Sciences of India. He has supervised PhD, postgraduate, and undergraduate students for their research work and has authored over 40 peer-reviewed articles, 20 book chapters, and 11 edited books with international publishers. He is a member of many scientific communities as well as a reviewer for many international journals. He has presented many papers at various national and international conferences. Dr. Kesharwani received his PhD from the Indian Institute of Information Technology, Allahabad, India, and worked at NIT Warangal. He has been a recipient of the Ministry of Human Resource Development (India) Fellowship and a Senior Research Fellowship from the Indian Council of Medical Research, India. His research fields of interest are medical informatics, protein structure and function prediction, computer-aided drug designing, structural biology, drug delivery, cancer biology, nano-biotechnology, and biomedical sciences.

Sippy Singh, DPhil
Assistant Professor, Department of Zoology, S.S. Khanna Girls' Degree College (A Constituent College of University of Allahabad, Prayagraj), India

Sippy Singh, DPhil, is an Assistant Professor at the Department of Zoology, S.S. Khanna Girls' Degree College, Prayagraj (A Constituent College of the University of Allahabad, Prayagraj). She has more than seven years of academic experience in various institutes in India. Dr. Singh has completed her DPhil in Zoology with a specialization in Neurobiology from the University of Allahabad. She has also served as Project Co-coordinator of the Societal Research Fellowship (SoRF) Scheme of the Department of Science and Technology, GoI at NASI, Prayagraj (The Oldest Science Academy). Dr. Singh has edited one book and has 20 publications in the form of research papers, book chapters, and popular articles in national and international journals of high repute. She is a life member of scientific bodies: Indian Science Congress Associatation, Indian Academy of Neurology (IAN), Indian Society of Cell Biology, Asian Biological Research Foundation, and IFEE. Dr. Singh is a member of IQAC (Internal Quality Assurance Cell) in the college. She is the recipient of two international awards: The "S.S. Parmar Foundation Award" of IAN, "Young Scientist Award 2019" the of International Foundation for Environment and Ecology (IFEE), and three national awards. Her research interest includes neurobiology and cell biology.

Kumar Sandeep, PhD

Scientist, All India Institute of Medical Sciences, New Delhi, India

Kumar Sandeep, PhD, has more than 10 years of research experience at various institutes in India. He obtained his doctoral degree in Life Science from the School of Life Sciences, Jawaharlal Nehru University, New Delhi, India. He has worked as Research Scientist at the Indian Council of Medical Research, Hqrs., New Delhi. Currently, he is working at the All India Institute of Medical Sciences, New Delhi as a Scientist. He was awarded a CSIR fellowship and a prestigious fellowship for Summer Curriculum in Cancer Prevention held at National Cancer Institute, USA. He has various projects from national funding agencies. His current research interests are prostate cancer and cancer awareness programs.

Contents

Contributors

Prerana Aditi
Department of Medicinal Chemistry, Banaras Hindu University, Varanasi, Uttar Pradesh, India

Eknath D. Ahire
Department of Pharmaceutics, MET's Institute of Pharmacy, Bhujbal Knowledge City (BKC), Affiliated to Savitribai Phule Pune University, Adgaon, Nashik, Maharashtra – 422003, India, Tel.: +91-9588431965, Fax: (02552)-303-515; (0253)-2303203, E-mail: eknathahire05@gmail.com

Rajesh A. Ahirrao
P. G. College of Pharmaceutical Science and Research, Chaupale, Nandurbar, Maharashtra – 425412, India

Rajeshwar Kamal Kant Arya
Department of Pharmaceutical Sciences, Bhimtal Campus, Kumaun University Nainital, Bhimtal, Uttarakhand, India, E-mail: rajeshwararya@gmail.com

Ankit Awasthi
Central University of Punjab, Bathinda, Punjab, India; Tel: +91-164-2864122; Tele Fax: +91-164-2864106; E-mail: awasthi0005@gmail.com

Dheeraj Bisht
Department of Pharmaceutical Sciences, Bhimtal Campus, Kumaun University Nainital, Bhimtal, Uttarakhand, India

Nayan A. Gujarathi
Shri Vile Parle Kelvani Mandal's Institute of Pharmacy, Dhule – 424001, Maharashtra, India

Charu Gupta
Amity Institute of Herbal Research and Studies, Amity University, Sector-125, Noida – 201313, Uttar Pradesh, India, E-mail: charumicro@gmail.com

Ashish S. Jain
Shri D. D. Vipsute College of Pharmacy and Research Center, Panvel, Maharashtra – 410206, India

Urmila Jarouliya
School of Studies in Biochemistry, Jiwaji University, Gwalior, Madhya Pradesh – 474011, India, E-mail: urmi_gudia@rediffmail.com

Meena Kausar
Department of Pharmaceutical Sciences, Bhimtal Campus, Kumaun University Nainital, Bhimtal, Uttarakhand, India

Raj K. Keservani
Faculty of B. Pharmacy, CSM Group of Institutions, Prayagraj, Uttar Pradesh – 212111, India

Shubham J. Khairnar
Department of Pharmacology, MET's Institute of Pharmacy, Bhujbal Knowledge City (BKC), Affiliated to Savitribai Phule Pune University, Nashik, Maharashtra – 422003, India

Sanjay J. Kshirsagar
Department of Pharmaceutics, MET's Institute of Pharmacy, Bhujbal Knowledge City (BKC), Affiliated to Savitribai Phule Pune University, Adgaon, Nashik, Maharashtra – 422003, India

Deepak Kumar
Department of Pharmaceutical Chemistry, School of Pharmaceutical Sciences, Shoolini University, Solan, Himachal Pradesh, India

Sunil K. Mahajan
Department of Pharmaceutical Chemistry, MGV's Pharmacy College, Panchvati, Nashik, Maharashtra, India

Shanti Bhushan Mishra
Department of Pharmacognosy, United Institute of Pharmacy, Prayagraj – 211010, Uttar Pradesh, India, Tel.: +917905848793, E-mail: shantipharma15@gmail.com

Pratiksha M. Pandagale
Department of Pharmacology, Oriental College of Pharmacy, Mumbai, Maharashtra, India

Afsar S. Pathan
Department of Pharmacology, RC Patel Institute of Pharmaceutical Education and Research, Dhule, Maharashtra, India

Daiana Romina Peralta
Laboratory of Antioxidants Oxidative Processes, Institute of Chemical Sciences, Faculty of Agronomy and Agro-Industries, National University of Santiago del Estero-CONICET, RN 9 Km 1125, Villa El Zanjón-Santiago del Estero, Argentina

Dhan Prakash
Amity Institute of Herbal Research and Studies, Amity University, Sector-125, Noida – 201313, Uttar Pradesh, India

Bhushan R. Rane
Head Department of Pharmaceutics, Shri D. D. Vipsute College of Pharmacy and Research Center, Panvel, Maharashtra – 410206, India, Tel.: +919421534437, E-mail: rane7dec@gmail.com

Mohammad Rashid
Department of Medicinal Chemistry and Pharmacognosy, College of Dentistry and Pharmacy, Buraydah Colleges, Buraydah, Al-Qassim – 31717, Saudi Arabia

Amit Kumar Singh
Department of Pharmaceutics, United Institute of Pharmacy, Prayagraj – 211010, Uttar Pradesh, India

Anil Kumar Singh
Department of Pharmaceutics, United Institute of Pharmacy, Prayagraj – 211010, Uttar Pradesh, India

Durgesh Singh
Department of Zoology, Sadanlal Sanwaldas Khanna Mahila Mahavidyalay (A Constituent College of University of Allahabad), Attarsuiya, Prayagraj, Uttar Pradesh – 211003, India, E-mail: ddurgesh12@gmail.com

Shradhanjali Singh
Department of Pharmaceutical Chemistry, United Institute of Pharmacy, Prayagraj – 211010, Uttar Pradesh, India

Vijayraj N. Sonawane
Department of Pharmaceutical Chemistry, Shreeshakti Shaikshanik Sanstha, Divine College of Pharmacy, Satana, Nashik, Maharashtra, India

Shivani Srivastava
Department of Physiology and Cell Biology, Davis Heart and Lung Research Institute,
The Ohio State University Wexner Medical Center, Columbus, OH, USA

Khemchand R. Surana
Department of Pharmaceutical Chemistry, Shreeshakti Shaikshanik Sanstha,
Divine College of Pharmacy, Satana, Nashik, Maharashtra, India

Hemant P. Suryawanshi
PSGVPMS, College of Pharmacy, Shahada, Maharashtra – 425409, India

Gokul S. Talele
Department of Pharmaceutics, Matoshri College of Pharmacy, Near Odhagaon, Eklahare, Nashik,
Maharashtra – 422105, India

Swati G. Talele
Department of Pharmaceutics, Sandip Institute of Pharmaceutical Sciences, Nashik,
Maharashtra – 422213, India

Duong Van Thanh
Department of Pharmacology and Clinical Pharmacy, VNU University of Medicine and Pharmacy,
Vietnam National University Ha Noi, 144 Xuan Thuy, Cau Giay, Ha Noi, Vietnam

Nguyen Phuong Thanh
Department of Pharmacology and Clinical Pharmacy, VNU University of Medicine and Pharmacy,
Vietnam National University Ha Noi, 144 Xuan Thuy, Cau Giay, Ha Noi, Vietnam

Nilima A. Thombre
Department of Pharmaceutics, MET's Institute of Pharmacy, Bhujbal Knowledge City (BKC),
Affiliated to Savitribai Phule Pune University, Adgaon, Nashik, Maharashtra – 422003, India

Bui Thanh Tung
Department of Pharmacology and Clinical Pharmacy, VNU University of Medicine and
Pharmacy – Vietnam National University, Ha Noi, 144 Xuan Thuy, Cau Giay, Ha Noi, Vietnam,
Tel.: +84-4-85876172, Fax: +84-0437450188, E-mail: tungasia82@gmail.com

Durgavati Yadav
Scientist B, Department of Urology, AIIMS, New Delhi, India, E-mail: durgavati45yadav@gmail.com

Srivastava Yashi
Central University of Punjab, Bathinda, Punjab, India, Tel: +91-164-2864122;
Tele Fax: +91-164-2864106; E-mail: yashicupb@gmail.com

Nguyen Thi Yen
Department of Pharmacology and Clinical Pharmacy, VNU University of
Medicine and Pharmacy – Vietnam National University, Vietnam

Abbreviations

AACE	American Association of Clinical Endocrinologists
ABTS	2,2'-azino-bis (3-ethylbenzothiazoline-6-sulphonic acid)
ACE	angiotensin-converting enzyme
ACEIs	angiotensin-converting enzyme inhibitors
AD	Alzheimer's disease
ADA's	American Diabetes Association
AgNPs	silver nanoparticles
AHA	American Heart Association
ALA	α-linolenic acid
AMDR	acceptable macronutrient distribution ranges
AMI	acute myocardial infarction
apo B	apolipoprotein B
APP-III	adult treatment panel III
ARBs	angiotensin receptor blockers
ASCEND	a study of cardiovascular events in diabetes
ASD	autism spectrum disorder
AT-II	angiotensin II
ATP	adult treatment panel
AuNPs	silver nanoparticles
BHA	butylated hydroxy anisole
BHT	butylated hydroxytoluene
BMI	body mass index
BW	body weight
CGs	cyanogenic glycosides
circRNAs	circular RNA
CKD	chronic kidney disease
COX	cyclo-oxygenases
Cr	chromium
CRP	C-reactive protein
Cu	copper
CVDs	cardiovascular diseases
DASH	dietary approaches to stop hypertension
Dfcy	de-ferri-ferri-chrysin
DHA	docosahexaenoic acid

DIO	diet-brought on obese
DM	diabetes mellitus
DMB	3,3-dimethyl-1-butanol
DPPH	2,2-diphenyl-1-picrylhydrazyl
EAE	ethyl acetate extract
EC	equivalent concentration
ECB	endocannabinoid system
ED	endothelial dysfunction
EDD	endothelium-dependent dilation
EGIR	European Group for Insulin Resistance Research
eNOS	endothelial nitric oxide synthase
EPA	eicosapentaenoic acid
EPA	enhanced physical activity
FADS	fatty acid desaturases
FAs	fatty acids
FDA	Food and Drug Administration
FFA	free fatty acid
FM	fat mass
FMO3	flavin monooxygenase 3
FOS	fructooligosaccharides
FUW	*Fritillariauni bracteata* var. *wabuensis*
GAE	gallic acid equivalents
GC	glucocorticoid
GDM	gestational diabetes mellitus
GI	glycemic index
GLUT2	glucose transporter 2
GLUT4	glucose transporter type 4
GnRH	gonadotropin-releasing hormone
GSH	glutathione
H_2O_2	hydrogen peroxide
HBAs	hydroxybenzoic acids
HCAs	hydroxycinnamic acids
HDL	high-density lipoprotein
HFD	high-fat diet
HIV	human immunodeficiency virus
HPB	health protection branch
HSH	hydrogenated starch hydrolysate
HSL	hormone-sensitive lipase
IBD	inflammatory bowel diseases

ICDRA	International Conference on Drug Regulatory Authorities
IDF	international diabetes federation
IL-6	interleukin-6
IOTF	international obesity task force
IR	insulin receptor
IR	insulin resistance
LA	linoleic acid
LAB	lactic acid bacteria
LCD	low-calorie diets
LDL	low-density lipoprotein
LDLR	LDL receptor
LNS	lipid-based nutrient supplements
LOX	lipoxygenases
LPS	lipopolysaccharide
LRP	LDL receptor-related protein
MA	media area
MAAs	mycosporine-like amino acids
MetS	metabolic syndrome
miRNAs	microRNAs
MS	metabolic syndrome
MT	media thickness
MUFA	monounsaturated fatty acids
NADH	nicotinamide adenine dinucleotide hydrogen
NAFLD	non-alcoholic fatty liver disease
NASH	non-alcoholic steatohepatitis
NCEP	national cholesterol education program
NCEP-ATP-III	National Cholesterol Education Program's-Adult Treatment Panel-III
NEFA	non-esterified fatty acid
NFHS-5	National Family Health Survey-5
NHANES	National Health and Nutrition Examination Survey
NO	nitric oxide
O/W	oil-in-water
PAI-1	plasminogen activator inhibitor-1
PAL	phenylalanine-ammonia-lyase
PCs	phenolic compounds
PHA	poly(hydroxamic acid)
piRNAs	piwi-interacting RNA
PP	polypropylene

PPAR	peroxisome proliferator-activated receptor
PRR	pathogen recognition receptors
PS	*Piper sarmentosum*
PUFAs	polyunsaturated fatty acids
PV	peroxide value
RAS	renin-angiotensin system
Rb	retinoblastoma
RCD	restricted calorie diet
RDA	recommended dietary allowance
RDI	reference daily intake
ROS	reactive oxygen species
SAM	S-adenosylmethionine
SAMR1	senescence-accelerated mouse resistance 1
SCFA	short-chain fatty acids
Se	selenium
sHE	soluble epoxide hydrolase
snRNPs	small nuclear ribonucleoproteins
SOD	superoxide dismutase
T1DM	diabetes mellitus type 1
T2D	type 2 diabetes
T2DM	diabetes mellitus type 2
TAE	tannic acid equivalent
TBARS	thiobarbituric acid reactive substances
TBHQ	tert-butyl hydroquinone
TC	total cholesterol
TEAC	Trolox equivalent antioxidant capacity
TFC	total flavonoid content
TGs	triglycerides
TLRs	toll-like receptors
TMAO	trimethylamine-N-oxide
TNF-α	tumor necrosis factor-α
tPA	tissue plasminogen activator
TPC	total phenolic content
TSC	total saponin content
UPR	unfolded protein response
VLCD	very-low-calorie diets
WC	waist circumference
WHO	World Health Organization

Preface

The term "metabolic syndrome" represents the imbalance of energy metabolism in our body. It is the overall derangement of carbohydrate, protein, and lipid biochemical metabolism in our body. Human energy comes from the foods that we consume; the three essential nutrients used for energy are carbohydrates, proteins, and fats. The human body can also use protein and fat for energy once carbohydrates are depleted.

Once the food is fed, the human body breaks down vitamins into smaller components and absorbs them to apply as power or fuel. This method is referred to as the metabolism. This book discusses the physiological factors of the metabolic syndrome within the human body and spotlights how nutraceuticals/functional ingredients, botanicals, natural dietary supplements, structurally numerous antioxidants, B vitamins, and diverse amino acids and vital nutrients have an effect the metabolic syndrome.

Preserving a balance between energy delivery and strength expenditure may be essential for the complete frame to preserve the right body mass index (BMI). Better BMI results in obesity and a different way of life and metabolic disorders like Diabetes Type II, cardiovascular illnesses, etc. Recent research has confirmed that proper vitamins, including antioxidants, nutrients, micronutrients, and selected amino acids, can enhance the body's metabolism and defend it from inflammatory insults.

There are some chapters on caffeine so that readers may also understand its professionals and cons. Also, a chapter highlighting the damaging interactions of nutraceuticals and functional foods with generally used prescription medicines is covered in this book. The observation and future guidelines section summarizes the earlier chapters, highlights paradigm shifts inside the discipline, and envisions future direction.

This book thus describes the basic biochemistry and pathophysiology of the human body and demonstrates working out, the right nutrition, and selected ingredients can assist people in staying healthy and satisfied. A developing frame of proof demonstrates that nutraceutical and functional meals may additionally clearly normalize this sensitive balance.

The book, *Metabolic Syndrome: Dietary Supplements and Food Ingredients* comprises 15 chapters that deal with pathophysiology, management, prevention, and alertness of nutritional and meal substances. This book demonstrates how the right nutrient balance is beneficial to reinforce the proper metabolism of the body to keep it disease-free. The book covers nutraceuticals, dietary, food, herbs, polyphenols, natural microbes, antioxidants, trace elements, vitamins, etc.

A healthy living system and a properly nutritious diet can improve the patient's condition. Exercise and weight control are generally prescribed with some medicines for treating patients with metabolic syndrome. Chapter 1, "An Overview of Metabolic Syndrome," written by Rajeshwar Kamal Kant Arya and coworkers, gives a general overview of epidemiology and diseases related to metabolic syndrome. In addition, the chapter discusses treatments for metabolic diseases.

There are numerous medicinal plants that have been reported to possess strong antioxidant activity along with free radical scavenging activity. These have been presented in Chapter 2, "Role of Microbes as Natural Antioxidants of Pharmaceutical Importance," by Charu Gupta and associates. The authors describe the use of natural antioxidants in food products and discuss that besides plants, various microbes, including algae, mushrooms, bacteria, and fungi, also possess powerful antioxidant activity.

Chapter 3, "Role of Bioactive Compounds in Metabolic Syndrome," written by Urmila Jarouliya, has addresses the possible beneficial effect of bioactive compounds in the management and cure of metabolic syndrome and its complications.

Chapter 4, "Preventive Measures of Type 2 Diabetes via Nutrition," written by Eknath D. Ahire and colleagues concentrate on healthy nutritional styles that emphasize the general diet and calorie desires for weight management and diabetes prevention and control.

Chapter 5, "Potential of Trace Elements and Vitamins in Management of Oxidative Stress Linked Diabetes," written by Shanti Bhushan Mishra and his coworkers, discusses that micronutrient deficiency is extra frequent amongst diabetic adults compared to the regular adult population. In addition, vitamins and trace elements play a vital role in the management and prevention of disease.

Microbiota symbiotically living inside our gut plays a crucial role in the maintenance of human body physiology, metabolism, and function. As a result, both mutually benefit from each other. These have been described in Chapter 6, "Gut-Microbiome and Metabolic Syndrome," by Durgavati Yadav and her colleagues.

Chapter 7, "Role of Dietary in Prevention and Management of Obesity," written by Eknath D. Ahire et al., briefly discusses about the benefits of dietary fibers in the prevention and management of obesity.

Chapter 8, "Effects of Phenolic Compounds in Plants for Treatment of Type 2 Diabetes," written by Bui Thanh Tung and Nguyen Thi Yen, gives updated information about Type 2 Diabetes (T2D) and phenolic compounds, strength against T2D.

A healthy diet, including antioxidant and anti-inflammatory vitamins, along with proper physical activity, is key to avoiding and managing metabolic syndrome. Chapter 9, "Role of Vitamins in Metabolic Syndrome," written by Durgesh Singh, describes various vitamins used in the management and prevention of metabolic syndrome.

Chapter 10, "Pathophysiology and Applications of Herbs in Metabolic Syndrome," written by Bhushan R. Rane and associates, provides a summary, pathophysiology, and literature related to the syndrome's definition and herbal remedy techniques for each of the danger elements comprising metabolic syndrome.

Chapter 11, "Metabolic Syndrome a Concerning Area for Future Research," *w*ritten by Eknath D. Ahire and Associates, focuses on valuable insights as well as critical thoughts regarding MS, which will guide future research.

Chapter 12, "Beneficial Effect of Dietary Supplements in Metabolic Induced Obesity," written by Bhushan R. Rane and coworkers, speaks about the useful consequences of different dietary supplements for the management and prevention of obesity.

Chapter 13, "Nutraceuticals in the Prevention and Management of Type II Diabetes Mellitus," written by Urmila Jarouliya, focuses on functional herbal foods, various micronutrients, and physical activity, which usually help in preventing and managing diabetes mellitus (DM) and its associated complications.

Chapter 14, "The Effect of Medicinal Plants in Metabolic Syndrome," written by Bui Thanh Tung et al., presents particular phytochemical, preclinical, and clinical concerns of plant species related to MS using examples of commonly recognized and partially defined plant species.

Chapter 15, "Role of Plant-Based Bioactive Nutritional or Dietary Supplements in Metabolic Syndrome of Women and Child During COVID-19," written by Yashi Srivastava and Ankit Awasthi, discusses plant-based nutrients for women and children with special emphasis on COVID-19.

CHAPTER 1

An Overview of the Metabolic Syndrome

RAJESHWAR KAMAL KANT ARYA,[1] MEENA KAUSAR,[1] DHEERAJ BISHT,[1] DEEPAK KUMAR,[2] MOHMMAD RASHID,[3] and RAJ K. KESERVANI[4]

[1]Department of Pharmaceutical Sciences, Bhimtal Campus, Kumaun University Nainital, Bhimtal, Uttarakhand, India, E-mail: rajeshwararya@gmail.com (R. K. K. Arya)

[2]Department of Pharmaceutical Chemistry, School of Pharmaceutical Sciences, Shoolini University, Solan, Himachal Pradesh, India

[3]Department of Medicinal Chemistry and Pharmacognosy, College of Dentistry and Pharmacy, Buraydah Colleges, Buraydah, Al-Qassim – 31717, Saudi Arabia

[4]Faculty of B. Pharmacy, CSM Group of Institutions, Prayagraj, Uttar Pradesh, India

ABSTRACT

Metabolic syndrome (MetSyn) is affecting the largest population worldwide, and the International Diabetes Federation says approximately 25% of adults worldwide are suffering from metabolic syndrome. MetSyn is considered a group of metabolic disorders that are associated with cardiovascular diseases (CDVs). In this, metabolic dysfunction causes more than one closely associated disease, e.g., cardiac disease is associated with obesity and diabetes. Disease like atherosclerosis, hypertension, and diabetes is the most common metabolic disorder diseases. A sedentary lifestyle and an unhealthy diet are mainly responsible for obesity. Neurohormones also play a crucial in the development of the MetSyn. The deposition of fat in adipose tissue induces

The Metabolic Syndrome: Dietary Supplements and Food Ingredients. Raj K. Keservani, MPharm, Durgavati Yadav, PhD, Rajesh K. Kesharwani, PhD, Sippy Singh, DPhil & Kumar Sandeep, PhD (Eds.)

the generation of leptin, a neurohormone that causes atherosclerosis, inflammation, and smooth muscle proliferation, and eventually increases the risk of cardiac disorder. Adipose tissue also releases inflammatory indicator C-reactive protein (CRP) that is found elevated in diabetes and cardiac disease. The etiology of MetSyn is not well established. A patient with a high triglyceride level, apple shape body, a waist size of more than 35 inches, and an increased blood glucose level is diagnosed as a MetSyn patient. Sleep disturbance and psoriasis are also associated with heart disease. Genetic framework, gender, ethnological, and racial background are also responsible for the development of MetSyn. A healthy living system and a properly nutritious diet can improve the patient's condition. Exercise and weight control are generally prescribed with some medicines for treating the patients of MetSyn. In this chapter, we have discussed MetSyn and associated disease, the etiology, and pathophysiology of metabolic syndrome, the diagnosis method, and the line of treatment.

1.1 INTRODUCTION

Earlier cardiac patients were noticed with some other diseases like diabetes and obesity, and the researcher has found the link between cardiac disorders and other diseases; the main culprit was considered metabolic syndrome (MetSyn) or syndrome (Rochlani et al., 2017). MetSyn is considered to be a group of metabolic abnormalities linked with the development of cardiovascular disorders and diabetes (Rochlani et al., 2017). In 1923 Kylin gave the concept of MetSyn (Levesque and Lamarche, 2008). The most commonly known metabolic risk factors are atherosclerosis, hypertension, and diabetes (Alberti et al., 2009). The etiology of the syndrome is not been well disclosed till now. The prime culprits of the syndrome in most patients are thought to be an unhealthy diet, minimum body movement or workout, and a sedentary lifestyle that increases body weight (BW) or obesity (Rochlani et al., 2017). Obesity is also considered as the main causative factor for triggering hyperglycemia and other diseases, interconnected to metabolic syndrome. However, numerous inter-connected factors are considered as significant components for developing the syndrome (Standl, 2005). If patients suffering from two or more factors among hyperglycemia, hypoalphalipoproteinemia, high triglyceride level, apple-shaped body, and hypertension, then he or she is more prone to be diagnosed with MetSyn disorder; recently, the interest of the scientific community developed in MetSyn due to the high prevalence of obesity worldwide (Rochlani et al., 2017). The chances of developing MetSyn are very similar to the chances of obesity or diabetes (Saklayen, 2018). The patients with psoriasis were also observed with myocardial infarction and heart failure; these incidences also established a relationship between psoriasis and heart disease (Zhou et al., 2020). Sleep disturbance is also considered a significant factor in developing hypertension and diabetes, or a risk factor for metabolic syndrome.

1.2 EPIDEMIOLOGY

Globally, about 10 to 84% of incidences of MetSyn are reported, which depends on the age, gender, ethnological, and racial background of the studied populace, geographical environment (rural or urban) (Desroches and Lamarche, 2007; Kolovou et al., 2007). According to the International Diabetes Federation (IDF), about 25% of adults in the world are suffering from MetSyn (Cameron, 2004). The main factors related to MetSyn include high body mass index (BMI), a sedentary lifestyle, and higher socioeconomic status. It has been proved that the unhealthy lifestyle that causes an imbalance in our dietary habits, smoking, low body movement or workout, education, and inherited hyperglycemia are the substantial factors causing this syndrome (Park et al., 2003). National Health and Nutrition Examination Survey (NHANES) stated the observed incidence of MetSyn was found to be only 5% for standard weight subjects, 22% for overweighed subjects, and 60% for subjects with high obesity (Ford et al., 2002). Studies performed based on age revealed that the incidence was found to be 10% in subjects with the age group of 20–29, 20% in subjects with the age group of 40–49, and 45% in subjects with the age group of 60–69 (Ponholzer et al., 2008). According to the National Cholesterol Education Program's-Adult Treatment Panel II (NCEP-ATP-III) 2001, the occurrence of this syndrome varied from 7% to 56% in females and from 8% to 43% in males around the world. This indicates that females are more prone to this, and also observed the age of the subject is also important for the occurrence of this syndrome, the chances enhanced with age. Men at the age of 60 and women at 70 are more susceptible to this syndrome (Gobal et al., 2011; Dai et al., 2013). Postmenopausal women have approximately a 30–42% chance of occurring MetSyn (Wilson et al., 1999). The report suggests gaining weight of more than 2.25 kg in 16 years enhances the possibility of getting metabolic syndrome (Palaniappan et al., 2004). A relationship between waist circumference and MetSyn was also established, which suggests an increment of 11 cm in waist circumference increases the chances of up to 80% of the syndrome arising within 5 years (Reilly and Rader, 2003). The frequent change in metabolic function and other factors take place simultaneously, which was not anticipated earlier to increase the risk of cardiac disorder (Andreadis et al., 2007).

1.3 PATHOPHYSIOLOGY OF METABOLIC SYNDROME

MetSyn is also varied with the geographical condition; recent worldwide studies underlined obesity as the prominent factor for MetSyn, and for which our living style and environmental aspects such as improper diet, sleep, and insufficient body movement or workout are mainly responsible (Rochlani et al., 2017). Visceral adiposity is considered to be a primary agent for activating the functioning employed in MetSyn. The rich calorie diet is considered as the main causative agent (Boden and Shulman, 2002). The key culprit for the instigation, progression and transformation of MetSyn to the cardiac disorder is diabetes, inflammatory response, and neurohormonal function. Recent studies disclosed psoriasis (Zhou et al., 2020), sleep disorder (Cappuccio et al., 2008), and hereditary (Poulsen and Vaag, 2003) as key players in developing cardiac disorder and insulin resistance or diabetes and considered as risk factors for metabolic syndrome (Figure 1.1) (Rochlani et al., 2017).

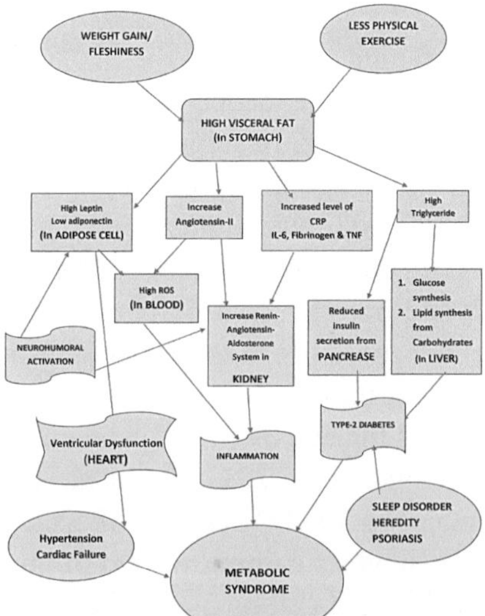

FIGURE 1.1 Pathophysiology of MetSyn (Rochlani et al., 2017).

1.4 DISEASES RELATED TO METABOLIC SYNDROME

1.4.1 ROLE OF TYPE 2 INSULIN INSENSITIVITY IN METABOLIC SYNDROME

Insulin insensitivity or type-2 diabetes is related to a bunch of risk factors for the cardiac disorder (Frayn, 2000). Insulin insensitivity is a key component in the MetSyn genesis. In type-2 diabetes, the body becomes insensitive to insulin, which affects the process of glucose consumption in the liver and muscle (Wilcox et al., 2005) and also suppresses hepatic gluconeogenesis and lipolysis in adipose tissue (Sears and Perry, 2015). Insulin insensitivity impairs the inhibitory function of lipolysis in adipose tissues and enhances the hydrolysis of triglycerides. As a result, the fatty acid is released into the systemic circulation (Tooke and Hannemann, 2000). The circulating free fatty acids (FFAs) reduce the glucose uptake in muscles by inhibiting the protein kinase induction (Ragheb et al., 2009) and stimulate gluconeogenesis and lipogenesis by activating protein kinase in the liver, and a hyper-insulinemic condition arises to compensate for the blood sugar level by minimizing insulin release. Free FAs are harmful to pancreatic β-cells and lead to low insulin release (Tripathy et al., 2003). Fat deposits in visceral parts also contribute to insulin insensitivity; the breakdown of visceral fat enhances the hepatic availability of FFA. FFA induces triglyceride synthesis and apolipoprotein-B production, which contain triglycerides rich in very low-density lipoprotein, specifically in the liver (Wallace et al., 2001). The alteration in hepatic metabolism of lipids, the insulin insensitivity also show a secondary function. It enhances the level of low-density lipoprotein and minimizes the high-density lipoprotein (Ormazabal et al., 2018). Visceral adipose tissues participate in metabolic function (Kershaw and Flier, 2004) and produce a biologically active protein plasminogen activator inhibitor which is associated with cardiac disorder (Rochlani et al., 2017). Insulin insensitivity enhances the overexpression of plasminogen activator inhibitors in arteries, vessels, and blood that, increase the risk of cardiac disorder, plaque formation, and atherosclerotic lesion (Juhan-Vague et al., 2003; Matsuzawa et al., 2011). Studies suggest insulin insensitivity causes atherosclerosis due to the induction of lipid genesis on arteries, and the rapid growth of smooth muscle (Howard et al., 1996). Insulin insensitivity impairs vasodilatation and raises blood pressure, and FFA also causes vasoconstriction (Juhan-Vague et al., 2003), salt accumulation, and fluid retention (De Fronzo, 1975), suppresses glycogen synthesis, and enhances lipid oxidation that induces oxidative stress resulted in the development of inflammatory response and destroy the cell (Ormazabal et al., 2018).

1.4.2 ROLE OF NEUROHORMONES IN METABOLIC SYNDROME

The central nervous system releases neurohormones having both hormonal as well as neurotransmitter functions (Srivastava et al., 2019). They play a crucial role in the development of MetSyn. Some studies exhibit the stimulation of neurohormones was found in congenital heart disease (Bolger et al., 2002). The adipose cell or tissue shows immunological and endocrine function, this help in understanding the genesis of the metabolic syndrome (Rochlani et al., 2017). Visceral adipose tissue releases adipokines that are connected with cardiovascular diseases (CDVs) and metabolic syndrome (Lindsay et al., 2002). An important adipokine, leptin, help maintain the energy balance in the hypothalamus and activate the immune system through the Th1 pathway. In obese subjects, a high amount of leptin is found to increase the chances of cardiac disease (Rochlani et al., 2017) because leptin causes atherosclerosis, inflammation, and smooth muscle proliferation. Adiponectin, another adipokine, is considered a savior of hypertension due to the antagonist function of adiponectin. Adiponectin improves the hypertensive condition, hyperglycemia, and myocardial infarction due to the antagonizing its function (Pischon et al., 2004; Vaněčková et al., 2014), but the increased adipose tissue in obese conditions enhances leptin levels and also reduces adiponectin, eventually enhances the chances of cardiovascular disease. The Renin-angiotensin-aldosterone system is a neurohormonal pathway, exhibits a crucial role in the controlling of hypertension and mineral level; it is an important component of MetSyn and connected co-morbidity (De Kloet et al., 2010). On stimulation, the angiotensin-converting enzyme (ACE) converts angiotensin-I into II, adipose tissue is a reservoir of angiotensin-I. The angiotensin released from the adipose tissue causes vasoconstriction and elevates blood pressure (Mehta and Griendling, 2007) in obese people (Ailhaud et al., 2000). It also exhibits ROS-generating activity (Gobal et al., 2011). ROS generation can aggravate smooth muscle injuries, inflammation, fibrosis, and oxidative modification of low-density lipoproteins (LDL) that result in atherogenesis (Matsuura et al., 2008). Angiotensin II (AT-II) also activates Lectin-like oxidized LDL receptor-1 (LOX-1) (Limor et al., 2005), which is also considered as the main causative factor for developing hypertension, hyperglycemia, atherosclerosis, cardiac hypertrophy, and it is found in an endothelial cell (Pant et al., 2014; Hotamisligil et al., 1994; Barreto et al., 2021). LOX-1 generates ROS that stimulates vascular cell death and cellular damage (Sharma, 2011).

1.4.3 ROLE OF INFLAMMATORY RESPONSE IN METABOLIC SYNDROME

International Diabetic Federation emphasized the importance of inflammatory conditions concerning MetSyn and other physiological changes (De Luca and Olefsky, 2008). The stimulation of the atherosclerosis process is believed as an accepted process of inflammatory response that is finally connected with MetSyn. Oxidative stress generated in the body causes insulin insensitivity in adipose tissue (Mohammadi et al., 2017), atherogenesis, and tissue fibrosis of smooth muscles in obese people (Matsuura et al., 2008). Studies show the relation between insulin insensitivity and stimulation of inflammatory responses. The diabetic and insulin-insensitive subject was found with an enhanced level of tumor necrosis factor-α (TNF-α), interleukin, interleukin-6 (IL-6), and C-reactive protein (CRP) (Rochlani et al., 2017), and the inflammation also participates in etiology of cardiac disorder and MetSyn (Hotamisligil et al., 1994). The inflammatory indicator CRP was found elevated in cardiac disorder and MetSyn patients (De Luca and Olefsky, 2008). The visceral adipose tissues not only release the adipokines it also releases inflammation indicator IL-6, TNF-α, and CRP (Tsigos et al., 1999). The macrophages present in adipocyte releases the TNF-α, and the production of TNF-α depends on obesity. The TNF-α is responsible for deactivating insulin receptor in smooth muscle cells and adipocytes, TNF-α also induce the breakdown of lipid and enhances the FFA level and also ceases the release of adiponectin (Fried et al., 1998). The high serum level of TNF-α is related to obesity and insulin insensitivity, and both of the components are interconnected to MetSyn (Ferreira-Hermosillo et al., 2015). Adipose cells also release cytokines and CRP. In insulin insensitivity, a high level of cytokines is observed with obesity (Bastard et al., 2000). Obesity produces cytokines (IL-6), which cause insulin insensitivity (Wisse et al., 2004). The production of cytokine and their immunity regulation mechanism from the adipose cell is quite complex (Doi et al., 2005). The cytokines activate the process of generation of an acute phase reactant, CRP, in hepatocytes by acting on hepatocytes, bone marrow, and endothelial cells (Rochlani et al., 2017). The higher concentration of CRP in the blood is associated with cardiovascular disorder and insulin sensitivity. Studies revealed a close relation between CRP levels in blood and diabetes (Doi et al., 2005; Grundy et al., 2004). The enhanced concentration of IL-6 causes the prothrombic condition, and participates in the pathogenesis of Ang-II-stimulated blood pressure and chronic kidney disease (CKD), kidney dysfunction, kidney fibrosis, and their progression. IL-6 also stimulates the expression of adhesion molecules in the endothelium that is responsible for vascular inflammation (Zhang et al., 2012; Poulsen et al., 2001).

1.4.4 ROLE OF PSORIASIS IN METABOLIC SYNDROME

For a long time, Psoriasis (Jarouliya and Keservani, 2018), a disease characterized by skin inflammation, is also associated with the cardiac disorder; the MetSyn is a prevalent cause of psoriasis that enhances plaque formation and occlusion in the arteries. Psoriasis affects 2–3% of the young generation and expedites the rate of atherosclerosis and, eventually, the formation of hard plaque in the arteries that lead to a heart attack. NIH performed a study on psoriasis patients, and the study revealed all the patients with psoriasis were found with MetSyn and also have insulin resistance, and high cholesterol levels (Zhou et al., 2020).

1.4.5 ROLE OF GENETICS IN METABOLIC SYNDROME

As we know, diabetes is a genetic disorder, and there is a requirement to explore the connection between our lifestyle and genetics. A study on monozygotic and dizygotic twins revealed that monozygotic twins are more susceptible to developing diabetes and hypertension (Roche et al., 2005). The gender of twins is also responsible for the prevalence of disease; the difference in the intrauterine environment of monozygotic or dizygotic twins makes monozygotic twins that are more prone to develop MetSyn (Pollex and Hegele, 2006). The siblings are at high risk of developing type-2 diabetes in comparison to others (Levesque and Lamarche, 2008). A new concept of studying the correlation between nutrition and genetics or nutrigenomics came into the light, where the relation of nutrition on genetics is explored, and nutrigenomics also gives a new perception in the investigation of MetSyn. With this concept, we can understand the influence of dietary habits on individuals or a specific community, and the genetic framework-based specific diet can be used utilized to protect from diseases. The human genomic framework has not been modified for many years, but we can see MetSyn is dramatically changing, and that supports the genomic and environmental relationship. The nutritious diet and proper living style ameliorate the MetSyn (Pollex and Hegele, 2006), which exhibits a poor nutritious diet works as an environmental factor for MetSyn, and it is related to the heredity (Hirshkowitz et al., 2015). The variability in the ethnic background also affects MetSyn and obesity due to the variation in the genetic framework (Hirshkowitz et al., 2015). The genetic and nutritive relation can also be altered by physical work, a non-nutritive environmental aspect (Yaggi et al., 2006). As we know, less physical work is responsible for obesity, less physical work, and high fatty diet also caused type-2 diabetes, because both the component interacts to PPARg Pro12Ala polymorphism (Yaggi et al., 2006).

1.4.6 ROLE OF SLEEP DISTURBANCE IN METABOLIC SYNDROME

For the last few years, the scientific community has been searching for the relation of sleep with MetSyn, and there is various kind of sleep disorders like insomnia (Keservani et al., 2016), change in the sleep cycle, sleep deprivation, sleep apnea, or other kinds of disturbance in sleep linked with obesity and cardiac disorder (Cappuccio et al., 2008). Insufficient sleep affects cardiac health and metabolic syndrome. The prescribed sleeping time for children 10–13 hrs. and for adults is 7–9 hrs. (Yaggi et al., 2006). But a survey done by the National Sleep Foundation, exhibit about 41% of the adult was found with less than 6 hr. sleep time, 11% with less than 5 hr., 29% with 7 hr. and rest has taken 8–10 sleep, in school going children's it was found that about 27% and in teenagers it was found that about 45% with under the prescribed limit. A global meta-analysis study on adults and children disclosed the participants with less sleep were found with high chances of obesity (Cappuccio et al., 2008). The altered sleeping system enhanced the chances of causing cardiac disorder. Sleep for a short duration (less than five hrs.) can cause hypertension, insulin insensitivity, or type-2 diabetes; the sleep duration of more than 9 hrs. in adults also causes the chances of developing insulin sensitivity (Reaven et al., 1988). These data establish the relation between sleep and other related risk factors for metabolic syndrome.

1.5 CLINICAL CONSEQUENCES ASSOCIATED METSYN

CDVs and diabetes are considered as the foremost clinical implication of MetSyn (NCEP, 2000). People with CDVs have the chances of developing a multifacet disease like dyslipidemia, elevated blood pressure, and elevated blood sugar; closely interconnected to each other (Liberopoulos et al., 2005) makes the patient susceptible to various other abnormalities like fatty liver, asthma, gallstones, sleep disorder, and sometimes carcinoma (NCEP, 2000).

1.6 DIAGNOSIS METHODS

Based on the clinical investigation National Cholesterol Education Program (NCEP) Adult Treatment Panel III (ATP-III) suggested a system for diagnosis of MetSyn that include measurement of waistline, triglyceride level, blood pressure, high-density lipid, and fasting blood sugar level, among these five components if a subject has at least three components in the higher side would be considered as a MetSyn patient (Scott, 2003). This diagnosis system is a widely accepted system for diagnosing the MetSyn. The baseline for the waistline is recommended >40 inches for males, and >35 inches for females, respectively. The triglyceride level is recommended ≥ 150 mg dL^{-1}. The cholesterol level is recommended <40 mg dL^{-1} for males, and <50 mg dL^{-1} for females, respectively. The blood pressure limit is recommended ≥130/85 mmHg, whereas the glucose level in fasting condition is recommended ≥110 mg dL^{-1} (6.1 mmol L^{-1}) (Scott, 2003; Wagh and Stone, 2004). AHA and NHLBI validated the ATP-III indicator and made some amendments to the ATP-III indicators (Scott, 2003). They laid down some conditions to the previous indicators; the baseline for waistline measurement was adjusted low for the people of the ethnic groups susceptible for insulin sensitivity was adjusted, the triglyceride level, high blood pressure, high-density lipid would be considered as unusual for the patients who are under treatment for these diseases. The glucose level was also reduced to 100 mg/dL from 110 mg/dL. The American Diabetes Association (ADA) also redrafted the definition of impaired fasting glucose. The waistline measurement was considered as the most important factor, the patient with increased waistline from the baseline having two or more above-mentioned disorders would be considered as MetSyn patient. The waistline measurement was also amended as per the ethnicity. For Europeans, the waistline was amended to ≥37 inches for males, and ≥31.5 inches for females, respectively. For the people of South Asia and China, it is prescribed as ≥35.4 inches for males, and ≥31.5 for females respectively (Wagh and Stone, 2004). For the patient receiving the treatment of hyperlipidemia, the baseline for triglyceride is amended to ≥150 mg/dL. For the patients who are under medication, the high-density lipoproteins (HDL) level was also reduced ≤to 40 mg/dL for males and ≤50 mg/dL for female patients, respectively. For the patients who are under blood pressure treatment, the value is increased to ≤130 mmHg systolic, and ≤85 mmHg diastolic. The blood glucose value also rose higher than 100 mg/dL for previously diagnosed diabetic patients. We have earlier discussed the relationship between psoriasis and MetSyn, and the clinical investigation of psoriasis-related MetSyn becomes necessary for the management of the interconnected disease. The clinical investigation includes regular assessment of skin conditions and other biochemical analyzes. The assessment of psoriasis and cardiac disorder includes measurement of blood pressure, waistline girth, and biochemical investigation like elevated level plasma lipid content, uric acid, blood sugar, and hepatic enzyme (Gisondi, 2018).

1.7 TREATMENT

We know MetSyn is a bunch of closely interconnected diseases; specific sort of treatment strategies is required for MetSyn therapy. Therapy will help in preventing the chances of cardiac-related disorders, diabetes, and other closely related diseases. Obesity is considered the main root cause of MetSyn. The treatments include three approaches, including a low carbohydrate diet, low fat, maintaining a proper weight or no more weight gain by doing regular exercise and reducing 10 kg weight per year, and change in living style. Being a multifacet disease, the other component of MetSyn must be taken into account and should be recognized first and targeted (Grundy et al., 2005). The therapeutic system is prescribed depending on the disease's condition, as given below.

1.7.1 OBESITY AND CARDIO-VASCULAR DISEASE

Obesity and cardiovascular treatment are generally performed simultaneously because obesity is the main causative factor for the development of cardiovascular disease (Keservani et al., 2010a, b; Keservani and Sharma, 2016). The treatment includes a weight management exercise regimen, and diet plan. For weight management, BMI should be <25 kg/m^2), and approximately 10% weight loss within one year is recommended. Regular workouts for 30 min for at least 5 days a week, brisk walks, and aerobics are prescribed to achieve this level. For higher-risk subject exercise programmed with experts, a restricted diet should be followed, including low saturated fat, trans-fat, and cholesterol, and sugar is prescribed to achieve a waist circumference of <40 inches in men and <35 inches in women (Grundy et al., 2005; Tripathi, 2018). Recently AHA prescribed a low dose of aspirin to a subject with a ≥10% score of Framingham risk scoring for the management of cardiovascular disorder (Scott, 2003). Some medicines used for weight management are lipase inhibitors (Xenical), appetite suppressants (meridian), and anti-obesity (rimonabant) (Scott, 2003; Grundy et al., 2005). Other medication includes fibrates, ezetimibe, nicotinic acid, or combinations. Statins are also prescribed to control cholesterol levels, e.g., lovastatin, pravastatin, simvastatin, and pravastatin, etc. (Wagh and Stone, 2004). The fibrate medication shows the reversal effect on hyperlipidemia related to MetSyn, and this includes clofibrate, gemfibrozil, fenofibrate, etc. (Gisondi, 2018).

The ideal parameter for blood pressure should be <140/90 mmHg for hypertensive, <130/80 mmHg for diabetic and kidney patients, for maintaining this level regular exercise regimen is prescribed with a low sodium diet, weight management, a low-fat diet, and cooking oils. Medication including anti-hypertensive which includes ACE inhibitors, vasodilators, renin inhibitors, calcium channel blockers, etc., is generally prescribed for blood pressure maintenance (Grundy et al., 2005; Tripathi, 2018; Mallare et al., 2005). Omega-3 fatty acids (fish oils) are very helpful in decreasing triglyceride levels and ameliorating insulin insensitivity.

1.7.2 DIABETES

For the management of diabetes proper exercise regimen and change in living style should be followed for weight management and modification of risk factors, e.g., obesity, lack of body movement, and lipid malfunction. The anti-diabetic medication is also prescribed for the patients (e.g., thiazolidinediones, metformin, meglitinides, acarbose) (Grundy et al., 2005; Tripathi, 2018; Mallare et al., 2005).

1.7.3 PSORIASIS

The treatment of psoriasis includes anthralin topical application with UV light therapy (Basavaraj et al., 2010). Other medical treatment includes imiquimod, cimetidine for psoriasis, and some cardiac drugs, e.g., ACE inhibitors, digoxin, and some psychotropic drugs like benzodiazepine are also given in the treatment of psoriasis-related MetSyn. Weight management with the help of exercise and a low-calorie diet is also recommended for treating MetSyn (Gisondi, 2018).

KEYWORDS

- **diabetes**
- **insulin insensitivity**
- **leptin**
- **metabolic syndrome**
- **neurohormone**
- **obesity**
- **psoriasis**

REFERENCES

Ailhaud, G., Fukamizu, A., Massiera, F., Negrel, R., Saint-Marc, P., & Teboul, M., (2000). Angiotensinogen, angiotensin II and adipose tissue development. *Int. J. Obes. Relat. Metab. Disord., 24*(4), S33–35. https://doi.org/10.1038/sj.ijo.0801501.

Alberti, K. G. M. M., Eckel, R. H., Grundy, S. M., Zimmet, P. Z., Cleeman, J. I., Donato, K. A., Fruchart, J. C., et al., (2009). Harmonizing the metabolic syndrome. *Circulation, 120*(16), 1640–1645.https://doi.org/10.1161/CIRCULATIONAHA.109.192644.

Andreadis, E. A., Tsourous, G. I., & Tzavaraetal, C. K., (2007). MetSyn and incident cardiovascular morbidity and mortality in a Mediterranean hypertensive population. *The American Journal of Hypertension, 20*(5), 558–564.

Barreto, J., Karathanasis, S. K., Remaley, A., & Sposito, A. C., (2021). Role of LOX-1 (lectin-like oxidized low-density lipoprotein receptor 1) as a cardiovascular risk predictor. *Arteriosclerosis, Thrombosis, and Vascular Biology, 41*(1), 153–166. https://doi.org/10.1161/ATVBAHA.120.315421.

Basavaraj, K. H., Ashok, N. M., Rashmi, R., & Praveen, T. K., (2010). The role of drugs in the induction and/or exacerbation of psoriasis. *Int J Dermatol, 49*(12), 1351–1361. https://doi.org/10.1111/j.1365-4632.2010.04570.x.

Bastard, J. P., Jardel, C., Bruckert, E., Blondy, P., Capeau, J., Laville, M., Vidal, H., & Hainque, B., (2000). *Elevated Levels of Interleukin 6 are Reduced in Serum and Subcutaneous Adipose Tissue of Obese Women After Weight Loss, 85*(9), 5.

Boden, G., & Shulman, G. I., (2002). Free fatty acids in obesity and type 2 diabetes: Defining their role in the development of insulin resistance and beta-cell dysfunction. *Eur. J. Clin. Invest., 32*(Suppl. 3), 14–23.

Bolger, A. P., Sharma, R., Li, W., Leenarts, M., Kalra, P. R., Kemp, M., Coats, A. J. S., et al., (2002). Neurohormonal activation and the chronic heart failure syndrome in adults with congenital heart disease. *Circulation, 106*(1), 92–99. https://doi.org/10.1161/01.CIR.0000020009.30736.3F.

Cameron, A. J., Shaw, J. E., & Zimmet, P. Z., (2004). The metabolic syndrome: Prevalence in worldwide populations. *Endocrinology and Metabolism Clinics of North America., 33*(2), 351–375.

Cappuccio, F. P., Taggart, F. M., Kandala, N. B., Currie, A., Peile, E., Stranges, S., & Miller, M. A., (2008). Meta-analysis of short sleep duration and obesity in children and adults. *Sleep, 31*(5), 619–626.

Dai, Y., Mercanti, F., Dai, D., et al., (2013). LOX-1, a bridge between GLP-1R and mitochondrial ROS generation in human vascular smooth muscle cells. *Biochim Biophys Res Commun, 437*(1), 62–66.

De Kloet, A. D., Krause, E. G., & Woods, S. C., (2010). The renin-angiotensin system and the metabolic syndrome. *Physiol Behav, 100*(5), 525–534. https://doi.org/10.1016/j.physbeh.2010.03.018.

De Luca, C., & Olefsky, J. M., (2008). Inflammation and insulin resistance. *FEBS Lett., 582*(1), 97–105. https://doi.org/10.1016/j.febslet.2007.11.057.

DeFronzo, R. A., Cooke, C. R., Andres, R., Faloona, G. R., & Davis, P. J., (1975). The effect of insulin on renal handling of sodium, potassium, calcium, and phosphate in man. *J. Clin. Invest., 55*(4), 845–855. https://doi.org/10.1172/JCI107996.

Desroches, S., & Lamarche, B., (2007). The evolving definitions and increasing prevalence of metabolic syndrome. *Applied Physiology, Nutrition and Metabolism, 32*(1), 23–32.

Doi, Y., Kiyohara, Y., Kubo, M., Tanizaki, Y., Okubo, K., Ninomiya, T., Iwase, M., & Iida, M., (2005). Relationship between C-reactive protein and glucose levels in community-dwelling subjects without diabetes: The hisayama study. *Diabetes Care, 28*(5), 1211–1213. https://doi.org/10.2337/diacare.28.5.1211.

Expert Panel on Detection, Evaluation, and Treatment of High Blood Cholesterol in Adults, (2001). Executive summary of the third report of the national cholesterol education program (NCEP) expert panel on detection, evaluation, and treatment of high blood cholesterol in adults (adult treatment panel III). *JAMA, 285*(19), 2486–2497. https://doi.org/10.1001/jama.285.19.2486.

Ferreira-Hermosillo, A., Molina-Ayala, M., Ramírez-Rentería, C., Vargas, G., Gonzalez, B., Isibasi, A., Archundia-Riveros, I., & Mendoza, V., (2015). Inflammatory cytokine profile associated with metabolic syndrome in adult patients with type 1 diabetes. *J. Diabetes Res., 2015*. https://doi.org/10.1155/2015/972073.

Ford, E. S., Giles, W. H., & Dietz, W. H., (2002). Prevalence of the metabolic syndrome among US adults: Findings from the third national health and nutrition examination survey. *Journal of the American Medical Association., 287*(3), 356–359.

Frayn, K. N., (2000). Visceral fat and insulin resistance--causative or correlative? *Br. J. Nutr., 83*(1), S71–77. https://doi.org/10.1017/s0007114500000982.

Fried, S. K., Bunkin, D. A., & Greenberg, A. S., (1998). Omental and subcutaneous adipose tissues of obese subjects release interleukin-6: Depot difference and regulation by glucocorticoid. *J Clin. Endocrinol. Metab., 83*(3), 847–850. https://doi.org/10.1210/jcem.83.3.4660.

Gisondi, P., Fostini, A. C., Fossà, I., Girolomoni, G., & Targher, G., (2018). Psoriasis and the metabolic syndrome. *Clin. Dermatol., 36*(1), 21–28. https://doi.org/10.1016/j.clindermatol.2017.09.005.

Gobal, F., Deshmukh, A., Shah, S., & Mehta, J. L., (2011). Triad of metabolic syndrome, chronic kidney disease, and coronary heart disease with a focus on microalbuminuria death by overeating. *J. Am. Coll. Cardiol., 57*(23), 2303–2308. https://doi.org/10.1016/j.jacc.2011.02.027.

Gobal, F., Deshmukh, A., Shah, S., et al., (2011). Triad of metabolic syndrome, chronic kidney disease, and coronary heart disease with a focus on microalbuminuria: Death by overeating. *J. Am. Coll. Cardiol., 57*(23), 2303–2308.

Grundy, S. M., Brewer, H. B., Cleeman, J. I., Smith, S. C., Lenfant, C., & American Heart Association; National Heart, Lung, and Blood Institute, (2004). Definition of metabolic syndrome: Report of the national heart, lung, and blood institute/American heart association conference on scientific issues related to definition. *Circulation, 109*(3), 433–438. https://doi.org/10.1161/01.CIR.0000111245.75752.C6.

Grundy, S. M., Cleeman, J. I., Daniels, S. R., Donato, K. A., Eckel, R. H., Franklin, B. A., Gordon, D. J., et al., (2005). Diagnosis and management of the metabolic syndrome. *Circulation, 112*(17), 2735–2752. https://doi.org/10.1161/CIRCULATIONAHA.105.169404.

Hirshkowitz, M., Whiton, K., Albert, S. M., Alessi, C., Bruni, O., DonCarlos, L., Hazen, N., et al., (2015). National sleep foundation's sleep time duration recommendations: Methodology and results summary. *Sleep Health, 1*(1), 40–43. https://doi.org/10.1016/j.sleh.2014.12.010.

Hotamisligil, G. S., Murray, D. L., Choy, L. N., & Spiegelman, B. M., (1994). Tumor necrosis factor-alpha inhibits signaling from the insulin receptor. *Proc. Natl. Acad. Sci. U S A, 91*(11), 4854–4858. https://doi.org/10.1073/pnas.91.11.4854.

Howard, G., O'Leary, D. H., Zaccaro, D., Haffner, S., Rewers, M., Hamman, R., Selby, J. V., et al., (1996). Insulin sensitivity and atherosclerosis. *Circulation, 93*(10), 1809–1817. https://doi.org/10.1161/01.CIR.93.10.1809.

Jarouliya, U., & Keservani, R. K., (2018). Nutraceutical components in the treatment of psoriasis and psoriatic arthritis. In: Siba, P. R., Smriti, R., & Debasis, B., (eds.), *Psoriasis and Psoriatic Arthritis Pathophysiology, Therapeutic Intervention, and Complementary Medicine*. CRC Press, Taylor & Francis Group, Book chapter. ISBN 9781498756068.

Juhan-Vague, I., Alessi, M. C., Mavri, A., & Morange, P. E., (2003). Plasminogen activator inhibitor-1, inflammation, obesity, insulin resistance and vascular risk. *Journal of Thrombosis and Haemostasis, 1*(7), 1575–1579. https://doi.org/10.1046/j.1538-7836.2003.00279.x.

Kershaw, E. E., & Flier, J. S., (2004). Adipose tissue as an endocrine organ. *The Journal of Clinical Endocrinology & Metabolism, 89*(6), 2548–2556. https://doi.org/10.1210/jc.2004-0395.

Keservani, R. K., Kesharwani, R. K., Sharma, A. K., Vyas, N., & Chadoker, A., (2010b). Nutritional supplements: An overview. *Int. J. of Curr. Pharm. Rev. and Res., 1*(1), 59–75.

Keservani, R. K., Kesharwani, R. K., Vyas, N., Jain, S., Raghuvanshi, R., & Sharma, A. K., (2010a). Nutraceuticals and functional food as future food. *Der Pharm. Lett., 2*(1), 106–116.

Keservani, R. K., Sharma, A. K., & Kesharwani, R. K., (2016). An overview and therapeutic applications of nutraceutical & functional foods. In: Keservani, R. K., Sharma, A. K., & Kesharwani, R. K., (eds.), *Recent Advances in Drug Delivery Technology* (pp. 160–201). IGI Global Publisher, USA, Ch 6. ISBN:9781522507543.

Keservani, R. K., Sharma, A. K., & Kesharwani, R. K., (2016). Nutraceutical and functional foods for cardiovascular health. In: Meghwal, M., & Goyal, M. R., (eds.), *Food Process Engineering: Emerging Technologies and Innovations in Food Engineering* (pp. 257–278). Apple academic press, CRC Press, Taylor and Francis, Ch 10. ISBN: 9781771884020.

Kolovou, G. D., Anagnostopoulou, K. K., Salpea, K. D., & Mikhailidis, D. P., (2007). The prevalence of metabolic syndrome in various populations. *The American Journal of Medical Sciences, 333*(6), 362–371.

Levesque, J., & Lamarche, B., (2008). The metabolic syndrome: Definitions, prevalence and management. *J. Nutrigenet. Nutrigenomics, 1*(3), 100–108. https://doi.org/10.1159/000112457.

Liberopoulos, E. N., Mikhailidis, D. P., & Elisaf, M. S., (2005). Diagnosis and management of the metabolic syndrome in obesity. *Obesity Reviews, 6*(4), 283–296. https://doi.org/10.1111/j.1467–789X.2005.00221.x.

Limor, R., Kaplan, M., Sawamura, T., Sharon, O., Keidar, S., Weisinger, G., Knoll, E., Naidich, M., & Stern, N., (2005). Angiotensin II Increases the expression of lectin-like oxidized low-density lipoprotein receptor-1 in human vascular smooth muscle cells via a lipoxygenase-dependent pathway. *American Journal of Hypertension, 18*(3), 299–307. https://doi.org/10.1016/j.amjhyper.2004.09.008.

Lindsay, R. S., Funahashi, T., Hanson, R. L., Matsuzawa, Y., Tanaka, S., Tataranni, P. A., Knowler, W. C., & Krakoff, J., (2002). Adiponectin and development of type 2 diabetes in the Pima Indian population. *Lancet, 360*(9326), 57–58. https://doi.org/10.1016/S0140-6736(02)09335-2.

Mallare, J. T., Karabell, A. H., Velasquez-Mieyer, P., Stender, S. R. S., & Christensen, M. L., (2005). Current and future treatment of metabolic syndrome and type 2 diabetes in children and adolescents. *Diabetes Spectrum, 18*(4), 220–228. https://doi.org/10.2337/diaspect.18.4.220.

Matsuura, E., Hughes, G. R. V., & Khamashta, M. A., (2008). Oxidation of LDL and its clinical implication. *Autoimmun Rev, 7*(7), 558–566. https://doi.org/10.1016/j.autrev.2008.04.018.

Matsuzawa, Y., Funahashi, T., & Nakamura, T., (2011). The concept of metabolic syndrome: Contribution of visceral fat accumulation and its molecular mechanism. *J. Atheroscler. Thromb., 18*(8), 629–639. https://doi.org/10.5551/jat.7922.

Mehta, P. K., & Griendling, K. K., (2007). Angiotensin II cell signaling: Physiological and pathological effects in the cardiovascular system. *Am J Physiol Cell Physiol, 292*(1), C82–97. https://doi.org/10.1152/ajpcell.00287.2006.

Mohammadi, M., Gozashti, M. H., Aghadavood, M., Mehdizadeh, M. R., & Hayatbakhsh, M. M., (2017). Clinical significance of serum IL-6 and TNF-α levels in patients with metabolic syndrome. *Rep. Biochem. Mol. Biol, 6*(1), 74–79.

Ormazabal, V., Nair, S., Elfeky, O., Aguayo, C., Salomon, C., & Zuñiga, F. A., (2018). Association between insulin resistance and the development of cardiovascular disease. *Cardiovascular Diabetology, 17*(1), 122. https://doi.org/10.1186/s12933-018-0762-4.

Palaniappan, L., Carnethon, M. R., & Wang, Y., (2004). Predictors of the incident metabolic syndrome in adults: The insulin resistance atherosclerosis study. *Diabetes Care, 27*(3), 788–793.

Pant, S., Deshmukh, A., Gurumurthy, G. S., Pothineni, N. V., Watts, T. E., Romeo, F., & Mehta, J. L., (2014). Inflammation and atherosclerosis--revisited. *J. Cardiovasc. Pharmacol. Ther., 19*(2), 170–178. https://doi.org/10.1177/1074248413504994.

Park, Y. W., Zhu, S., Palaniappan, L., Heshka, S., Carnethon, M. R., & Heymsfield, S. B., (2003). The metabolic syndrome: Prevalence and associated risk factor findings in the US population from the third national health and nutrition examination survey, 1988–1994. *Archives of Internal Medicine, 163*(4), 427–436.

Pischon, T., Girman, C. J., Hotamisligil, G. S., Rifai, N., Hu, F. B., & Rimm, E. B., (2004). Plasma adiponectin levels and risk of myocardial infarction in men. *JAMA, 291*(14), 1730–1737. https://doi.org/10.1001/jama.291.14.1730.

Pollex, R. L., & Hegele, R. A., (2006). Genetic determinants of the metabolic syndrome. *Nat. Clin. Pract. Cardiovasc. Med., 3*(9), 482–489. https://doi.org/10.1038/ncpcardio0638.

Ponholzer, A., Temml, C., Rauchenwald, M., Marszalek, M., & Madersbacher, S., (2008). Is the metabolic syndrome a risk factor for female sexual dysfunction in sexually active women? *International Journal of Impotence Research, 20*(1), 100–104.

Poulsen, P., & Vaag, A., (2003). The impact of genes and pre- and postnatal environment on the metabolic syndrome. Evidence from twin studies. *Panminerva Med., 45*(2), 109–115.

Poulsen, P., Vaag, A., Kyvik, K., & Beck-Nielsen, H., (2001). Genetic versus environmental aetiology of the MetSyn among male and female twins. *Diabetologia, 44*(5), 537–543. https://doi.org/10.1007/s001250051659.

Ragheb, R., Shanab, G. M. L., Medhat, A. M., Seoudi, D. M., Adeli, K., & Fantus, I. G., (2009). Free fatty acid-induced muscle insulin resistance and glucose uptake dysfunction: Evidence for PKC activation and oxidative stress-activated signaling pathways. *Biochem. Biophys. Res. Commun., 389*(2), 211–216. https://doi.org/10.1016/j.bbrc.2009.08.106.

Reaven, G. M., (1988). Banting lecture 1988. Role of insulin resistance in human disease. *Diabetes, 37*(12), 1595–1607. https://doi.org/10.2337/diab.37.12.1595.

Reilly, M. P., & Rader, D. J., (2003). The metabolic syndrome: More than the sum of its parts? *Circulation, 108*(13), 1546–1551. https://doi.org/10.1161/01.CIR.0000088846.10655.E0.

Roche, H. M., Phillips, C., & Gibney, M. J., (2005). The metabolic syndrome: The crossroads of diet and genetics. *Proc. Nutr. Soc., 64*(3), 371–377. https://doi.org/10.1079/pns2005445.

Rochlani, Y., Pothineni, N. V., Kovelamudi, S., & Mehta, J. L., (2017). Metabolic syndrome: Pathophysiology, management, and modulation by natural compounds. *Ther. Adv. Cardiovasc. Dis, 11*(8), 215–225. https://doi.org/10.1177/1753944717711379.

Saklayen, M. G., (2018). The global epidemic of the metabolic syndrome. *Curr. Hypertens. Rep., 20*(2), 12. doi: 10.1007/s11906-018-0812-z.

Scott, C. L., (2003). Diagnosis, prevention, and intervention for the metabolic syndrome. *Am. J. Cardiol., 92*(1A), 35i–42i. https://doi.org/10.1016/s0002-9149(03)00507-1.

Sears, B., & Perry, M., (2015). The role of fatty acids in insulin resistance. *Lipids in Health and Disease, 14*, 121. https://doi.org/10.1186/s12944-015-0123-1.

Sharma, P., (2011). Inflammation and the metabolic syndrome. *Indian J. Clin. Biochem, 26*(4), 317–318. https://doi.org/10.1007/s12291-011-0175-6.

Srivastava, A., Dedhia, P., & Thakar, C. V., (2019). Chapter 112 – cardiorenal syndrome type 2. In: Ronco, C., Bellomo, R., Kellum, J. A., & Ricci, Z., (eds.), *Critical Care Nephrology* (3rd edn., pp. 690–695.e2); Elsevier: Philadelphia. https://doi.org/10.1016/B978–0-323-44942-7.00112-6.

Standl, E., (2005). Aetiology and consequences of the metabolic syndrome. *European Heart Journal Supplements*, *7D*, 10–13.

Tooke, J. E., & Hannemann, M. M., (2000). Adverse endothelial function and the insulin resistance syndrome. *J. Intern. Med., 247*(4), 425–431. https://doi.org/10.1046/j.1365-2796.2000.00671.x.

Tripathi, K. D., (2018). *Essentials of Medical Pharmacology* (8th edn.). Jaypee Brothers Medical.

Tripathy, D., Mohanty, P., Dhindsa, S., Syed, T., Ghanim, H., Aljada, A., & Dandona, P., (2003). Elevation of free fatty acids induces inflammation and impairs vascular reactivity in healthy subjects. *Diabetes, 52*(12), 2882–2887. https://doi.org/10.2337/diabetes.52.12.2882.

Tsigos, C., Kyrou, I., Chala, E., Tsapogas, P., Stavridis, J. C., Raptis, S. A., & Katsilambros, N., (1999). Circulating tumor necrosis factor-alpha concentrations are higher in abdominal versus peripheral obesity. *Metabolism, 48*(10), 1332–1335. https://doi.org/10.1016/s0026-0495(99)90277–9.

Vaněčková, I., Maletínská, L., Behuliak, M., Nagelová, V., Zicha, J., & Kuneš, J., (2014). Obesity-related hypertension: Possible pathophysiological mechanisms. *J. Endocrinol., 223*(3), R63–78. https://doi.org/10.1530/JOE-14-0368.

Wagh, A., & Stone, N. J., (2004). Treatment of metabolic syndrome. *Expert. Rev. Cardiovasc. Ther., 2*(2), 213–228. https://doi.org/10.1586/14779072.2.2.213.

Wallace, A. M., McMahon, A. D., Packard, C. J., Kelly, A., Shepherd, J., Gaw, A., & Sattar, N., (2001). Plasma leptin and the risk of cardiovascular disease in the west of Scotland coronary prevention study (WOSCOPS). *Circulation, 104*(25), 3052–3056. https://doi.org/10.1161/hc5001.101061.

Wilcox, G., (2005). Insulin and insulin resistance. *Clin. Biochem. Rev., 26*(2), 19–39.

Wilson, P. W. F., Kannel, W. B., Silbershatz, H.,'Agostino, R. B. D., (1999). Clustering of metabolic factors and coronary heart diseases. *Archives of Internal Medicine, 159*(10), 1104–1109.

Wisse, B. E., (2004). The inflammatory syndrome: The role of adipose tissue cytokines in metabolic disorders linked to obesity. *J. Am. Soc. Nephrol., 15*(11), 2792–2800. https://doi.org/10.1097/01.ASN.0000141966.69934.21.

Yaggi, H. K., Araujo, A. B., & McKinlay, J. B., (2006). Sleep duration as a risk factor for the development of type 2 diabetes. *Diabetes Care, 29*(3), 657–661. https://doi.org/10.2337/diacare.29.03.06.dc05-0879.

Zhang, W., Wang, W., Yu, H., Zhang, Y., Dai, Y., Ning, C., Tao, L., et al., (2012). Interleukin 6 underlies angiotensin ii–induced hypertension and chronic renal damage. *Hypertension, 59*(1), 136–144.

Zhou, W., Abdelrahman, K. M., Dey, A. K., Reddy, A., Uceda, D. E., Lateef, S. S., et al., (2020). Association among noncalcified coronary burden, fractional flow reserve, and myocardial injury in psoriasis. *Journal of the American Heart Association, 9*(22), e017417.

CHAPTER 2

Role of Microbes as Natural Antioxidants of Pharmaceutical Importance

CHARU GUPTA,[1] DHAN PRAKASH,[1] and DAIANA ROMINA PERALTA[2]

[1]*Amity Institute of Herbal Research and Studies, Amity University, Sector-125, Noida – 201313, Uttar Pradesh, India, E-mail: charumicro@gmail.com (C. Gupta)*

[2]*Laboratory of Antioxidants Oxidative Processes, Institute of Chemical Sciences, Faculty of Agronomy and Agro-Industries, National University of Santiago del Estero-CONICET, RN 9 Km 1125, Villa El Zanjón-Santiago del Estero, Argentina*

ABSTRACT

Antioxidants are of significant importance to offer protection against various deteriorating ailments similar to malignancy, diabetes mellitus (DM), inflammatory diseases, neurodegenerative disorders, and aging. There are several therapeutic floras that have been testified to have resilient antioxidant action alongside with free radical scavenging activity. The part of antioxidants is to hunt the free radicals intricate in an extensive array of lethal oxidative reactions, as well as to inhibit pro-oxidant enzymes that catalyze them and also to complex dynamic metal ions able to yield oversensitive oxygen species, most of them free radicals, that initiate oxidative process. A great variety of antioxidant phytochemicals are widely distributed in grains, fruits, vegetables, and herbs. Their consistent intake lessens the danger of chronic illnesses connected with oxidative damage. Polyphenols are considered to be the utmost active antioxidants; they can also strengthen the action of added antioxidants. Besides plants, various microbes, including algae,

The Metabolic Syndrome: Dietary Supplements and Food Ingredients. Raj K. Keservani, MPharm, Durgavati Yadav, PhD, Rajesh K. Kesharwani, PhD, Sippy Singh, DPhil & Kumar Sandeep, PhD (Eds.)

mushrooms, bacteria, and molds also have potent antioxidant action. Some of the microbes belong to the probiotic group that has likely to defend the body from hazardous free radicals. Thus, the above-mentioned section would focus on the usage of various microbial sources as natural antioxidants and their prospective to be used as pharmaceuticals, thereby preventing many diseases.

2.1 INTRODUCTION

Free radicals are involved in the start and spread of a variety of toxic oxidative reactions (Kesharwani et al., 2016) such as peroxidation of membrane lipids that leads to the production and accumulation of lipid peroxides. They also hinder the activity of mitochondrial respiratory chain enzymes, causes disintegration or haphazard cross-connecting of particles, including DNA, enzymes, and proteins, which ultimately primes to cell loss. The free radicals are so potent that they also destroy the naturally occurring antioxidant enzymes like superoxide dismutase, catalase, and peroxidase; instigating damage and fatal cellular effects (e.g., apoptosis) through oxidization of membrane lipids, cellular proteins, DNA, and enzymes and shut down cellular processes (Zhang and Tsao, 2016).

Antioxidant' (Keservani and Sharma, 2014) is the term for any composite that is capable to hunk or postpone the oxidative injury initiated by the oxidants through giving electrons and/or hydrogen atoms and thus faltering the series reactions (Apak et al., 2016). The approach of action of an antioxidant is through numerous ways, such as preventing the production of free radicals, disturbing, and inhibiting the autoxidation chain reaction, slaking the singlet oxygen, decreasing the reactive oxygen species (ROS) into steady complexes, chelating metal pro-oxidants, curbing other antioxidant enzymes and impeding pro-oxidative enzymes (Zhang and Tsao, 2016). The development of free radicals can be prohibited inside the cell through the use of indigenous antioxidant enzymes, for example, glutathione peroxidase, catalase, and the superoxide dismutase (SOD) enzymes. These antioxidant enzymes though extremely competent in reducing the free radicals and maintaining them at low concentrations, still are inadequate to hostage oxidative stress and persons. They depend solely on their endogenous antioxidant systems that can be obtained only through sufficient dietary intake or other health supplements (Carocho and Ferreira, 2013; Rane and Keservani, 2020). Fruitlets and taters are a significant portion of a proportionate and fit food for individuals due to their high antioxidant amount (Tan et al., 2015). The ingesting of an antioxidant-rich food has been revealed to be well-related epidemiologically with a condensed danger of evolving cardiovascular ailments and certain sorts of tumors.

There are various types of phytochemicals with good antioxidant activity that are obviously existing in diet and they are recognized to exercise positive effects on humanoid fitness. They provide protection against various oxidative deterioration (Tiwari, 2001). Numerous epidemiological and clinical trial studies proved that the regular consumption of fresh seasonal fruits, vegetables, and whole grains, reduces the risk of chronic diseases associated with oxidative damage (Nichenametla et al., 2006). Some of the examples of natural antioxidants are carotenoids, tocopherols, ascorbates, lipoic acids, and polyphenols with free radical scavenging activity. Polyphenols are a big clan of naturally arising plant foods that are generally dispersed in plant foods, comprising fruitlets, spuds, nuts, kernels, florae, and bark. Synthetic antioxidants such as butylated hydroxy anisole (BHA) and butylated hydroxytoluene (BHT) play a useful part in diet and medicine trades (Scalbert et al., 2005) but due to their health hazard, they are used under strict regulation. Thus, interest in the use and development of natural antioxidants arises; for example, alpha-tocopherol, ascorbic acid as well as hydrolyzed proteins from numerous bodily and vegetal bases (Lee et al., 2010).

Thus far, foodstuff bases have been the chief emphasis of natural antioxidants (Tang et al., 2016); nevertheless, they are also found plentifully in metabolites created by microorganisms (Sadrati et al., 2013). There are varieties of group of microbes (bacteria, fungi, mushrooms, algae) that are reported to possess antioxidant activity. These microbes possess tremendous potential to be utilized as sources of antioxidant compounds which can be incorporated to develop a functional food or a nutraceutical (Gupta et al., 2013; Keservani et al., 2010a).

2.2 ANTIOXIDANTS FROM MICROBIAL SOURCE

Microorganisms (bacteria, fungi, micro-algae, mushrooms) are one of the most plentiful and varied classes originate on the globe. The utilization of microbes to yield diet constituents has been going on from the ancient times. Though, the separation of bacterial antioxidants originated into light only in the early 1980s, though researchers see recognized a link amongst antioxidants and microbes (Forbes et al., 2007; Meisinger et al., 2005). Since this initial work, a huge quantity of complexes and microbes have been reported that are able to produce them; some of them have been characterized. About 23,000 ancillary metabolites from microbes are identified, out of which approximately 42% are exclusively produced by actinomycetes, whereas molds form nearly equivalent volume (42%), and residual 16% is formed by eubacteria (Nannipieri et al., 2020). Microbial ancillary metabolites, comprising growing hormones, pigments, antibiotics, antitumor agents, etc., are not used for their growing and expansion, yet they have revealed an exceptional future for humanoid well-being. These bioactive complexes have developed noteworthy foundations for life-saving drugs, which comprise penicillin, amphotericin, streptomycin, erythromycins, tetracycline, vancomycin, etc. (Demain, 2014). Antibiotics testified from different microorganism act at particular target sites. These ancillary metabolites are mainly formed due to the commencement of a few enigmatic genetic factor groups, which are normally indolent under usual situations; henceforth, their manifestation would be substantial in using the biochemical diversity of microorganisms (Keservani et al., 2016a). Many of these ancillary metabolites embrace definite antioxidant and antimicrobial potential. Such bioactive molecules are the attentive source for biotechnological uses, especially for drugs, nutraceuticals, and cosmetics. This section would emphasis on the part of microbes as natural antioxidants of pharmaceutical importance.

2.2.1 ANTIOXIDANTS FROM MICRO-ALGAE

There are innumerable types of marine microbiota, including microalgae or marine algae that are the rich source of antioxidants. They possess diverse chemical structures with a number of health effects. Most of them are generally still unexploited and they possess potent bioactive compounds with great diversity in their structure. Nutraceuticals obtained from marine algae are used in various aspects of processing healthier foods and in developing new functional or preventive foods. Micro-algae including seaweeds have received great attention in previous years as probable natural antioxidants (Murugesan et al., 2015; Keservani et al., 2016b). They are known to comprise responsive antioxidant molecules, such as ascorbate ions or ascorbic acid and glutathione (GSH), carotenoids (and its types including – and -carotene, fucoxanthin, astaxanthin), mycosporine-glycine (mycosporine-like amino acids), catechin counting epigallocatechin, gallate, phlorotannins such as phloroglucinol, eckol, and tocopherols (including α-, χ-, δ-tocopherols). Besides this, many other prenyl toluquinones were isolated from the brown alga Cystoseira crinite; these compounds exhibited potent radical scavenging effects (Keservani et al., 2010a, b). The brown alga *Ecklonia stolonifera* collected from South Korea yielded a new phlorotannin, eckstolonol that possessed a potent 2,2-diphenyl-1-picrylhydrazyl (DPPH) radical-scavenging activity. Brown alga *Sargassum thunbergii* produced a novel antioxidant compound that is chemically a chromene called sargothunbergol A. Similarly, two other bioactive compounds monogalactosyl-diacyl-glycerols were also extracted from alga *S. thunbergii* (Kim et al., 2007). Another alga *Ecklonia* cava was used to isolate the compound named fucodiphlorethol G that is chemically a tetrameric phlorotannin. It also possesses a strong free radical scavenging activity. Some other macroalgae have also been reported to produce some potent antioxidants, such as terpenoids, phlorotannins, polyphenols, phenolic acids, anthocyanins, hydroxycinnamic acid derivatives, and flavonoids. Phytochemicals such as flavonoids exhibit strong anti-allergic, anti-viral properties, and free radical scavenging activities. They also provide protection against cardiovascular diseases (CDVs). There are some drawbacks of these synthetic antioxidants as they can cause liver damage and carcinogenesis. Some other examples of marine algae that were reported to possess antioxidant activities are fucoxanthin in *Hijikia fusiformis*; phycocyanin and phycocyanobilin in *Spirulina platensis* and *Aphanizomenon flos-aquae*; phenolic acids, tocopherols, and beta-carotene in Spirulina maxima; fatty acids, polyphenols, and phlorotannin in *Sargassum kjellamanianum*, *S. siliquastrum*, *Rhodomela confervoides*, *Symphjocladia latiuscula*, and *Kappaphycus alvarezzi*; lutein in *Botryococcus braunii*; and carotenoids in *Dunaliella salina* (Vadlapudi et al., 2012).

The various types of antioxidants that are produced by microalgae and cyanobacteria are carotenoids, polyunsaturated fatty acids, polysaccharides, and mycosporine-like amino acids. The antioxidant activity of these algae has been determined by Folin-Ciocalteu test (Li et al., 2007; Heller et al., 2016). By definition, polyphenolic secondary metabolites are those compounds that exhibit a wide diversity of chemical structures and are generally present in earthly plant life and water macrophytes without microalgae.

Research studies have shown that both microalgae and cyanobacteria also comprise phenolic and cinnamic acid products at gram levels on a dry mass base (Onofrejová et al., 2010). There are numerous studies that have proved the presence of microalgal polyphenols. Thus, the need of an hour is to develop or improve the existing methodologies for the detection and quantification of phenolic compounds in both microalgae and cyanobacteria (López et al., 2015). Thus, the natural antioxidants from marine sources (microalgae and cyanobacteria) represent attractive alternative sources, with potential health benefits for consumers (Chacón-Lee and González-Mariño, 2010).

In a recent study, the antioxidant and anti-proliferative activity of *Asparagopsis taxiformis* was evaluated. The outcomes revealed that red algae hold solid antioxidant and cytotoxic action that recommends their likely usage in the growth of medicinal remedies (Neethu et al., 2017).

In hitherto additional new research, the aqueous and methanolic excerpts of some selected microalgae such as *Ankistrodesmus* sp., *Spirogyra* sp., *Euglena cantabrica*, and *Caespitella pascheri* and cyanobacteria such as Nostoc sp., *Nostoc commune*, *Nodularia spumigena*, *Leptolyngbya protospira*, *Phormidiochaete* sp., and *Arthrospira platensis* were studied and their extracts were screened for radical scavenging action by DPPH assay. The methanol extract was found to possess more efficient in extracting the antioxidant rich compounds as compared to their aqueous counterpart. Besides, some other bioactive compounds such as gallic acid, syringic acid, protocatechuic acid and chlorogenic acid (chemically phenolics) and catechin and epicatechin (chemically flavonoids) were also synthesized by *Euglena cantabrica*. They formed a maximum amount of together gallic and protocatechuic acids amounting to 5.87 and 2.97 mg per gram of dry biomass, separately. Both their aqueous and methanolic extracts also displayed the maximum antioxidant action, due to the occurrence of the great quantity of phenolic compounds (Idaira et al., 2017).

There is an excessive shop call of natural antioxidants nowadays all over the world. Research was done to decide the relation between the two bioactive complexes viz. carotenoids and phenolic acids from a fucoxanthin-producing algae (Foo et al., 2017). Besides this, total carotenoids, phenolic acid, fucoxanthin contents and fatty acid profile of five types of microalgae and one type of macroalga were also quantified, followed by their bioactivity determination using different antioxidant assays. The maximum antioxidant action was stated in *Chaetoceros calcitrans* and *Isochrysis galbana*, followed by *Odontella sinensis* and *Skeletonema costatum* that exhibited a moderate one. *Phaeodactylum tricornutum* and Saccharina japonica showed the smallest antioxidant activity midst the algae species examined. The results were analyzed by using Pearson correlation and manifold linear regression that proved that both carotenoids and phenolic acids were knowingly associated ($p<0.05$) with the antioxidant actions, thus indicating the influence of these bioactive compounds on these algal antioxidant capacities (ABTS) (Singh et al., 2017).

In another research, cell-free extracts of 20 earthly cyanobacteria were assessed for their antioxidant properties in terms of free-radical scavenging activity (DPPH and 2,2'-azino-bis (3-ethylbenzothiazoline-6-sulphonic acid) (ABTS) assays), metal chelating activity and deoxyribose protection. Extract of *Anabaena constricta* was the most noticeable antioxidant agent (IC_{50} for DPPH and ABTS scavenging activities of 0.91 and 0.231 mg ml^1, respectively, for DNA protection 0.631 mg ml^1 and Fe^{2+}-ion chelating 0.91 mg ml^1). The strong antioxidant activity of cyanobacterial species was due to the occurrence of a great quantity of total phenol and total flavonoid. The cyanobacteria species also exhibited relatively high phenylalanine-ammonia-lyase (PAL) activity. It was found due to the presence of polyphenolics such as gallic acid, chlorogenic acid, caffeic acid, vanillic acid and ferulic acid and flavonoids such as rutin, quercetin, and kaempferol. The presence of polyphenols is directly correlated with their free radical scavenging, metal chelating and anti-oxidative damage protecting properties. Thus, these studies proved cyanobacteria to be the most feasible, promising, and attractive alternatives for natural antioxidants. Thus, they are of immense value for industrial applications in pharmaceuticals and nutraceuticals companies. Moreover, the presence of polyphenols as dominant phytochemicals in cyanobacteria also reflects their alteration approaches in contradiction of abiotic strains for their environmental accomplishment in diverse habitations (Singh et al., 2017).

2.2.2 ANTIOXIDANTS FROM BACTERIA

Sentience on antioxidants and their importance in humanoid healthcare has improved numerous folds in the modern period. Amplified request required on friendly novel and substitute assets for natural antioxidants. Some obligate aerobic microorganisms have been established to have the ability to synthesize a type of non-photosynthetic colored pigments. For example, Gram positive bacteria Staphylococcus aureus produces carotenoids that are well known antioxidants. Carotenoids are chiefly originated in fruitlets and taters and exhibit strong antioxidant activity by the quality of their free-radical scavenging properties and remarkable capability to quench singlet oxygen. *S. aureus* could exploit its golden carotenoid tint to fight oxidant-based approval modes of the host native immune structure (Gupta et al., 2013).

In a study, about nine genera of bacteria, including both Gram-positive and Gram-negative were found to produce the compound with antioxidant activity. The maximum activity was found in the methanol fractions of three Pseudomonas species (Smith and John, 1970).

In a research, pigmented bacteria, *Pseudomonas koreensis* strain that is attached with seaweed Sargassum were screened for its antioxidant potentials due to the presence of yellowish-brown pigment. The study concluded that 55.6% of the organisms were gifted to manufacture such antioxidant compounds and DPPH assay revealed that 20% of the creatures touched an antioxidant zone of 1.1 cm and 8.31% of the strains reached more than 3 cm. This bacterium showed better activity than many known commercially available synthetic antioxidants, such as butylated hydroxyl toluene towards DPPH scavenging. Similarly, *Serratia rubidaea* strain was found to be usually associated with seaweed *Ulva* sp. and *Pseudomonas argentinensis* strain as an epiphyte of Chaetomorpha. They both exhibit strong ABTS scavenging and FRAP activities, respectively. Moreover, the 16S rRNA gene sequence analysis showed that those bacteria that have greater antioxidant activity usually belong to a class Gamma proteobacteria. These observations were statistically analyzed for their phenolic substances in relative to other factors like DPPH, ABTS, reducing power, and FRAP and are all well correlated. Thus, the inference from the above study was that the seaweed-related pigmented bacteria have vast perspective as the foundation of potent antioxidant compounds, and efforts should be developed for their large-scale extraction for clinical applications (Pawar et al., 2015).

Research revealed that a strain of *Streptomyces* sp. isolated from mangrove soil in Kuala Selangor, Malaysia, as a probable home of antioxidants. The actinomycetes strain was identified as *Streptomyces* sp. by means of 16S rRNA gene phylogenetic study. Their extract exhibited potent antioxidant action through DPPH, ABTS, and superoxide radical scavenging assays and also metal-chelating activities. Moreover, their extract was also found to inhibit lipid peroxidation and also restore the existence of Vero cells from H_2O_2-induced oxidative damage. This was due to the occurrence of good antioxidant activities from its total phenolic contents. The results were confirmed from the gas chromatography–mass-spectrometry examination which showed the occurrence of phenolic complexes that were responsible for their antioxidant activities. Besides phenolics, other chemical constituents were also detected, like hydrocarbons, alcohols, and cyclic dipeptides. These compounds also contribute to the total antioxidant capability of the actinomycete extract. Thus, the study concluded that *Streptomyces* sp. possess good perspective as a home of new bioactive molecules with anti-oxidative healing result active against several oxidative stress-related diseases (Tan et al., 2017).

In modern times, several results have shed new light on the empathetic of the antioxidant capability of probiotics. The fitness welfares of probiotic bacteria are lowers diarrhea; kindle humoral and cellular immunity; thwart tumor; reduce negative metabolites, comprising ammonium and pro-cancerogenic enzymes in the colon; encourage antitumor immunity and get rid of irritable bowel syndrome. In a recent study by scientists (Wang et al., 2017) the culture supernatant of *Bifidobacterium animalis* were found to hunt hydroxyl radicals and superoxide anion *in vitro* while increasing the antioxidant activities of mice *in vivo* (Shen et al., 2011). In addition, the oxidative stress in patients with type 2 diabetes (T2D) can be enhanced by multispecies probiotics (Asemi et al., 2013). Lactic acid bacteria (LAB) strains have been studied broadly together in faunas and the humanoid. It is shown that LAB can fight ROS, including peroxide radicals, superoxide anions, and hydroxyl radicals. Rats fed high-fat diets augmented with Lactobacillus plantarum P-8 offered raised antioxidant capability, as revealed by restricting the increase of liver lipids and defending strong liver function. In people, *Lactobacillus rhamnosus* exercised sturdy antioxidant activity in conditions of high-bodily strain. Sportspersons open to oxidative stress might profit from the capability of *Lactobacillus rhamnosus* to raise antioxidant levels and counteract the properties of reactive oxygen species (Wang et al., 2017).

Another animal study has revealed that the probiotic bacteria Streptococcus thermophilus helps in recovery from malnutrition due to short-term fasting and reduces the associated intestinal atrophy (Dock et al., 2004). The probiotic bacteria also possess strong antioxidant activity and protect the body from dangerous free radicals. The rise in free radicals in the body depends upon several factors such as aging, stress, sugar intake or high sugar level in blood, antibiotics, and other chemicals and toxins.

In yet another study, the strains of Lactobacillus fermentum were isolated from miang and their antibacterial and antioxidant activities were investigated and it was found that its two strains produced certain bioactive complexes that display good antioxidant action that was conveyed in terms of Trolox equivalent antioxidant capacity (TEAC) and equivalent concentration (EC) values for free radical scavenging and reducing mechanisms, respectively. Thus, both the strains showed good alternative as a source of promising health-promoting probiotics together with antimicrobial and antioxidant properties (Klayraung and Okonogi, 2009).

It is well known that aquatic microbes yield natural colorings; however, the capacity of aquatic bacterial symbionts to yield natural colorants has been less considered. In a recent study, marine bacteria associated with soft-coral *Sarcophyton* sp. were collected from Karimunjawa Island and separated to produce the carotenoids. These strains were found to be closely associated to *Pseudoalteromonas shioyasakiensis*, *Pseudoalteromonas rubra*, *Virgibacillus salaries*, and *Pseudoalteromonas spongiae*. The antioxidant activity of the extracts was determined by calculating the inhibitory concentration (IC_{50}) of pigment extract against DPPH. The antioxidant activity measurement of CBSCP 1-1, CBSCP 2-2, CBSCP 2-3, and CBSCP 2-4 extract, and -carotene was 2,015, 5,017, 2,520, 4,213, and 1,980 mg l^1, correspondingly (Kusmita et al., 2017).

2.2.3 ANTIOXIDANTS FROM FUNGI

Molds are the greatest varied and plentiful collection on the globe and are an outstanding aspirant for bioactive metabolite making. It has been projected that there are as many as 5.2 million species of fungi, out of which only around 1 lakh species have been reported in literature (Kirk et al., 2011). Between these fungi, 14,000 species are mushrooms, about 5,000 are macrofungi; and only around 1,800 are those fungal species that possess pharmacological, therapeutic, and medicinal properties (Kirk et al., 2011).

The fungi belonging to the genera of Ascomycota and Basidiomycota mostly possess the pharmacological, therapeutic, and medicinal properties as they produce the secondary metabolites, while the fungi from group Zygomycota have been less studied (Cui et al., 2015; Zhong and Xiao, 2009). Few members of pharmaceutically important fungi also belong to the group Mucor and Umbelopsis of Zygomycota (Cui et al., 2015).

The two genera of zygomycetes (Mucor and Rhizopus) can be a great source of natural antioxidants. The production of antioxidants from microbial source has several benefits, such as being cost-effective; saves time due to the high multiplication rate; easy to handle large amount of culture at the same time and non-competitive with food sources. Notable examples of antioxidant producing fungi are *Mucor circinelloides*. The fungi can easily be mass multiplied in fermenters or bioreactors using various cheap carbon sources under controlled environmental conditions. Some latest technical information elucidates the antioxidant capacity of group *Rhizopus* (e.g., *Rhizopus oryzae*) along with food used for fermentation (Sugiharto et al.,

2016). *Mucor circinelloides f. lusitanicus* (CBS 277.49), *Mucor circinelloides* (WJ11), and *Mucor circinelloides* (CBS 108.16) strains are excellent producers of -carotene, lycopene, lipids, and bioactive component (Hameed et al., 2017).

In a study, the antioxidant activity of extracts of some *Penicillium*, *Aspergillus*, and *Rhizopus oryzae*, was calculated by the thiocyanate technique. Their extracts were found to show decent antioxidant action as compared to the control. *Aspergillus candidus* strain protected both the oil and BHA. In yet another study, it was found that when sucrose/lactose and ammonium sulfate was added in the culture media, it enhanced the growth of the fungus and indirectly the production of antioxidants (Arora and Chandra, 2010).

In yet another study, some of the filamentous fungi *Grifola frondosa*, *Monascus purpureus*, *Pleurotus* spp., *Lentinula edodes*, and *Trametes versicolor* were assessed for their antioxidant activity and the results were expressed in terms of -carotene/linoleic acid bleaching, radical scavenging, reduction of metal ions and chelating capabilities against ferrous ions. The fungus was developed in plunged flask cultures and different organic solvents were used to obtain the antioxidant compounds. The methanolic extract of each fungal strain was found to be more effective in antioxidant properties. The higher value of the total phenolic content was related to the high antioxidant activity and production by submerged cultivation was found to be more actual method for the manufacture of natural antioxidants (Smith et al., 2015).

In another research, the anti-oxidative activity of an ethyl acetate extract (EAE) from *Aspergillus terreus* LS01 was assessed by means of numerous *in vitro* tests such as DPPH, hydroxyl-scavenging free radical scavenging, and -carotene-linoleate model assay. The antioxidants in the extract were secluded, separated by silica gel column chromatography, and identified by UV-vis spectra, MS, and NMR analyses. The complexes were recognized as terreic acid and terremutin, and they also exhibited the highest level of DPPH free radical scavenging action. Thus, the research findings proved that *A. terreus* is a potential source of natural antioxidant products (Dewi et al., 2012).

Similarly, fungal endophytes also possess significant therapeutic potential. In a research study, a polyphenol producing endophytic fungus *Aspergillus austroafricanus* was secluded from Zingiber officinale rhizome. They were tested both for their antimicrobial (by disc diffusion method) and antioxidant activities (by DPPH, hydroxyl, and nitric oxide radical-scavenging methods). Their DNA defense from the OH radicals was also determined by agarose gel electrophoresis. HPLC was used to decide the total yield of polyphenols. Their EAE showed noteworthy antioxidant, antimicrobial activity and DNA injury defense capability. The HPLC study displayed the occurrence of polyphenols such as p-coumaric acid, ferulic acid and cinnamic acid. Thus, it was concluded that the fungus *Aspergillus austroafricanus* strain isolated from *Z. officinale* is a favorable prospective therapeutic agent and can be used as a substitute basis of polyphenols like p-coumaric acid, ferulic acid, and cinnamic acid (Danagoudar et al., 2017).

In another study on fungal endophytes, about 53 mycological endophytes were extracted from the bulbs of *Fritillariauni bracteata var. wabuensis* (FUW) to study their antioxidant activity. The majority of them belonged to the genus Fusarium and all the fungiform fermented filtrates displayed antioxidant activities. The DPPH activity, total ABTS, reduction power (FRAP), total phenol activity (TPC), total flavonoid content (TFC), and total saponin content (TSC) were assessed using different organic solvents such as petroleum ether, ethyl acetate, n-butyl alcohol and ethanol fractions removed from five representative fungoid cultures. Noteworthy positive connections were established amongst the structures (TPC, TFC, and TSC) and ABTS (DPPH, ABTS, and FRAP). In totaling, many natural antioxidant components were recognized from the fungiform extracts, comprising gallic acid, rutin, phlorizin, 2,4-di-tert-butylphenol and 2,6-di-tert-butyl hydroquinone (TBHQ); these were determined preliminarily by TLC-bioautography, HPLC, and GC-MS analysis. This study displayed plentiful mycological resources in *Fritillariauni bracteata var. wabuensis*. phenolics, flavonoids, and saponins are critical bioactive constituents in these plentiful fungiform endophytes and can be observed as novel perspective antioxidant assets (Pan et al., 2017).

In a different reading, the antioxidant action of methanolic extract of endophytic molds was investigated that was isolated from *Calotropis procera* plant. The methanolic extracts of these molds were prepared and evaluated for the antioxidant action by DPPH and reducing power assays. The extracts were characterized for the existence of phenolics and flavonoids. A whole of 8 endophytic molds were isolated from *C. procera* explants. Out of these 8 fungal isolates, the raw extracts of two molds, i.e., CPIMR-2 and CPIL-1 exhibited affirmative antioxidant action. From the morphological features, the isolates CPIMR-2 and CPIL-1 were recognized as *Penicillium* spp. and *Aspergillus* spp., individually. Both the isolates were found to have antioxidant potential with a percentage inhibition value of 81.16% and 80.97% in the DPPH radical scavenging assay. The phytochemical selection of the methanolic extracts displayed the manifestation of phenolics and flavonoids. The entire flavonoid content in *Penicillium* species and *Aspergillus* species were found to be 130.50 μg/mg and 94.91 μg/mg of extracts, respectively. The TPC was found to be 9.16 μg/mg and 12.13 μg/mg of extracts in *Penicillium* species and *Aspergillus* species, correspondingly. Thus, it was established from this research that endophytic fungi possess medicinal value in *C. procera*. The probable bioactive components for antioxidant action produced by the fungi are flavonoids and phenolics. These metabolites produced by endophytic molds from *C. procera* must be further explored as potential source of novel natural antioxidant compounds (Nagda et al., 2017).

In another recent study, the activity evaluation of 100 fungoid strains (26 endophytes, 42 entomopathogens, and 32 saprophytes) were carried out to decide the antioxidant ability of the fungal ethanolic extracts using three different assays (reducing power, TCP, and flavonoid contents). The results suggested that the verified strains, particularly those of endophytes, possess a good potential of natural antioxidants that can find various applications in food and cosmetics industries (Zohri et al., 2017).

In an alternative investigation, five mycological endophytes related with the medicinal plant *Mussaenda luteola* L. (Rubiaceae) were evaluated for the presence of phytochemicals, antioxidant, and antimicrobial properties. The study concluded that the EAE of *Alternaria* sp. strain revealed maximum total phenolic and flavonoid content. It also exhibited strong DPPH scavenging activity at the concentration of about 300 μg/ml and high reducing power. The other fungal extracts also showed a good amount of total phenol and flavonoid content, antioxidant, and antibacterial activity. Thus, the endophytic fungi secluded from *M. luteola* can be a prospective source for antioxidant and antibacterial compounds (Gunasekaran et al., 2017).

2.2.4 ANTIOXIDANTS FROM MUSHROOMS

Mushrooms are also known to be good sources of antioxidant compounds. Apart from their dietary abilities, mushrooms are also homes of physiologically useful bioactive constituents that encourage good well-being. Consumption of mushroom has also improved remarkably because of their desired smell, palate, and great nutritive constituents. Fitness endorsing stuffs, e.g., antioxidant, antimicrobial, anticancer, cholesterol-lowering and immune-stimulatory effects, have been stated for some species of mushrooms (Kosani et al., 2012). Both fruiting bodies and the mycelium contain composites with varied range of antioxidant and antimicrobial activities (Oyetayo et al., 2009). Some common edible mushrooms have presently been found to have antioxidant action, which is well associated with their total phenolic content (TPC). Mushrooms have been well-known for their nutritional value. They contain a good amount of polyphenols which is correlated with their antioxidant activity. There are several studies that reported the total phenols and antioxidant activity of new and cooked; wild and commercial mushrooms (Keles et al., 2011).

In yet another study, the measurement and comparison was made amongst the total antioxidant action, TPC and TFC of *Agaricus bisporous* (white button mushroom) and *Agaricus brasiliensis* (Brazilian mushroom) in together aqueous and ethanol extracts. The results proved that button mushrooms had noteworthy greater TPC in aqueous extract, whereas Brazilian mushrooms had significant higher TPC in 60% ethanolic extracts. In terms of TFC, Brazilian mushroom had a greater amount than button mushrooms in both types of solvents. For FRAP assay, Brazilian mushroom had considerably greater total antioxidant activity than the button mushroom in 60% ethanol but reverse trend with aqueous. For DPPH radical scavenging activity, Brazilian mushroom (60% ethanol) had the lowest EC50 value, trailed by button mushroom (60% ethanol), Brazilian mushroom (aqueous), and button mushroom (aqueous). Thus, the research established that the ethanolic extracts of mushrooms exhibit good antioxidant activity due to the presence of phenolic content and thus can possibly be used as a source of natural antioxidants (Gan et al., 2013).

In yet another study, the ethanolic extracts of oyster mushrooms (*Pleurotus pulmonarius*, *P. ostreatus*, *P. djamor* var. *djamor* and *P. djamor* var. *roseus*) and the split gill mushroom (*Schizophyllum commune*) were studied for their antioxidant activities through DPPH assay. The reducing ability of the mushrooms were determined by the ferricyanide complex (FRAP assay). The highest ability to decrease ferricyanide complex to ferrous form was detected in mushroom *Pleurotus djamor* var. *djamor*. Meanwhile, the highest TPC was determined in *P. djamor* var. *djamor* extract whereas *S. commune* gave maximum reading of TFC. All the mushroom testers displayed significant antioxidant properties, which, therefore, can be endorsed as natural antioxidant favorite in food and therapeutic productions (Arbaayah and Umi, 2013).

In a scientific study, the calculation of antioxidant perspective of lyophilized mushroom mycelia from 5 strains of the species *Pleurotus ostreatus* and *Coprinus comatus* (acquired by plunged cultivation in batch system) was studied as ethanolic extracts by estimating ABTS and the hydroxyl scavenging activity, FRAP method, chelating capacity, inhibition of human erythrocyte hemolysis, and inhibition of xanthine oxidase activity. Largely, outcomes established that the biologically dynamic materials content is controlled by the extraction process used. The utmost useful extract was that of *C. comatus* strain, followed by *the Pleurotus ostreatus* strain. A significant amount of -tocopherol was found to be present along with flavones such as rutin and apigenin. In the *P. ostreatus* strain extract, tocopherol acetate is known to play an important role as an antioxidant in skin defense against oxidative stress generated by UV rays. The various observations portrayed an important pharmacological perspective as well as the likelihood of practice in the improvement of novel purposeful produces (Vamanu, 2014).

In a recent scientific study, extractions were performed for the calculation of total phenolic content (TPC) and total flavonoid content (TFC) and antioxidant properties of five edible mushroom samples *Lentinus edodes*, *Volvariella volvacea*, *Pleurotus eous*, *Pleurotus sajor-caju* and *Auricularia auricular* using three diverse extractants. Among the three dissimilar extractants, 50% v/v ethanol was the most appropriate for antioxidant extraction from the mushroom testers. The 50% (v/v) ethanolic extract of dried *L. edodes* contained higher TPC and TFC than in the other mushroom extract samples. The antioxidant activities of 50% (v/v) ethanolic extract of dried *L. edodes* exhibited the studies tDPPH scavenging assay compared to BHA and -tocopherol. The ethanolic extract displayed a lesser reducing power compared to BHA and -tocopherol. Moreover, the *L. edodes* ethanolic extract also had the maximum chelating ability which was lower than for ethylene-diamine-tetra-acetic acid and displayed the highest superoxide radical-scavenging activity as compared to BHA and -tocopherol. Thus, it was concluded from the study that the 50% (v/v) ethanolic extract of *L. edodes* can be used as a probable natural anti-oxidant source or as a component in the fish and fishery produce trades (Boonsong et al., 2016).

In modern years, nanotechnology exploration is evolving as cutting-edge knowledge interdisciplinary with physics, chemistry, biology, material science and medicine. Gold and silver nanoparticles (AuNPs and AgNPs) are a significant type of nanomaterial for a varied array of biomedical uses. Production of AuNPs and AgNPs through natural way is favored owing to its environment-friendly and cost-effective features. A recent study demonstrated a recyclable and low-cost process of biosynthesis of AuNPs and AgNPs using basidiomycetes mushroom fungoid strains. The antioxidant probable of the basidiomycetes mushroom mycological strains was analyzed by TFC, FRAP assay, ABTS assay, metal chelating action, phospho-molybdenum assay, superoxide radical scavenging activity assay, free radical scavenging activity on DPPH along with the assessment of total phenolic and tannin contents in the mushroom mycelial extracts. These biologically synthesized AuNPs and AgNPs were tried for antioxidant action and showed good antioxidant activity. Thus, the research established that the mushrooms which are effectual manufacturers of AuNPs and AgNPs could also act as a harmless and cost-effective source of natural antioxidants (Madhanraj et al., 2017).

2.3 APPLICATIONS OF SIDEROPHORES AS ANTIOXIDANTS

Among metabolites with antioxidant capacity produced by microbes, siderophores have a remarkable role. Siderophores are low-molecular-weight iron-chelating agents, which the main function is to capture the insoluble iron from different habitats and to provide this essential metal as a nutrient (Khan et al., 2018). Beyond proving iron to cells, microbial siderophores can also act as antioxidant agents by sequestering free iron and preventing Haber-Weiss/Fenton reactions or through other varied mechanisms related to its chemical structure (Figure 2.1) that include the scavenging of free radicals. In spite of the possible of siderophores to turn as antioxidants, studies focusing on their usage as such have been very limited. In the nutrition business, siderophores could find a place as food-grade antioxidants, safeguarding food protection and keeping quality. For example, Todokoro et al. (2016) assessed the probable of de-ferri-ferri-chrysin (Dfcy), siderophore formed by *Aspergillus oryzae*, as an antioxidant food additive to reduce oxidation in food products. In the study, authors found that the siderophore was safe for consumption and stable at various pH and temperatures values after pasteurization and sterilization processes (Todokoro et al., 2016). Moreover, Moreover, Dfcy exhibited high antioxidant activity when compared to EDTA and -tocopherol in oil-in-water (O/W) emulsion model; where inhibition of lipid oxidation was assessed by peroxide value (PV) and thiobarbituric acid reactive substances (TBARS) assays to estimate both primary and secondary oxidation products, respectively. Dfcy reduced PV and TBARS by 83 and 75%, respectively and its antioxidant activity was equivalent to or greater than that of EDTA (75 and 72%, respectively) and -tocopherol (9.9 and 39%, respectively). These results strongly suggest that Dfcy can be a new food-grade antioxidant and the first natural substitute to the created iron chelator EDTA (Todokoro et al., 2016). Previously, other works had investigated the use of siderophores and hydroxamates to prevent lipid oxidation. Chen et al. (2012) studied the effect of deferoxamine (DFO) as an iron chelator to increase the oxidative stability of bulk oils containing reverse micelles. DFO was able to inhibit lipid oxidation in bare soybean oil (SSO) containing reverse micelles due to its ability to inhibit iron-promoted generation of free radicals through pathways such as hydroperoxide decomposition (Chen et al., 2012). In another work, it was demonstrated that a siderophore-mimetic active packaging film has the ability to prevent iron-promoted lipid oxidation in emulsified food products. This film is able to chelate iron and was developed by grafting poly(hydroxamic acid) (PHA) from the exterior of polypropylene (PP) films. They also detected that PP-g-PHA films achieved even better than EDTA in averting the development of unstable oxidation foodstuffs (Tian et al., 2013). These studies clearly exhibited the high potential of siderophores and their structure for inhibiting lipid oxidation. Then, these metabolites produced by microorganisms need to be further explored as potential source of novel natural antioxidant compounds.

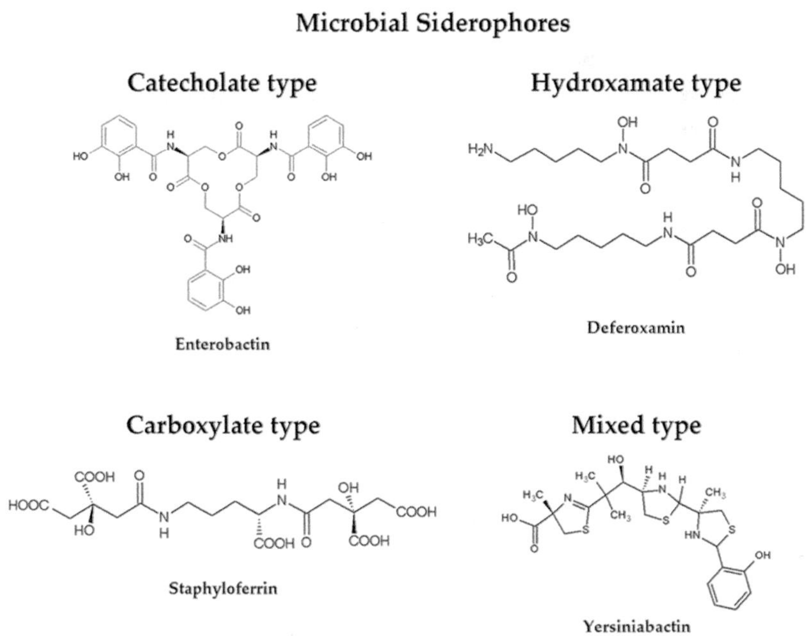

FIGURE 2.1 Representative chemical structure of microbial siderophores. Siderophores produced by microbes are classified into four main types: catecholates (enterobactin); hydroxamates (deferoxamine); carboxylate (staphyloferrin); and mixed type (yersiniabactin).

Source: Reprinted from with permission from Holden et al, 2015. © Oxford University Press.

2.4 APPLICATIONS OF NATURAL ANTIOXIDANTS IN FOOD PRODUCTS

The use of antioxidants in a vegetarian diet is less often as compared to non-vegetarian foods as the former contains some indigenous antioxidants such as tocopherols, carotenoids, or flavonoids with the exception of edible and essential oils. Synthetic antioxidants prevailed for the stabilization of foods of plant origin in earlier applications, but in the previous era or two, natural antioxidants have been intensively applied, following consumers' wishes and demands.

2.4.1 IN EDIBLE OILS

Eatable oils develop bad on storing, the kind of rank off-flavor dependent on their fatty acid structure. Edible oil makers attempt to extend the keeping quality of eatable oils by altered methods, comprising the toting of antioxidants. The existence of natural antioxidants should always be taken into justification, when proper amounts of extra antioxidants are deliberated. Alpha-tocopherol is the most common antioxidant that is used to prevent the rancidity in refined oils. Tocopherol (vitamin E) is effective as an antioxidant only after its hydrolysis in the body. The hydrolytic product is engrossed, clearances through the intestinal wall and then turn as an antioxidant in the plasma serum (Pokorn´y et al., 2001; Keservani et al., 2016).

2.4.1.1 STABILIZATION OF SUNFLOWER OIL

Sunflower oil is edible oil with a high sensory and nutritional value, but the major drawback is its unstablity during storage. This is owing to the occurrence of a relatively high percentage of linoleic acid (60–70%) and low amount of gamma- and delta-tocopherols. The later are more active antioxidants in oil than -tocopherol, which is the main tocopherol present in sunflower oil. Various phenolic compounds are more strong antioxidants as compared to tocopherols and citric acids. The majority of recent studies have suggested the use of natural plant extracts as antioxidants as compared to their synthetic counterpart (Pokorn´y et al., 2001).

2.4.1.2 STABILIZATION OF SOYBEAN OIL FROM RANCIDITY

Soybean oil is usually stable from getting rancid due to the presence of a high percentage of linoleic acid that is usually about 50% total fatty acids present in the oil. They also encompass a minor quantity of linolenic acid. However, soybean oil suffers from flavor reversion due to the presence of specific trace fatty acids. Synthetic antioxidants such as tert-Butyl hydroquinone (TBHQ) are also sometimes used to support tocopherols as stabilizers of edible oils against rancidification, but due to their side effects, natural antioxidants are preferred (Pokorn´y et al., 2001).

2.4.1.3 STABILIZATION OF RAPESEED OIL

Rapeseed oil is readily prone to oxidation, in spite of its relatively low linoleic acid content, as it contains higher amounts of extremely oxylabile linolenic acid. Nevertheless, it could be stabilized efficiently by adding natural antioxidants (Pokorn´y et al., 2001).

2.4.1.4 STABILIZATION OF OTHER VEGETABLE OILS

Sesame oil is moderately unsaturated and also contains numerous naturally current antioxidants (containing lignin assembly). Notable one is sesamol, which helps in its stabilization. Similarly, peanut oil also possesses a fair amount of linoleic acid, but it is comparatively less stable as it contains a very low amount of tocopherols, and no other endogenous antioxidants. However, their stability can be further improved by the use of natural antioxidants (Pokorn´y et al., 2001).

2.4.2 IN EMULSIFIED FAT PRODUCTS

2.4.2.1 STABILIZATION OF MARGARINES

Margarines are water-in-oil emulsions, containing carotene as a nature-identical additive, or related substances, imparting yellow color to the product. Carotene contributes to their stability against oxidation. The combination of 0.1% ascorbyl palmitate and 0.01% propyl gallate was the most efficient anti-oxidant in a dietetic margarine rich in (n–3) fatty acids. Among many natural antioxidants, rosemary extracts are also useful. All creamed fat goods can be stowed for a shorter duration than edible oils, because they are highly susceptible to microbial spoilage. So, they should always be stored under refrigeration conditions so as to prevent or lower the rate of autoxidation and naturally occurring antioxidant-tocopherols can stabilize them (Pokorn´y et al., 2001).

2.4.2.2 STABILIZATION OF MAYONNAISE

In contrast to margarine, mayonnaise is an oil-in-water emulsion. The oil phase (about 70% w/w) usually consists of edible oil.

Mayonnaise finds its application in the prevention of CDVs. These days, mayonnaise augmented with (n–3) polyunsaturated fatty acids and fortified with fish oil is used to develop ready-to-eat products. As fish oil is very oxygen-sensitive, they are highly susceptible to oxidation and causes rancidity. Tocopherol has been commonly used as a natural antioxidant to provide protection and stabilize mayonnaise. It has also been observed that the antioxidants also disturb the rheological things of the produce. The addition of synthetic antioxidant tert-Butyl hydroquinone (TBHQ) enhanced the keeping quality of the product (Pokorn´y et al., 2001).

2.4.2.3 STABILIZATION OF SALAD DRESSINGS

Salad dressings are oil-in-water suspensions, equivalent to mayonnaise; hence, their stabilization displays a good resemblance. Numerous kinds of salad dressings are habitually stowed after opening for an extended period in a fridge, so stabilization against rancidification is useful. Usually, zings or flavor extracts to be used for their stabilization. The use of BHA, BHT, TBHQ, and their combinations has been patented, with enhanced keeping quality of fish oil salad dressing (Pokorn´y et al., 2001).

2.4.2.4 STABILIZATION OF SWELTERING OILS AND SAUTÉED FOODSTUFFS

Sweltering oils must have non-volatile thermostable antioxidants; else, the antioxidants are steadily vanished by decomposition or vaporization at high cooking temperatures, generally in the array of 160–180°C and in a stream of water vapor made from H_2O existing in the cooked material. Numerous non-volatile manmade antioxidants have been made, but they are not on the list of compounds acceptable as foodstuff extracts. Several antioxidants with high inhibitory activity at storing temperatures quickly drop their efficacy with growing heat. Sesame oil, especially oil from cooked sesame kernels, was found to be very competent for boiling croutons, maybe due to the higher sesaminol amount. Such oil can be used in blend with other fewer steady fats. Research has shown that their stability can be significantly improved by the addition of natural antioxidants (Pokorn´y et al., 2001).

2.4.3 USE IN CEREAL FOODSTUFFS (GRAINS, FLOUR, FEAST MUESLIS)

Natural antioxidants also find application in increasing the storage of many cereal products for example de-hulled rice, white flour, grits, etc. In full grain flours, enzymes have to be disabled to upsurge storage capacity. Next cooking, natural antioxidants from fibers are generally incorporated to stabilize lipids.

Antioxidants are also incorporated to feast mueslis to upsurge their storage life. The nourishing importance of feast mueslis is improved by the addition of flavonoids and associated herbal antioxidants, which prolong keeping quality (Shukla, 1993).

2.4.4 APPLICATION IN EXTRUDED PRODUCTS

Cereal products such as flatbread come under the category of extruded products. The antioxidants are again used here to improve the shelf life of these products. They are usually added along with the flour and other additives directly to the extruder barrel so that they are uniformly distributed in the final product. Spices can also be used for stabilization as they impart good flavor to the final product. Normally, manmade antioxidants such as BHA are usually used (Pokorn´y et al., 2001). Natural antioxidants are more preferable by the modern consumers because of no side effects.

2.4.5 APPLICATION IN CAKES, CRACKERS, COOKIES, AND OTHER BAKERY PRODUCTS

Antioxidants also find application in prolonging the keeping quality of bakery products such as cakes, crackers, cookies, etc. Bakery items are usually made from cereals and their products that contain mostly added fats. Hydrogenated edible oil and/or fillings are also used that are rich in fat. The hydrogenated oils and fats are more susceptible to the development of off-flavors during storage. Both synthetic and natural antioxidants and mixtures of both additives are used to preserve these special products. Certain spices, Maillard products, and essential oils can also be considered as substitutes for natural antioxidants. For example, sugar-snap cookies are generally stabilized by using butylated hydroxyl anisole (BHA), but they are nowadays being replaced with natural antioxidants. Some additional instances of natural antioxidants comprise ferulic acid plus sodium phytate (Hix et al., 1997).

2.4.6 APPLICATION IN FRUITS AND VEGETABLES

The antioxidants from natural sources can also be used to increase the shelf life of fruits and vegetables as they are very prone to damage through fat oxidation. Thus, antioxidants can be used to prevent oxidation of polyphenols, which causes enzymatic browning in fruits and mushrooms (Nisperos-Carriedo et al., 1991). The quantity of fats in fruitlet and spuds is around 1% or less, so their spoilage by rancidity can be prevented by these bioactive constituents. Some citrus fruits contain natural essential oils mainly in their peel such as Citrus limetta and other Citrus species, which possess strong antioxidant activities but are also easily oxidized. They are usually protected by synthetic antioxidants as glyceridic oils. If their lipoxygenase enzymes are inactivated by blanching, the content of natural antioxidants (mostly flavonoids) would be sufficient enough to protect their lipid fraction against oxidation. Natural antioxidants are applied to protect carotenoids or anthocyanins against oxidation. Pigmented orange juice can be stabilized with ascorbic acid and phenolic acids followed by pasteurization (Maccarone et al., 1988). The antioxidants can also be used to stabilize the dried products.

2.4.7 USE IN SEASONING AGENTS, ZESTS, AND ESSENTIAL OILS

Seasoning agents, zests, and essential oils are most frequently incorporated to increase not only the flavor, but also the keeping quality, as several of them are strong antioxidants. They are frequently added in minor quantities, around 1%, so that the adding of the antioxidant portion is much lesser. Nonetheless, their result should not be overlooked, as they act in amalgamation with many native antioxidants or synergists. On occasion, these condiments entail stabilization, for example, essential oils from citrus fruits. Manmade antioxidants are more shared than natural antioxidants in this case because of their lesser expense, but natural antioxidants may be desirable in distinct cases (Vadlapudi et al., 2012).

2.4.8 APPLICATION IN ECO-FRIENDLY AGRO-BIOTECHNOLOGY PRODUCTS

Eco-friendly agro-biotechnology products are actually produced by special agricultural scientists, without utilizing the industrial fertilizers and synthetic pesticides. These kinds of florae are normally developed in life-threatening ecological situations, so they produce high amount of phenolic compounds and other stress factors. Besides, they contain less fat as compared to other conventional products so they can be used in food and other dietary supplements. They can be stabilized by adding tocopherol rich virgin edible oils, herbs, and spices (Pokorn´y et al., 2001; Rane et al., 2020).

2.5 CONCLUSION

From the foregoing discussion, it is imperative to state that the future of microbial-based antioxidants holds great potential as natural antioxidants. It may be anticipated that the starter of altered fresh materials and new technologies will improve the stability of foods and, thus, eliminate the necessity of using synthetic antioxidants. Steadily growing requirements to extend shelf life will create new possibilities for the application of antioxidants.

We believe that new natural antioxidants from microbial sources will be presented to the marketplace in the close prospect.

We believe that new natural antioxidants from microbial sources will be introduced to the market in the near future.

ACKNOWLEDGMENTS

The writer significantly recognizes the finance assistances MinCyT-Argentina and DST-India for their monetary funding and aid. The writers are also obliged to Dr. Ashok K. Chauhan, Founder President and Mr. Atul

Chauhan, Chancellor, Amity University UP, Noida, India for the inspiration, and research facilities.

Ethical Issues

There is nothing to be acknowledged.

Conflict of Interest

The writer(s) authorize that this piece has nope conflict of interest.

KEYWORDS

- **algae**
- **bacteria**
- **fungi**
- **mushrooms**
- **natural antioxidants**
- **pharmaceuticals**

REFERENCES

Apak, R. A., Özyürek, M., Güclü, K., & Capanoglu, C. E., (2016). Antioxidant activity/capacity measurement. 2. Hydrogen atom transfer (HAT)-based, mixed-mode (electron transfer (ET)/HAT), and lipid peroxidation assays. *Agric. Food Chem., 64*, 1028–1045.

Arbaayah, H. H., & Umi, K. Y., (2013). Antioxidant properties in the oyster mushrooms (*Pleurotus spp.*) and split gill mushroom (*Schizophyllum commune*) ethanolic extracts. *Mycosphere, 4*(4), 661–673.

Arora, D. S., & Chandra, P., (2010). Assay of antioxidant potential of two *aspergillus* isolates by different methods under various physio-chemical conditions. *Braz. J. Microbiol., 41*, 765–777.

Asemi, Z., Zare, Z., Shakeri, H., Sabihi, S. S., & Esmaillzadeh, A., (2013). Effect of multispecies probiotic supplements on metabolic profiles, hs-CRP, and oxidative stress in patients with type 2 diabetes. *Ann. Nutr. Metab., 63*, 1–9.

Boonsong, S., Klaypradit, W., & Wilaipun, P., (2016). Antioxidant activities of extracts from five edible mushrooms using different extractants. *Agric. Nat. Resources, 50*(2), 89–97.

Carocho, M., & Ferreira, I. C., (2013). A review on antioxidants, pro-oxidants and related controversy: Natural and synthetic compounds, screening and analysis methodologies and future perspectives. *Food Chem. Toxicol., 51*, 15–25.

Chacón-Lee, T. L., & González-Mariño, G. E., (2010). Microalgae for healthy foods-possibilities and challenges. *Comprehensive Rev. Food Sc. Food Safety., 9*(6), 655–675.

Chen, B., Panya, A., McClements, D. J., & Decker, E. A., (2012). New insights into the role of iron in the promotion of lipid oxidation in bulk oils containing reverse micelles. *J. Agric. Food Chem., 60*(13), 3524–3532.

Cui, J. L., Guo, T. T., Ren, Z. X., Zhang, N. S., & Wang, M. L., (2015). Diversity and antioxidant activity of culturable endophytic fungi from alpine plants of *Rhodiola XXXndusiate, R. angusta* and *R. sachalinensis*. *PloS One, 10*(3), e0118204.

Danagoudar, A., Joshi, C. G., Kumar, R. S., Poyya, J., Nivya, T., Hulikere, M. M., & Appaiah, K. A. A., (2017). Molecular profiling and antioxidant as well as anti-bacterial potential of polyphenol producing endophytic fungus-*Aspergillus austroafricanus* CGJ-B3. *Mycology, 8*(1), 28–38.

Demain, A. L., (2014). Importance of microbial natural products and the need to revitalize their discovery. *J. Industrial Microbiol. Biotechnol., 41*(2), 185–201.

Dewi, R. T., Tachibana, S., Itoh, K., & Ilyas, M., (2012). Isolation of antioxidant compounds from *Aspergillus terreus* LS01. *J. Microbial. Biochem. Technol., 4*, 010–014.

Dock, D. B., Aguilar-Nascimento, J. E., & Latorraca, M. Q., (2004). Probiotics enhance the recovery of gut atrophy in experimental malnutrition. *Biocell.,* 28, 143–150.

Foo, S. C., Yusoff, F. M., Ismail, M., Basri, M., Yau, S. K., Khong, N. M., Chan, K. W., & Ebrahimi, M., (2017). Antioxidant capacities of fucoxanthin-producing algae as influenced by their carotenoid and phenolic contents. *J. Biotechnol., 241*, 175–183.

Forbes, B. A., Sahm, D. F., & Weissfeld, A. S., (2007). Laboratory methods and strategies for antimicrobial susceptibility testing. In: *Bailey and Scott's Diagnostic Microbiology* (12th edn., pp. 187–214). Mosby Elsevier, St. Louis.

Gan, C. H., Amira, B. N., & Asmah, R., (2013). Antioxidant analysis of different types of edible mushrooms (*Agaricus bisporous* and *Agaricus brasiliensis*). *Intern. Food Res. J., 20*(3), 1095–1102.

Gunasekaran, S., Sathiavelu, M., & Arunachalam, S., (2017). *In vitro* antioxidant and antibacterial activity of endophytic fungi isolated from *Mussaenda luteola*. *J. Appl. Pharmaceut. Sc., 7*(08), 234–238.

Gupta, C., Prakash, D., & Gupta, S., (2013). Functional foods enhanced with microbial antioxidants. *Academic J. Nutr., 2*(2), 10–18.

Hameed, A., Hussain, S. A., Yang, J., Ijaz, M. U., Liu, Q., Suleria, H. A. R., & Song, Y., (2017). Antioxidants potential of the filamentous fungi (*Mucor circinelloides*). *Nutr., 9*, 1101.

Heller, M. I., Wuttig, K., & Croot, P. L., (2016). Identifying the sources and sinks of CDOM/FDOM across the Mauritanian shelf and their potential role in the decomposition of superoxide (O^{-2}). *Front. Marine Sc., 3*, article 132.

Hix, D. K., Klopfenstein, C. F., & Walker, C. E., (1997). Physical and chemical attributes and consumer acceptance of sugar-snap cookies containing naturally occurring antioxidants. *Cereal Chem., 74*, 281–283.

Holden, V.I., Bachman, M.A. Diverging roles of bacterial siderophores during infection, Metallomics, Volume 7, Issue 6, June 2015, Pages 986–995, https://doi.org/10.1039/c4mt00333k

Jerez-Martel, I., García-Poza, S., Rodríguez-Martel, G., Rico, M., Afonso-Olivares, C., & Gómez-Pinchetti, J. L., (2017). Phenolic profile and antioxidant activity of crude extracts from microalgae and cyanobacteria strains. *J. Food Quality, 2017*, 8. Article ID 2924508.

Keles, A., Koca, I., & Gençcelep, H., (2011). Antioxidant properties of wild edible mushrooms. *J. Food Process Technol.,* 2, 130.

Keservani, R. K., & Sharma, A. K., (2014). Flavonoids: Emerging trends and potential health benefits. *J. Chin. Pharm. Sci, 23*(12), 815–822.

Keservani, R. K., Kesharwani, R. K., & Sharma, A. K., (2016a). Naturally occurring toxicants as etiologic agents of food-borne disease. In: Bagchi, D., & Swaroop, A., (eds.), *Food Toxicology* (p 245–262). CRC Press, Taylor and Francis, Ch 12. ISBN: 9781498708746.

Keservani, R. K., Kesharwani, R. K., Sharma, A. K., Vyas, N., & Chadoker, A., (2010b). Nutritional supplements: An overview. *Int. J. of Curr. Pharm. Rev. and Res., 1*(1), 59–75.

Keservani, R. K., Kesharwani, R. K., Vyas, N., Jain, S., Raghuvanshi, R., & Sharma, A. K., (2010a). Nutraceuticals and functional food as future food. *Der Pharm. Lett., 2*(1), 106–116.

Keservani, R. K., Sharma, A. K., & Kesharwani, R. K., (2016b). An overview and therapeutic applications of nutraceutical & functional foods. In: Keservani, R. K., Sharma, A. K., & Kesharwani, R. K., (eds.), *Recent Advances in Drug Delivery Technology* (pp. 160–201). IGI Global Publisher, USA, Ch 6. ISBN:9781522507543.

Kesharwani, R. K., Singh, P., & Keservani, R. K., (2016). Green coffee bean extract and chlorogenic acids: Chemistry and novel antioxidant benefits. In: Bagchi, D., Moriyama, H., & Swaroop, A., (eds.), *Green Coffee Bean Extract in Human Health* (pp. 1–17). Boca Raton, Taylor and Francis, USA., Ch 1. ISBN: 9781498716376.

Khan, A., Singh, P., & Srivastava, A., (2018). Synthesis, nature and utility of universal iron chelator – siderophore: A review. *Microbiol. Res., 212, 213*, 103–111.

Kim, Y. H., Kim, E. H., Lee, C., Kim, M. H., & Rho, J. R., (2007). Two new monogalactosyl diacylglycerols from brown alga *Sargassum XXXndusiateXXX. Lipids, 42*, 395–399.

Kirk, P. M., Stalpers, J. A., Minter, D. W., & Cannon, P. F., (2011). *Dictionary of the Fungi*. CAB International: Wallingford, UK.

Klayraung, S., & Okonogi, S., (2009). Antibacterial and antioxidant activities of acid and bile resistant strains of *Lactobacillus fermentum* isolated from Miang. *Braz. J. Microbiol., 40*, 757–766.

Kosanić, M., Ranković, B., & Dašić, M., (2012). Mushrooms as possible antioxidant and antimicrobial agents. *Iranian J. Pharmaceut. Res., 11*(4), 1095–1102.

Kusmita, L., Mutiara, E. V., Nuryadi, H., Pratama, P. A., Wiguna, A. S., & Radjasa, O. K., (2017). Characterization of carotenoid pigments from bacterial symbionts of soft-coral *Sarcophyton* sp. From North Java Sea. *Int. Aquat. Res., 9*, 61–69.

Lee, J. K., Yun, J. H., Jeon, J. K., Kim, S. K., & Byun, H. G., (2010). Effect of antioxidant peptide isolated from *Brachionus calciflorus*. *J. Korean Society for App. Bio. Chem., 53*, 192–197.

Li, H. B., Cheng, K. W., Wong, C. C., Fan, K. W., Chen, F., & Jiang, Y., (2007). Evaluation of antioxidant capacity and total phenolic content of different fractions of selected microalgae. *Food Chem., 102*(3), 771–776.

López, A., Rico, M., Santana-Casiano, J. M., González, A. G., & González-Dávila, M., (2015). Phenolic profile of *Dunaliella tertiolecta* growing under high levels of copper and iron. *Environ. Sc. Pollution Res., 22*(19), 14820–14828.

Maccarone, E., Longo, M. L., Leuzzi, U., Maccarone, A., & Passerini, A., (1988). Stabilization of pigmented orange juice with physical treatments and phenolic additives. *Chim Industria., 70*, 95–98.

Madhanraj, R., Eyini, M., & Balaji, P., (2017). Antioxidant assay of gold and silver nanoparticles from edible basidiomycetes mushroom fungi. *Free Radicals Antioxidants, 7*(2), 137–142.

Meisinger, D., Engel, A. S., Lee, N., Porter, M. L., Stern, L. A., & Bennett, P. C., (2005). Molecular and functional diversity of anaerobic metabolic guilds in aphotic, redox stratified microbial mats from Lower Kane Cave, Wyoming, international symposia for environmental biogeochemistry and subsurface microbiology: Jackson, Wyo. *Am. Society Microbiol.*, 120.

Murugesan, S., Bhuvaneswari, S., & Thamizh, S. N., (2015). Evaluation of antioxidant activity of methanol extracts of red algae *Chondrococcus hornemannii* and *Spyridia fusiformis*. *J. Chem. Pharmaceut. Res.*, *7*(1), 333–337.

Nagda, V., Gajbhiye, A., & Kumar, D., (2017). Isolation and characterization of endophytic fungi from *Calotropis procera* for their antioxidant activity. *Asian J. Pharm Clin. Res.*, *10*(3), 254–258.

Nannipieri, P., Ascher-Jenull, J., Ceccherini, M. T., Pietramellara, G., Renella, G., & Schloter, M., (2020). Beyond microbial diversity for predicting soil functions: A mini-review. *Pedosphere, 30*(1), 5–17.

Neethu, P. V., Suthindhiran, K., & Jayasri, M. A., (2017). Antioxidant and antiproliferative activity of Asparagopsis taxiformis. *Pharmacognosy Res.*, *9*(3), 238–246.

Nichenametla, S. N., Taruscio, T. G., Barney, D. L., & Exon, J. H., (2006). A review of the effects and mechanism of polyphenolics in cancer. *Crit. Rev. Food Sc. Nutr.*, *46*, 161–183.

Nisperos-Carriedo, M. O., Baldwin, E. A., & Shaw, P. E., (1991). Development of an edible coating for extending the postharvest life of selected fruits and vegetables. *Proc. Fla. State Hortic. Soc.*, *104*, 122–125.

Onofrejová, L., Vašíčková, J., Klejdus, B., et al., (2010). Bioactive phenols in algae: The application of pressurized-liquid and solid-phase extraction techniques. *J. Pharmaceut. Biomed. Anal.*, *51*(2), 464–470.

Oyetayo, V. O., Dong, C. H., & Yao, Y. J., (2009). Antioxidant and antimicrobial properties of aqueous extract from *Dictyophora XXXndusiate*. *The Open Mycology, J.*, *3*, 20–26.

Pan, F., Su, T. J., Cai, S. M., & Wu, W., (2017). Fungal endophyte-derived *Fritillaria unibracteata* var. Wabuensis: Diversity, antioxidant capacities *in vitro* and relations to phenolic, flavonoid or saponin compounds. *Sc. Reports, 7*, 42008.

Pawar, R., Mohandass, C., Sivaperumal, E., Sabu, E., Rajasabapathy, R., & Jagtap, T., (2015). Epiphytic marine pigmented bacteria: A prospective source of natural antioxidants. *Braz. J. Microbiol.*, *46*(1), 29–39.

Pokorn´y, J., Trojáková, L., & Takácsová, M., (2001). The use of natural antioxidants in food products of plant origin. In: Pokorny, J., Yanishlieva, N., & Gordon, M., (eds.), *Antioxidants in Food Practical Applications.* Woodhead Publishing Limited, Abington Hall, Abington, Cambridge CB1 England.

Rane, B. R., & Keservani, R. K., (2020). Introduction to nutraceutical & dietary supplements. In: Keservani, R. K., Sharma, A. K., & Kesharwani, R. K., (eds.), *Nutraceutical and Dietary Supplements* (pp. 35–55). Apple Academic Press, Taylor & Francis Group, Chapter 3. ISBN: 9781771888738.

Rane, B. R., Bharath, M. S., Patil, R. R., Keservani, R. K., & Jain, A. S., (2020). Novel approaches in nutraceuticals. In: Kesharwani, R. K., Keservani, R. K., & Sharma, A. K., (eds.), *Enhancing the Therapeutic Efficacy of Herbal Formulations Through Novel Drug Delivery Systems* (pp. 241–266). IGI Global International Publisher, Pennsylvania, USA. Chapter 11. ISBN13: 9781799844532.

Sadrati, N., Daoud, H., Zerroug, A., Dahamna, S., & Bouharati, S., (2013). Screening of antimicrobial and antioxidant secondary metabolites from endophytic fungi isolated from wheat (*Triticum durum*). *J. Plant Prot. Res.*, *53*, 128–136.

Scalbert, A., Manach, C., Morand, C., & Remesy, C., (2005). Dietary polyphenols and the prevention of diseases. *Crit. Rev. Food Sc. Nutr., 45*, 287–306.
Shen, Q., Shang, N., & Li, P., (2011). *In vitro* and *in vivo* antioxidant activity of *Bifidobacterium animalis* 01 isolated from centenarians. *Curr. Microbiol., 62*, 1097–1103.
Shukla, T. P., (1993). High nutrition breakfast cereals. *Cereal Foods World, 38*, 437, 438.
Singh, D. P., Prabha, R., Verma, S., Meena, K. K., & Yandigeri, M., (2017). Antioxidant properties and polyphenolic content in terrestrial cyanobacteria. *Biotech., 7*(2), 134.
Smith, H., Doyle, S., & Murphy, R., (2015). Filamentous fungi as a source of natural antioxidants. Food Chem., 185, 389–397.
Smith, J. L., & John, A. A., (1970). Presence of antioxidant materials in bacteria. *Lipids, 5*(10), 795–799.
Sugiharto, S., Yudiarti, T., & Isroli, I., (2016). Assay of antioxidant potential of two filamentous fungi isolated from the Indonesian fermented dried cassava. *Antioxidants.,* 5.
Tan, L. T. H., Chan, K. G., Khan, T. M., Bukhari, S. I., Saokaew, S., Duangjai, A., Pusparajah, P., et al., (2017). *Streptomyces* sp. MUM212 as a source of antioxidants with radical scavenging and metal chelating properties. *Front. Pharmacol., 8*, 276.
Tan, L. T. H., Lee, L. H., Yin, W. F., Chan, C. K., Abdul, K. H., Chan, K. G., et al., (2015). Traditional uses, phytochemistry, and bioactivities of *Cananga odorata* (ylang-ylang). *Evid. Based Complement. Alternat. Med.,* 896314.
Tang, C., Hoo, P. C. X., Tan, L. T. H., Pusparajah, P., Khan, T. M., Lee, L. H., et al., (2016). Golden needle mushroom: A culinary medicine with evidenced-based biological activities and health promoting properties. *Front. Pharmacol., 7*, 474.
Tian, F., Decker, E. A., & Goddard, J. M., (2013). Controlling lipid oxidation via a biomimetic iron chelating active packaging material. *J Agric. Food Chem., 61*(50), 12397–1404.
Tiwari, A. K., (2001). Imbalance in antioxidant defense and human diseases: Multiple approach of natural antioxidants therapy. *Current Sc., 81*, 1179–1187.
Todokoro, T., Fukuda, K., Matsumura, K., Irie, M., & Hata, Y., (2016). Production of the natural iron chelator deferriferrichrysin from *Aspergillus oryzae* and evaluation as a novel food-grade antioxidant. *J. Sc. Food Agric., 96*(9), 2998–3006.
Vadlapudi, V., Kaladhar, D. S. V. G. K., Paul, M. J., Kumar, S. V. N., & Behara, M., (2012). Antioxidant activities of marine algae: A review. *Inter. J. Recent Scientific Res., 3*(7), 574–580.
Vamanu, E., (2014). Antioxidant properties of mushroom mycelia obtained by batch cultivation and tocopherol content affected by extraction procedures. *BioMed Res. Intern., 2014*, Article ID 974804: 8.
Wang, Y., Wu, Y., Wang, Y., Xu, H., Mei, X., Yu, D., Wang, Y., & Li, W., (2017). Antioxidant properties of probiotic bacteria. *Nutrients, 9*(521), 1–15.
Zhang, H., & Tsao, R., (2016). Dietary polyphenols, oxidative stress and antioxidant and anti-inflammatory effects. *Curr. Opin. Food Sci.,* 8, 33–42.
Zhong, J. J., & Xiao, J. H., (2009). Secondary metabolites from higher fungi: Discovery, bioactivity, and bioproduction. *Adv. Biochem. Eng. Biotechnol., 113*, 79–150.
Zohri, A. A., Moharram, A. M., & Abd El-Ghani, O. A., (2017). Antioxidant potentialities of some strains belonging to endophytic, entomopathogenic and saprophytic fungi. *Eur. J Biol. Res., 7*(1), 76–85.

CHAPTER 3

Role of Bioactive Compounds in the Metabolic Syndrome

URMILA JAROULIYA

School of Studies in Biochemistry, Jiwaji University, Gwalior, Madhya Pradesh – 474011, India, E-mail: urmi_gudia@rediffmail.com

ABSTRACT

Metabolic syndrome is a health situation that forces individuals at a higher threat of coronary heart disorder, diabetes, and atherosclerosis. The underlying causes of metabolic syndrome consist of weight problems, insulin resistance, bodily state of being inactive, eating behavior, way of life, and genetic factors. Bioactive compounds found in purposeful ingredients have proven ability as preventive and therapeutic options within the treatment of metabolic syndrome and without any side effects. Those bioactive compounds (polyphenols, flavonoids, prebiotics, fibers, and vitamins) are received from fruits, veggies, and entire grains, and play an essential function in treating metabolic syndrome complications. Having plenty of pharmaceutical properties, these bioactive compounds have proven to offer effective results in treating diabetes mellitus (DM), high blood pressure, obesity, hyperlipidemia, and cardiovascular illnesses. This chapter focuses on the feasible useful impact of bioactive compounds in the prevention and remedy of metabolic syndrome and its complications.

3.1 INTRODUCTION

Metabolic syndrome (MetS) additionally referred to as the insulin resistance syndrome refers to the gathering of numerous metabolic irregularities,

The Metabolic Syndrome: Dietary Supplements and Food Ingredients. Raj K. Keservani, MPharm, Durgavati Yadav, PhD, Rajesh K. Kesharwani, PhD, Sippy Singh, DPhil & Kumar Sandeep, PhD (Eds.)

inclusive of dyslipidemia [elevated serum triglycerides and reduced high-density lipoprotein (HDL) cholesterol], dysglycemia, high blood pressure, and primary obesity, it is a main public health and medical assignment worldwide (Reaven, 1988). The term syndrome, originating from Greek and exactly meaning "running together," which mention to a set of signs or symptoms occurring together where the central pathophysiology leading to the concordance is unknown (Jablonski, 1991). Initially, in 1998, World Health Organization (WHO) counseled a definition for MetS (Alberti, 1998). During the last 22 years, different groups have evolved specific definitions for this circumstance. In 2001, the countrywide cholesterol schooling program adult remedy Panel III blanketed inside the MetS standards the fasting blood plasma glucose stage (Jama, 2001) (Figure 3.1).

Following this, the International Diabetes Federation proposed some other new definitions for MetS. To outline significant adiposity, they indexed the specific ethnic institution with thresholds for waist circumference (Ford, 2005). The American Heart Affiliation/countrywide heart, Lung, and Blood Institute (AHA/NHLBI) showed that MetS is recognized while three or greater of the subsequent standards are available: waist circumference >102 cm in men, and >88 cm in Women, blood pressure ≥130/≥85 mmHg, triglyceride (TG) plasma level ≥ 150 mg/dL (1.7 mmol/L), HDL cholesterol <40 mg/dL (1.03 mmol/L) in men, and <50 mg/dL (1.29 mmol/L) in women, and fasting plasma glucose ≥ 100 mg/dL (5.6 mmol/L) or any records of diabetes mellitus (DM) or taking any type of antidiabetic drugs (Grundy, 2004).

In general, we are able to say that the MetS represent a set of metabolic abnormalities that consist of high blood pressure, abdominal obesity, insulin resistance, dyslipidemia, irritation, and oxidative stress which might be firmly associated with an increased degree of diabetes, atherosclerosis, and non-atherosclerotic cardiovascular ailment (CVD).

3.2 EPIDEMIOLOGY OF METS

Metabolic syndrome (MetS) is an example of disease where both genetic and environmental factors lead to CVD. Genetic factors itself or its each component (obesity, type 2 diabetes (T2D), insulin resistance) are not simplest answerable for the improvement of the disorder, but the environmental factors or non-genetic elements associated with higher intake of meals wealthy in fat, carbohydrates, and salt, much less intake of fruits and veggies, and decrease tiers of physical activities additionally performs an essential function in inflicting MetS. The worldwide prevalence of MetS varies between 10 and 84% for urban populations based on the region, structure (age, sex, race) of the population. The international Diabetes Federation (IDF) predicted that about one-region of the worldwide adult populace has MetS, of which 28% have been men and 34% were women belonging to the atherosclerosis threat in communities (ARIC) (Desroches and Lamarche, 2007; Kolovou et al., 2007).

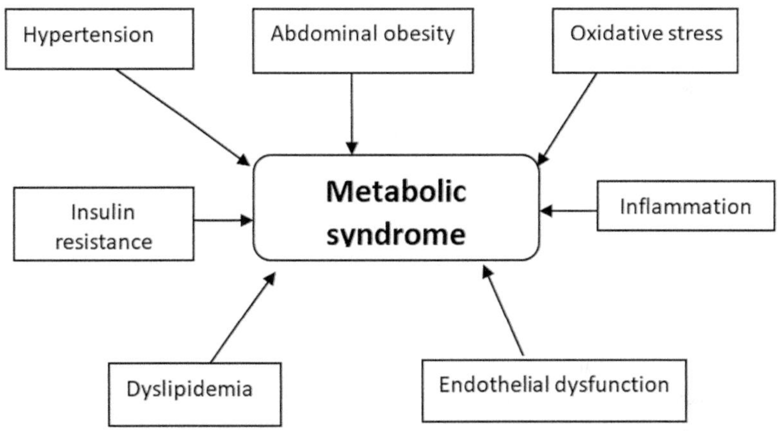

FIGURE 3.1 Schematic representation of metabolic syndrome.

Globally MetS has received tremendous significance currently because of the exponential growth in obesity. Early analysis is essential in order to adopt the way of life and to change risk factors. Many studies articles mentioned that the fundamental reasons of MetS is persistent and subclinical inflammation, as well as the morphological modifications of protein receptors (Urmila and Raj, 2019). The most important sites which have been concerned in beginning of persistent inflammation in MetS is liver, intestine, and adipose tissue (Henao-Mejia et al., 2012).

3.2.1 MAIN COMPOUNDS OF METS

The four essential compounds (insulin resistance, visceral adiposity, dyslipidemia, and high blood pressure), all groups collectively for causing MetS and are stated to be interconnected. In the present conditions, metabolic syndrome may be expressed with four principal traits they're: insulin resistance, visceral obesity, atherogenic dyslipidemia and endothelial disorder.

Sufferers affected by metabolic syndrome, if their weight reduces, it is able to enhance the multiple features, so a certain degree of adiposity seems to be required to take place for the odd pathophysiology.

Contrarily, there are a lot of sufferers who are obese; however, they no longer obvious any of the other functions of metabolic syndrome, predisposition to insulin resistance, and obesity is essential for the metabolic syndrome. So, insulin resistance and weight problems are two major compounds for metabolic syndrome. Insulin resistance and visceral obesity generate atherogenic dyslipidemia, expression of excessive serum TG levels and occasional HDL levels may be blanketed in this definition endothelial dysfunction (ED) is generally generated by hypertension also generates from insulin resistance and from adipokines and free fatty acids (FFAs) that are released from visceral adipose tissue. Each atherogenic dyslipidemia and ED make a contribution mechanistically to the improvement of atherosclerosis and CVD. The four predominant compounds – insulin resistance, visceral adiposity, atherogenic dyslipidemia, and endothelial disorder makes the exact meaning for the metabolic syndrome. Within the given below segment we will discuss these compounds.

3.2.1.1 INSULIN RESISTANCE

Insulin is produced through the beta cells of the pancreas in response to increasing level of blood glucose and stimulates the enzymes responsible for glucose metabolism in distinctive tissues. Insulin resistance, which may be described as a condition of decrease reactiveness to regularly circulating stages of insulin, performs a main function inside, in happening of type 2 DM. Insulin will increase glucose uptake in skeletal muscle, liver, and adipose tissue. Within the skeletal muscle and adipose tissue, insulin stimulates glucose uptake through translocation of the GLUT4 glucose transporter and in liver GLUT2 glucose transporter to the cell surface. Inside the skeletal muscle and liver, insulin stimulates the synthesis of glycogen from glucose and inhibits glycogenolysis. In adipose tissue, insulin inhibits fats breakdown, or lipolysis, and stimulates glucose uptake, leading to a boom in circulating free Fatty Acids (FFAs) that further inhibit the antilipolytic impact of insulin. FFAs inhibit protein kinase activation within the muscle leading to decreased glucose uptake. They increase protein kinase activation inside the liver that promotes gluconeogenesis and lipogenesis. The general impact of all of these modifications is to elevate glucose uptake, reduce circulating glucose levels and increase the regeneration of glucose into the form of glycogen or fat (Kim et al., 2006). In insulin resistance, adipose tissue, skeletal muscle, and liver cells do not respond well to insulin, and circulating glucose levels continue to be excessive, which results in pathology. Insulin resistance additionally reasons an elevation in serum viscosity, inducting a prothrombotic state, and release of pro-inflammatory cytokines from the adipose tissue that come up with the elevated risk of cardiovascular illnesses.

3.2.1.2 VISCERAL ADIPOSITY

Visceral weight problems result in insulin resistance greater than subcutaneous fats, as visceral fat lipolysis delivers FFAs to the liver. Extra accumulation of FFAs results in increased oxidative stress, will increase triglyceride synthesis and the manufacturing of apolipoprotein B (apo B) containing triglyceride, very low-density lipoprotein (LDL) within the liver. Similarly, altered lipid metabolism in liver causes elevated level of small dense LDL cholesterol and decrease in HDL cholesterol level, are indirect consequences of insulin resistance. Further to those FFAs, adipokines, that are made through adipose tissue regulates hyperlink among metabolism and vascular function (Kershaw and Flier, 2004). Visceral adipose tissue also taken into consideration more metabolically active and synthesizes notably higher quantities of bioactive secretory proteins, which includes plasminogen activator inhibitor, which promotes a prothrombotic state, and heparin-binding epidermal growth component like growth factor, which encourages smooth muscle cell proliferation and vascular transforming, and the induction of infection through the production of inflammatory cytokines (IL-6, TNF-α).

3.2.1.3 ATHEROGENIC DYSLIPIDEMIA

Atherogenic dyslipidemia, an element of metabolic syndrome, is indicated via increased levels of apo B-containing lipoproteins, consisting of very-low-density lipoprotein (VLDL) remnants and small low-density lipoprotein (LDL) particles, and decreased levels of excessive-density lipoprotein cholesterol (HDL-C) is commonly observed in patients with weight problems, insulin resistance, and type 2 DM. All of the factors of atherogenic dyslipidemia were mentioned as unbiased threat element for cardiovascular disease (Kathiresan et al., 2006).

3.2.1.4 ENDOTHELIAL DYSFUNCTION (ED)

Endothelial dysfunction (ED), is the initial step in the process of vascular disease, is frequently present in patients with diabetes or metabolic syndrome (MetS). Endothelial cells line the internal surface of blood vessels and function an essential mechanical and biological feature. The endothelium senses and responds to physiological and pathological stimuli, and produces vasoactive substances, consisting of nitric oxide (NO), prostacyclin, and endothelins. Normal endothelial function protects against inflammation, circulating platelets, hemostasis, and thrombosis. ED can be defined as the partial or entire lack of stability among vasoconstrictors and vasodilators, increase promoting and inhibiting factors, pro-atherogenic, and anti-atherogenic factors, and pro-coagulant and anticoagulant factors (Després and Brewer, 2008). ED, happens whilst the endothelium fails to serve its regular physiological and protecting mechanisms. Various probabilities exist that hyperglycemia, hyperinsulinemia, increased oxidative stress, and diabetic dyslipidemia can all contribute to ED individually or in co-relation with one another. ED in the brachial artery highly correlates with ED in the coronary circulation, which is coming up as an independent threat factor for cardiovascular disease.

3.2.2 MOLECULAR MECHANISM OF METABOLIC SYNDROME (METS)

Various pathophysiologic factors that may lead to the development of MetS, as mention, circulating fatty acids accumulation and adiposity are the main factors for insulin resistance (Montague and O’rahilly, 2000; Taniguchi et al., 2006; Barazzoni et al., 2018). Insulin resistance is a state of reduced receptivity to insulin action. Fats accumulation in abdominal adipose tissue, liver, skeletal muscle, coronary heart, and pancreas has been proven to release elevated quantities of tumor necrosis factor-alpha (TNF)-α, free fatty acids, as well as adipocytokines or adipokines such as plasminogen activator inhibitor-1 (PAI-1), resistin, and leptin (Friedman and Halaas, 1998; Leyva et al., 1998; Matsuzawa et al., 1999; Niemann et al., 2017; Reho and Rahmouni, 2017; Guzmán-Gerónimo et al., 2017; Louwen et al., 2018). Which without delay have an effect on insulin signaling, decreases glucose uptake within the muscle, drives exaggerated triglyceride synthesis and results in gluconeogenesis. Present-day excessive incidence of obesity can be defined through evolutionary stress for choice of genes promoting fat accumulation to survive in hunger. Insulin resistance regularly coexists with vital obesity, high blood pressure, and dyslipidemia, which have together been denoted as metabolic syndrome. These manifestations constitute robust threat factors for DM type 2 and cardiovascular ailment.

3.3 BIOACTIVE COMPOUNDS IN METABOLIC SYNDROME

It is stated that MetS and all its clinical implications are worldwide spread, and it is immensely necessary to find a logical therapeutic approach to prevent it. The preliminary healing approach implemented for the prevention of MetS is to change in eating behavior and way of life. It's miles clearly recognized that refinement in ingesting habits, with a depletion in caloric intake in the case of obese and weight problems, a reduction in the consumption of sodium, saturated fats, cholesterol, and simple sugars, can assist in the medical governance of MetS co-morbidities (Di Daniele et al., 2021). As there isn't always a single drug focused towards MetS, it's miles often essential a poly-pharmaceutical remedy that could have aspect outcomes and reduction in the affected person condition. Because of this, it is far crucial to discover an herbal method, freed from side outcomes, that can help the MetS patients to impede the beginning and development of co-morbidities for this syndrome (Rochlani et al., 2017). Now ever days, lots of research have called attention to how bioactive compounds can play a pivotal role in delaying the continuation of MetS, applying its long-term beneficial outcomes.

Bioactive compounds are plant based secondary metabolites present in foods able to regulate metabolic activities. Lots of clinical research on phytochemicals (such as polyphenols, protein or peptides, polysaccharides, saponins, and vitamins; found in vegetables, fruits, medicinal plants, and aromatic plants) recommends their possibility as healing and preventive sellers towards illnesses which include inflammation, oxidative stress, dyslipidemia, and insulin resistance (IR). In some other manner it is able to be said that, the term functional food encompasses the food which isn't always only nutritionally essential for our frame, however it's also concerned in resolving disease condition of the body (Jarouliya et al., 2012). It has diverse mechanisms of action, which include reducing elevated blood total cholesterol (TC) level and low-Density Lipoprotein (LDL) cholesterol level, beyond the delivery of simple nutrition. A fundamental precept of nutrients and metabolism is that body weight (BW) exchange is related to an imbalance among the energy gain and energy loss. On this basis, it's far regularly and simplistically theorized that a few human beings turn out to be obese genuinely due to the consumption of an excessive amount of food and workout too little (Urmila and Raj, 2019). Even though this is practically genuine, different individuals' energy balance have to be taken into consideration and want a higher expertise. As an example, diet composition, nutrient bioavailability, and bioactive ought to have a function in energy balance (Bordoni et al., 2019).

Widespread epidemiologic research recommends that eating regimen behavior and healthful life style could have a preventive impact on chronic ailments like cardiovascular disorder and however, bad behavior aggravates them. Consistent with the document by using WHO, those populations eating a massive quantity of fruits; vegetables in addition to sea meals, have been much less at risk of CVD cases (Meier et al., 2019; Raj et al., 2015). This finding introduced a revolution in human being's food regimen, they have become extra involved in their per day intake, consumption of that no longer most effective fulfill their dietary demand. However, additionally alter the level of cholesterol, as a result in the long run lowering the threat of cardiovascular illnesses (Boussageon et al., 2011). Food bioactives are purported to exert useful outcomes on abdominal obesity, glucose intolerance, dyslipidemia, and high blood pressure, consequently bioactives represents alternative remedy for the treatment of MetS. The bioactive compounds of plant (usually) and animal (fish) origin are mentioned in this chapter.

3.3.1 POLYPHENOLS

Polyphenols are natural bioactive material and are a large group of secondary metabolites of plants that are usually intricate in defense mechanisms against various activities such as antioxidant, anti-inflammatory, anticancer, antidiabetic, and antimicrobial properties (Arulselvan et al., 2014; Francesco et al., 2019). Flavonoids (Keservani and Sharma, 2014) are the important class of polyphenols segregated from natural plants that are helpful in managing metabolic syndrome-related complications. The flavonoids (phenolic compound) found in citrus fruits. These fruits are rich in diverse nutrients, consisting of vitamins A and C, folic acid, and dietary fibers; also, desirable supply of different bioactive compounds, which include coumarins, limonoids, and carotenoids (Ding et al., 2012; Turner and Burri, 2013). Flavonoids and other bioactive compounds act as antioxidants and can defend against oxidative stress associated inflammation process, as a consequence decreasing the threat of macromolecules harm due to the movement of reactive species (free radicals), giving out defense against several neurodegenerative diseases and decreasing the risk of evolving cardiovascular diseases (CDVs) (Benavente-García et al., 1997, 2013; Keservani and Sharma, 2014).

The grape juice in raw form can also confer a variety of healthful consequences, which include the cardio-protective impact and curing CVDs. The bioactive compounds in grape juice have an effect on blood pressure, oxidative stress, inflammation, formation of fatty streaks, endothelial functioning, and blood glucose level. Studies found out the impact of bioactive compounds in grape juice on LDL cholesterol, atherosclerosis, and blood lipid profile (Ahmadi et al., 2019). The polyphenols observed in red wine are gallic acid, caffeic acid, stilbenes additionally referred to as trans-resveratrol, p-coumaric acid and other flavonoids, e.g., quercetin, myricetin, rutin, epicatechin, and catechin (Ditano-Vázquez et al., 2019). Examine performed among the grape juice and red wine in terms of health promoting advantages (Barbalho et al., 2020), researchers observed that the grape juice and red wine decreased the body mass index (BMI), peroxidation of serum lipids, glucose level, waistline, blood pressure, TC, LDL-C, and many others (Figure 3.2).

FIGURE 3.2 Food rich in polyphenols, anthocyanins, and pomegranate.

3.3.2 ANTHOCYANINS

Anthocyanins are water-soluble bioactive compounds belongs to flavonoids family, broadly present in many cereals, vegetables, and fruits, along with grapes, plums, cherries, cranberries, strawberries, blueberries, blackberries (Keservani et al., 2016), elderberries, currants, beetroot, purple cabbage, and red onions. This compound has an antioxidant asset because of its capability to transport electrons or to donate the hydrogen atoms from numerous hydroxyl groups to free radicals (Wang et al., 1999; Huang et al., 2014). Numerous studies carried out on blueberries (anthocyanins) indicated its preventive and healing properties, which include the diminishing of oxidative stress and inflammatory responses (Lau et al., 2007), protective towards high blood pressure, diabetes, and cardiovascular problems (Figure 3.2).

3.3.3 POMEGRANATE

Pomegranate is an ancient fruit and in spite of a part of the food plan within the Mediterranean vicinity, the Middle East, and India. Pomegranate is wealthy in polyphenols together with ellagitannins and anthocyanin, and consequently it's been proposed within the prevention and remedy of MetS. Furthermore, *in vivo,* and *in vitro* research have proven that pomegranate exerts hypoglycemic results, which include elevated insulin sensitivity, inhibition of α-glucosidase, and effect on glucose transporter type 4 (GLUT4) receptor, however, is also liable for a reduction of total cholesterol, and the enhancement of blood lipid profiles, in addition to anti-inflammatory effects via the reconversion of peroxisome proliferator-activated receptor (PPAR) pathways. These outcomes can also provide an explanation for how pomegranate-derived compounds results in the improvement of destructive health effects because of metabolic syndrome (Svjetlana and Alois, 2013; Moazzen et al., 2017) (Figure 3.2).

3.3.4 FLAXSEED

Flaxseed, is also referred to as linseed. Flaxseed is cultivated all around the world, as it is the richest source of n–3, or omega-3 (α-linolenic acid (ALA)). ALA is an important polyunsaturated fatty acid depending on foods consumption that comprise it, because it can't be synthesized through the human body. Flaxseed is likewise composed of different functional bioactive compounds other than ALA, including lignans, cyclic peptides, cyanogenic glycosides (CGs), phytosterogenic lignans (secoisolariciresinol diglycoside-SDG), flax proteins, and dietary fibers. These compounds of flaxseed may also be liable for the cardiovascular outcomes (Parikh et al., 2018).

Studies on nutritional supplementation of flaxseed and its bioactive compounds had been carried out very well during the last many years for his or her beneficial activities on cardiovascular (Keservani et al., 2016) risk factors and other clinical conditions which includes hypertensive, cancer, and hormone alternative remedy (Khalesi et al., 2015). The cardiovascular results of dietary flaxseed have covered an antihypertensive motion through inhibition of soluble epoxide hydrolase (sEH) activity, antiatherogenic results by reducing inflammatory cytokine formation, antiplatelet movement via inhibition of antiaggregatory prostacyclin, antiarrhythmic agent by means of decreasing VF, and a development in cardiac cell survival by using upregulating antiapoptotic proteins (Bassett et al., 2009; Tomaz et al., 2011; Caligiuri et al., 2014). It is enriched with the omega-3 fatty acid (-linolenic acid) and the antioxidant (lignan secoisolariciresinol diglucoside), in addition to its high fiber content material and have been advantageous for the cardiovascular functions (Figure 3.3(a)).

3.3.5 SOY PROTEINS

Soy proteins, weight problems, and CVD are not unusual physiological issues in the world (Basharat, 2020). Food and Drug Administration (FDA) accepted soy protein as a useful meal to lessen the danger of cardiovascular illnesses an extended in the past. A meta-analysis based on 46 research through FDA proved that soy protein confers coronary heart defensive function via lowering notably total cholesterol and LDL within the blood of adult populace (Blanco et al., 2019). Soy-protein consists of plant materials like isoflavones, just like estradiol, is liable for reducing of cholesterol. Similarly, to isoflavones, soy-proteins comprise saponins, phytic acids, trypsin inhibitors, fibers, and globulins. These chemical compounds make soy protein a useful purposeful meal because it decreases blood lipids, progressed arterial compliances, lipid metabolism, and decrease the LDL-oxidation (Gil-Izquierdo et al., 2012). Food enriched with soybean has been suggested conferring lipid reducing impact and hence lowering CVDs and obesity as well. Soya bean and soy-based products have bioactive compounds which include, globulin proteins such as glycinin and beta-conglycinin, oligosaccharides, phytosterols (campesterol, sitostanol, and sitosterol), and antioxidant compounds. They have been mentioned to confer health-promoting useful outcomes on both humans and animals. The food supplements made from soybean protein, glycine max has been observed effective towards CVDs via enhancing triglycerides and lipids profile as well as obesity (Basharat, 2020) (Figure 3.3(b)).

3.3.6 OAT FIBER

Oat fiber is likewise referred to as viscous soluble fiber's, carries β-glucan (Joyce et al., 2019). It reduces the LDL within the blood and confers safety towards CVDs. It also consists of phenolic compounds, xyloglucan, arabinoxylan, insoluble fiber, lipids, protein, vitamins, and minerals. It reduces the blood lipid through decreasing the absorption of cholesterol and bile acids or via delaying the lipid digestion. Some other mode of reducing cholesterol in blood was observed upward thrust in thickness of gut contents (Wolever et al., 2010), which will increase the excretion of cholesterol and bile salts from the gut. Now a day, lots of research work has indicated that not only the viscous properties of β-glucan was effective in reducing cholesterol level. However, they may also helpful in preserving gut microbiota, the cholesterol homeostasis inside the gut (Jonsson and Bäckhed, 2017; Joyce et al., 2019). The Oatmeal or Oat β-glucan regulates the intestine microbiota as properly. The gut microbes metabolize the bile acids (precursor of that is cholesterol) and therefore will increase the excretion of cholesterol from the body together with the faces, therefore lowering the serum cholesterol level in experimental animals and human beings (Wolever et al., 2010) (Figure 3.3(c)).

3.3.7 FISH OIL

Fish oil, intake of fish and CVD have an opposite connection, and this was pronounced almost 30 years in the past while Greenland Eskimos had low rates of this CVDs, despite consuming a food regimen containing about 7 g/day of omega-3 (n–3) fatty acids from marine fish (Saravanan et al., 2010). Oily fish (consists of sardines, herring, salmon, trout, tuna, swordfish, and mackerel) and its oils are the primary resources of the omega-3 (n-three) fatty acids eicosapentaenoic acid (EPA) and docosahexaenoic acid (DHA) (Behera, 2019). Omega three fatty acid supplementation has been indicated to have beneficial results on lipid profile, cytokine cascade, oxidant, and antioxidant balance, parasympathetic, and sympathetic tone, and nitric oxide synthesis (Peter et al., 2013). Intake of 1 gram/day EPA and DHA is usually recommended to enhance cardiac feature through their anti-triglyceridemic, antihypertensive, hemostatic, antiarrhythmic (Di Minno et al., 2010; Kromhout et al., 2012), anti-atherogenic effects and improved endothelial function (De Caterina, 2011). Consumption of slight quantity of fish (one to two servings per week) has been linked with decreased CVD associated deaths, therefore it acts as a remedy (Einvik et al., 2010) and additionally decreases the metabolic risk elements for coronary heart diseases and the associated mortalities remarkably in a certain period of time (Ameen et al., 2020) (Figure 3.3(d)).

3.3.8 GARLIC

Garlic, is ranked as the top natural product utilized all around the world. As a long way, it was considered the excellent functional food. It is useful, has health promoting outcomes and effective in treating atherosclerosis and heart diseases. The intact garlic bulb consists of odorless amino acid called allicin, while when the garlic is crushed, the enzymes alliinase releases and convert the alliin into allicin. Allicin is the compound that offers a function odor to the garlic, then spontaneously, the allicin decomposes into many sulfur containing compounds (Beato et al., 2011). These compounds provide it a medicinal property by means of reducing the cholesterol level and blood stress in cardiovascular sufferers (Omar and Al-Wabel, 2010). The garlic has notable capacity to decrease the cholesterol level in blood, retaining lipid profile, lessening LDL oxidation, increases the fibrinolysis and reduce fibrinogen level within the blood (Sobenin et al., 2019) (Figure 3.3(e)).

3.3.9 GINGER

Ginger, has been cultivated for medicinal and cooking purposes for as a millennium. It is far established that ginger has the ability to lessen the coronary heart associated problems (Shoaib et al., 2016). It inhibits the formation of Nitric oxide (NO), inflammatory cytokines, cyclo-oxygenases

(COX) and Lipoxygenases (LOX) which oxidizes the arachidonic acid and causes the arterial inflammation (Mojani et al., 2014). Numerous studies have mentioned the lipid-lowering impact of ginger (Shoaib et al., 2016); a meta-analysis based examine noted the beneficial effect of ginger on Triglycerol level and LDL-C, at the same time as having no effect on TC and HDL-C. Furthermore, the low dose of ginger was determined to have greater lessening effect on total cholesterol and triglyceride levels (Pourmasoumi et al., 2018) (Figure 3.3(f)).

3.3.10 CURCUMIN

Curcumin, a natural yellow coloration phenol acquired from the rhizome of Curcuma longa. Curcumin is a multifunctional phytopharmaceutical compound, having beneficial effects on several conditions that include compounds of MetS and cardiovascular risk factors (Visioli and Davalos, 2011; Shureiqi and Baron, 2011; Panahi et al., 2014a, 2015a; Sahebkar et al., 2013; Mohammadi et al., 2013). The medical outcomes of curcuminoids supplementation on pulmonary feature and systemic inflammation had been investigated via Sahebkar, who investigated that curcuminoids had a great impact on several anti-inflammatory mediators under examine that incorporates: IL-6, IL-8, TNF, and hs-CRP (Sahebkar et al., 2013; Panahi et al., 2014b, 2015b). In a examine via Di Pierro et al. (2015), they found that a bioavailable form of curcumin ended in progressed weight control in obese subjects. Further, consumption of curcumin in conjunction with a low-calorie diet, elevates the loss of BW, decreasing fat mass (FM) %, BMI, and body circumferences (Mousavi et al., 2020; Akbari et al., 2019). Some other examine carried out by Panahi et al. (2016) resulted that curcumin improves the expression of adiponectin, a robust anti-inflammatory agent produced through adipose tissue involve in an anti-inflammatory pathway. After the consumption of 70 mg/day curcumin for 8 weeks among human beings with non-alcoholic fatty liver disorder (NAFLD), excellent results were observed, which showed a reduction in BMI and liver fat level (Rahmani et al., 2016). Even though, anthropometric parameters which includes BMI, weight, waist, and hip circumference, total body fats, and parameters of oxidative stress had been not affected following the consumption of curcumin at a dosage of 1 g/day in overweight (Mohammadi et al., 2013) as well as in MetS (Ghazimoradi et al., 2017) persons.

In a current *in vitro* examination, Wu et al. (2019) spotlighted that curcumin has the capability to disrupt the differentiation of adipocytes, and simultaneous induces apoptosis of pre-adipocytes. The bioavailability of curcumin after oral administration is very low, because of poor intestinal absorption, this interference can be corrected through the use of the particular inhibitors of the curcumin metabolism that enhance its absorption and sluggish down its removal (Dei Cas and Ghidoni, 2019). Furthermore, in certain Southeast Asian groups, the maximum dietary intake of 1.5 g per individual per day had recommended, without any harmful effect (Figure 3.3(g)).

3.3.11 CINNAMON

Cinnamon, called 'dalchini' in Hindi, farmed from historic instances within the Mediterranean area, Sri Lanka, and India, has been used as a spice and flavoring agent, and is frequently utilized in Turkish, Persian, Chinese, and Indian traditional to provide flavor to curries and different food objects (Gupta et al., 2017). On the basis of latest clinical documentation cinnamon and its biological active components which includes cinnamaldehyde, cinnamate, cinnamic acid and eugenol in the forms of aqueous and alcoholic extracts have a plenty of therapeutic functions (Mollazadeh and Hosseinzadeh, 2016). Many researchers mentioned that cinnamon's biologically active compounds exert antimicrobial, antiviral, antioxidant, antitumor, antihypertension, antilipemic, antidiabetic, gastroprotective, and immunomodulatory (Shen et al., 2012) activity and additionally effective towards multiple sclerosis, Alzheimer, and human immune deficiency (HIV) infection (Rao et al., 2014; Ranasinghe et al., 2013). Discrete factors of MetS consisting of high blood glucose, dyslipidemia, obesity, and high blood pressure are stepped forward with cinnamon extracts. Studies with cinnamon found out that this plant is a cardiovascular defensive agent and has a probable effect in diminishing MetS complications because of its antidiabetic, antioxidant, anti-inflammatory, and anti-lipidemic properties (Couturier et al., 2010; Shen et al., 2012). Even though, the water-soluble polyphenol compounds found in cinnamon which show insulin-potentiating, antioxidant, and associated activities are Type A doubly associated procyanidin oligomers of the catechins and epicatechin. Similarly, methyl chalcone polymer in cinnamon improves the lipase activity that hydrolyzes nutritional fat molecules, increases glycogen synthesis in liver, increase glucose uptake and phosphorylation of insulin receptor in skeletal muscle tissue and adipocytes (Gupta et al., 2017) (Figure 3.3(h)).

FIGURE 3.3 Sources of bioactive compounds: (a) flaxseed; (b) soy proteins; (c) oatmeal; (d) fish oil supplements; (e) Indian garlic; (f) Indian ginger; (g) Indian curcumin; (h) cinnamon.

3.4 CONCLUSION

Metabolic syndrome (MetS) is a group of situations that arise collectively, these situations encompass increased blood pressure, high blood sugar, extra body fats across the waist, and peculiar cholesterol or triglyceride levels. Preventive measures and therapeutic strategies, which includes modifications in way of life which include nutrition interventions and physical activity, are the feasible approaches to keep away from the onset of MetS. The improvement of nutraceuticals which are effectively available and with minimum side effects might also constitute an area of promise within the development of novel remedies. Consequently, research information indicates that bioactive compounds perform an essential role within the clinical control of MetS and its complications. Their long-time intake gives plenty of useful results including BW control, antidiabetic, antioxidant, anti-inflammatory, anti-lipidemic, anti-tumor, anti-hypertension, and immunomodulatory effect (Keservani et al., 2010a, b). Therefore, this proof indicates that the usage of food products wealthy in bioactive compounds might be a legitimate preventive and therapeutic complementary remedy for the treatment of MetS.

KEYWORDS

- **apolipoprotein B**
- **bioactive compounds**
- **cardiovascular diseases**
- **endothelial dysfunction**
- **functional foods**
- **metabolic syndrome**

REFERENCES

Ahmadi, L., El-Kubbe, A., & Roney, S. K., (2019). Potential cardio-protective effects of green grape juice: A review. *Curr. Nutr. Food Sci., 15*, 202–207.

Akbari, M., Lankarani, K. B., Tabrizi, R., Ghayour-Mobarhan, M., Peymani, P., Ferns, G., Ghaderi, A., & Asemi, Z., (2019). The effects of curcumin on weight loss among patients with metabolic syndrome and related disorders: A systematic review and meta-analysis of randomized controlled trials. *Front. Pharmacol., 10*, 649.

Alberti, K. G., & Zimmet, P. Z., (1998). Definition, diagnosis and classification of diabetes mellitus and its complications. Part 1: Diagnosis and classification of diabetes mellitus provisional report of a WHO consultation. *Diabet Med., 15*, 539–553.

Arulselvan, P., Ghofar, H. A. A., Karthivashan, G., Halim, M. F. A., Ghafar, M. S. A., & Fakurazi, S., (2014). Antidiabetic therapeutics from natural source: A systematic review. *Biomed. Prev. Nutr., 4*, 607–617.

Barazzoni, R., Cappellari, G. G., Ragni, M., & Nisoli, E., (2018). Insulin resistance in obesity: An overview of fundamental alterations. *Eat. Weight Disord., 23*, 149–157.

Barbalho, S. M., Bueno, O. A. M. M., Fiorini, A. M. R., Guiguer, É. L., Nicolau, C. C. T., Goulart, R. D. A., & Flato, U. A. P., (2020). Grape juice or wine: Which is the best option? *Crit. Rev. Food Sci. Nutr., 2020*, 1–14.

Basharat, S., (2000). Therapeutic effect of glycine max (soybean) bioactive components in Cvd and obesity. *J. Food Nutr., 6*, 1–7.

Bassett, C. M., Rodriguez-Leyva, D., & Pierce, G. N., (2009). Experimental and clinical research findings on the cardiovascular benefits of consuming flaxseed. *Appl. Physiol. Nutr. Metab., 34*, 965–974.

Beato, V. M., Orgaz, F., Mansilla, F., & Montaño, A., (2011). Changes in phenolic compounds in garlic (*Allium sativum* L.) owing to the cultivar and location of growth. *Plant Foods Hum. Nutr., 66*, 218–223.

Behera, S. S., (2019). Dietary fish oil concentrates associated health benefits: A recent development of cardiovascular risk reduction. *Curr. Pharm. Des., 25*, 4053–4062.

Benavente-García, O., Castillo, J., Marin, F. R., Ortuño, A., & Del, R. J. A., (1997). Uses and properties of citrus flavonoids. *J. Agric. Food Chem., 45*(12), 4505–4515.

Blanco, M. S., Messina, M., Li, S. S., Viguiliouk, E., Chiavaroli, L., Khan, T. A., Srichaikul, K., et al., (2019). A meta-analysis of 46 studies identified by the FDA demonstrates that soy protein decreases circulating LDL and total cholesterol concentrations in adults. *J. Nutr., 149*(6), 968–981.

Bordoni, A., Boesch, C., Malpuech-Brugère, C., Orfila, C., & Tomás-Cobos, L., (2019). The role of bioactives in energy metabolism and metabolic syndrome. *Proc. Nutr. Soc., 78*, 340–350.

Boussageon, R., Bejan-Angoulvant, T., Saadatian-Elahi, M., Lafont, S., Bergeonneau, C., Kassaï, B., Erpeldinger, S., et al., (2011). Effect of intensive glucose lowering treatment on all-cause mortality, cardiovascular death, and microvascular events in type 2 diabetes: Meta-analysis of randomized controlled trials. *BMJ., 26*, 343d4169.

Caligiuri, S. P., Aukema, H. M., Ravandi, A., & Pierce, G. N., (2014). Elevated levels of pro-inflammatory oxylipins in older subjects are normalized by flaxseed consumption. *Exp. Gerontol., 59*, 51–57.

Caligiuri, S. P., Edel, A. L., Aliani, M., & Pierce, G. N., (2014). Flaxseed for hypertension: Implications for blood pressure regulation. *Curr. Hypertens. Rep., 16*, 499–508.

Couturier, K., Batandier, C., Awada, M., Hininger-Favier, I., Canini, F., Anderson, R. A., Leverve, X., & Roussel, A. M., (2010). Cinnamon improves insulin sensitivity and alters the body composition in an animal model of the metabolic syndrome. *Arch. Biochem. Biophys., 501*, 158–161.

De Caterina, R., (2011). N–3 fatty acids in cardiovascular disease. *N. Engl. J. Med., 364*, 2439–2450.

Dei Cas, M., & Ghidoni, R., (2019). Dietary curcumin: Correlation between bioavailability and health potential. *Nutrients., 11*, 2147.

Després, J. P., & Brewer, H. B., (2008). Metabolic syndrome: The dysmetabolic state of dysfunctional adipose tissue and insulin resistance. *Eur. Heart J., 10*, B1–3.

Desroches, S., & Lamarche, B., (2007). The evolving definitions and increasing prevalence of the metabolic syndrome. *Appl. Physiol. Nutr. Metab., 32*, 23–32.

Di Daniele, N., Marrone, G., Di Lauro, M., Di Daniele, F., Palazzetti, D., Guerriero, C., & Noce, A., (2021). Effects of caloric restriction diet on arterial hypertension and endothelial dysfunction. *Nutrients, 13*, 274.

Di Minno, M. N., Tremoli, E., Tufano, A., Russolillo, A., Lupoli, R., & Di Minno, G., (2010). Exploring newer cardioprotective strategies: ω-3 fatty acids in perspective. *Thromb. Haemost., 104*, 664–680.

Di Pierro, F., Bressan, A., Ranaldi, D., Rapacioli, G., Giacomelli, L., & Bertuccioli, A., (2015). Potential role of bioavailable curcumin in weight loss and omental adipose tissue decrease: Preliminary data of a randomized, controlled trial in overweight people with metabolic syndrome. *Eur. Rev. Med. Pharmacol. Sci., 19*, 4195–4202.

Ding, X., Fan, S., Lu, Y., Zhang, Y., Gu, M., Zhang, L., Liu, G., et al., (2012). *Citrus ichangensis* peel extract exhibits anti-metabolic disorder effects by the inhibition of PPARγ and LXR signaling in high-fat diet-induced C57BL/6 mouse. *Evid. Based Complement. Alternat. Med.*, 678592.

Ditano-Vázquez, P., Torres-Peña, J. D., Galeano-Valle, F., Pérez-Caballero, A. I., Demelo-Rodríguez, P., Lopez-Miranda, J., Katsiki, N., et al., (2019). The fluid aspect of the Mediterranean diet in the prevention and management of cardiovascular disease and diabetes: The role of polyphenol content in moderate consumption of wine and olive oil. *Nutrients, 11*, 2833.

Einvik, G., Ole, K. T., Sandvik, L., & Hjerkinn, E. M., (2010). A randomized clinical trial on n–3 polyunsaturated fatty acids supplementation and all-cause mortality in elderly men at high cardiovascular risk. *Eur. J. Cardiovasc. Prev. Rehabil., 17*, 588–592.

Expert Panel on Detection, Evaluation, and Treatment of High Blood Cholesterol in Adults, (2001). Executive summary of the third report of the national cholesterol education program (NCEP) expert panel on detection, evaluation, and treatment of high blood cholesterol in adults (Adult Treatment Panel III). *JAMA, 285*, 2486–2497.

Ford, E. S., (2005). Prevalence of the metabolic syndrome defined by the international diabetes federation among adults in the. U.S. *Diabet. Care, 28*, 2745–2749.

Francini-Pesenti, F., Spinella, & Calò, L. A., (2019). Potential role of phytochemicals in metabolic syndrome prevention and therapy. *Diabetes Metab. Syndr. Obes., 12*, 1987–2002.

Friedman, J. M., & Halaas, J. L., (1998). Leptin and the regulation of body weight in mammals. *Nature, 395*, 763–770.

Garcia-Mora, P., & Ferreres, F., (2012). Soy isoflavones and cardiovascular disease epidemiological, clinical and -omics perspectives. *Curr. Pharm. Biotechnol., 13*, 624–631.

Ghazimoradi, M., Saberi-Karimian, M., Mohammadi, F., Sahebkar, A., Tavallaie, S., Safarian, H., Ferns, G. A., et al., (2017). The effects of curcumin and curcumin–phospholipid complex on the serum pro-oxidant–antioxidant balance in subjects with metabolic syndrome. *Phytother. Res., 31*, 1715–1721.

Gil-Izquierdo, A., Penalvo, J. L., Gil, J. I., Medina, S., Horcajada, M. N., Lafay, S., Silberberg, M., et al., (2016). Curcumin on serum adipokine concentrations: A randomized controlled trial. *Nutrition, 32*, 1116–1122.

Grundy, S. M., Brewer, H. B. Jr., Cleeman, J. I., Smith, S. C. Jr., & Lenfant, C., (2004). Definition of metabolic syndrome: Report of the national heart, lung, and blood institute/ American heart association conference on scientific issues related to definition. *Arterioscler. Thromb. Vasc. Biol., 24*(2), e13–18.

Gupta, J. S., Puri, S., Misra, A., Gulati, S., & Mani, K., (2017). Effect of oral cinnamon intervention on metabolic profile and body composition of Asian Indians with metabolic syndrome: A randomized double -blind control trial. *Lipids Health Dis., 16*, 113.

Guzmán-Gerónimo, R. I., Alarcon-Zavaleta, T. M., Oliart-Ros, R. M., Meza, A. J. E., Herrera-Meza, S., & Chavez-Servia, J. L., (2017). Blue maize extract improves blood pressure, lipid profiles, and adipose tissue in high-sucrose diet-induced metabolic syndrome in rats. *J. Med. Food, 20*, 110–115.

Henao-Mejia, J., Elinav, E., Jin, C., Hao, L., Mehal, W. Z., Strowig, T., Thaiss, C. A., et al., (2012). Inflammasome-mediated dysbiosis regulates progression of NAFLD and obesity. *Nature, 482*, 179–185.

Huang, W. Y., Liu, Y. M., Wang, J., Wang, X. N., & Li, C. Y., (2014). Anti-inflammatory effect of the blueberry anthocyanins malvidin-3-glucoside and malvidin- 3-galactoside in endothelial cells. *Molecules, 19*, 12827–12841.

Jablonski, S., (1991). Syndrome: Le mot de jour. *Am. J. Med. Genet., 39*, 342–346.

Jarouliya, U., & Keservani, R. K., (2019). Pathways leading to childhood obesity: an overview. *Global Perspectives on Childhood Obesity* (2nd edn.) Academic Press, USA.

Jarouliya, U., & Keservani, R. K., (2019). Protein function as cell surface and nuclear receptor in human diseases. In: Sanjay, B., (ed.), *Drug Research Reviews*. Apple Academic Press, CRC Press, Taylor & Francis Group.

Jarouliya, U., Zacharia, J. A., Kumar, P., Bisen, P. S., & Prasad, G. B., (2012). Alleviation of metabolic abnormalities induced by excessive fructose administration in Wistar rats by spirulina maxima. *Indian J. Med. Res., 135*, 422–428.

Jonsson, A. L., & Bäckhed, F., (2017). Role of gut microbiota in atherosclerosis. *Nat. Rev. Cardiol., 14*, 79–87.

Joyce, S. A., Kamil, A., Fleige, L., & Gahan, C. G. M., (2019). The cholesterol-lowering effect of oats and oat beta-glucan: Modes of action and potential role of bile acids and the microbiome. *Front. Nutr., 27*, 171.

Kathiresan, S., Otvos, J. D., Sullivan, L. M., Keyes, M. J., Schaefer, E. J., Wilson, P. W., D'Agostino, R. B., et al., (2006). Increased small low-density lipoprotein particle number: A prominent feature of the metabolic syndrome in the Framingham heart study. *Circulation, 113*, 20–29.

Kershaw, E. E., & Flier, J. S., (2004). Adipose tissue as an endocrine organ. *J. Clin. Endocrinol. Metab., 89*, 2548–2556.

Keservani, R. K., & Sharma, A. K., (2014). Flavonoids: Emerging trends and potential health benefits. *J. Chin. Pharm. Sci., 23*(12), 815–822.

Keservani, R. K., Kesharwani, R. K., & Sharma, A. K., (2016). *Medicinal Effect of Nutraceutical Fruits for the Cognition and Brain Health* (pp. 1–10). Scientifica. Article ID 3109254. Doi: 10.1155/2016/3109254.

Keservani, R. K., Kesharwani, R. K., Sharma, A. K., & Jarouliya, U., (2015). Dietary supplements, nutraceutical and functional foods in immune response (immunomodulators). In: Bagchi, D., Preuss, H. G., & Swaroop, A., (eds.), *Nutraceutical and Functional Foods in Human Life and Disease Prevention* (pp. 343–358). Boca Raton, Fl. Taylor and Francis; Ch 20. ISBN: 9781482237214.

Keservani, R. K., Kesharwani, R. K., Sharma, A. K., Vyas, N., & Chadoker, A., (2010b). Nutritional supplements: An overview. *Int. J. of Curr. Pharm. Rev. and Res, 1*(1), 59–75.

Keservani, R. K., Kesharwani, R. K., Vyas, N., Jain, S., Raghuvanshi, R., & Sharma, A. K., (2010a). Nutraceuticals and functional food as future food. *Der Pharm. Lett., 2*(1), 106–116.

Keservani, R. K., Sharma, A. K., & Kesharwani, R. K., (2016). Nutraceutical and functional foods for cardiovascular health. In: Meghwal, M., & Goyal, M. R., (eds.), *Food Process Engineering: Emerging Technologies and Innovations in Food Engineering* (pp. 257–278). Apple Academic Press, CRC Press, Taylor and Francis; Ch 10. ISBN: 9781771884020.

Khalesi, S., Irwin, C., & Schubert, M., (2015). Flaxseed consumption may reduce blood pressure: A systematic review and meta-analysis of controlled trials. *J. Nutr., 145*, 758–765.

Kim, J. A., Montagnani, M., Koh, K. K., & Quon, M. J., (2006). Reciprocal relationships between insulin resistance and endothelial dysfunction: Molecular and pathophysiological mechanisms. *Circulation., 113*, 1888–1904.

Kolovou, G. D., Anagnostopoulou, K. K., Salpea, K. D., & Mikhailidis, D. P., (2007). The prevalence of metabolic syndrome in various populations. *Am. J. Med. Sci., 333*, 362–371.

Kromhout, D., Yasuda, S., Geleijnse, J. M., & Shimokawa, H., (2012). Fish oil and omega-3 fatty acids in cardiovascular disease: Do they really work? *Eur. Heart J., 33*, 436-443.

Lau, F. C., Bielinski, D. F., & Joseph, J. A., (2007). Inhibitory effects of blueberry extract on the production of inflammatory mediators in lipopolysaccharide activated BV2 microglia. *J. Neurosci. Res., 85*, 1010–1017.

Leyva, F., Godsland, I. F., Ghatei, M., Proudler, A. J., Aldis, S., Walton, C., Bloom, S., & Stevenson, J. C., (1998). Hyperleptinemia as a component of a metabolic syndrome of cardiovascular risk. *Arterioscler. Thromb. Vasc. Biol., 18*, 928–933.

Medjakovic, S., & Jungbauer, A., (2013). Pomegranate: A fruit that ameliorates metabolic syndrome. *Food Funct., 4*, 19–39.

Meier, T., Gräfe, K., Senn, F., Sur, P., Stangl, G. I., Dawczynski, C., & Lorkowski, S., (2019). Cardiovascular mortality attributable to dietary risk factors in 51 countries in the WHO European region from 1990 to 2016: A systematic analysis of the global burden of disease study. *Eur. J. Epidemiol., 34*, 37–55.

Moazzen, H., & Alizadeh, M., (2017). Effects of pomegranate juice on cardiovascular risk factors in patients with metabolic syndrome: A double-blinded, randomized crossover-controlled trial. *Plant Foods Hum. Nutr., 72*, 126–133.

Mohammadi, A., Sahebkar, A., Iranshahi, M., Amini, M., Khojasteh, R., Ghayour-Mobarhan, M., & Ferns, G. A., (2013). Effects of supplementation with curcuminoids on dyslipidemia in obese patients: A randomized crossover trial. *Phytother. Res., 27*, 374–379.

Mojani, M. S., Ghasemzadeh, A., Rahmat, A., Loh, S. P., & Ramasamy, R., (2014). Assessment of bioactive compounds, nutritional composition and antioxidant activity of Malaysian young ginger (*Zingiber officinale* roscoe). *Int. Food Res. J., 21*, 1931–1935.

Mollazadeh, H., & Hosseinzadeh, H., (2016). Cinnamon effects on metabolic syndrome: A review based on its mechanisms. *Iran J. Basic Med. Sci., 19*, 1258–1270.

Montague, C. T., & O'Rahilly, S., (2000). The perils of portliness: Causes and consequences of visceral adiposity. *Diabetes, 49*, 883–888.

Mousavi, S. M., Milajerdi, A., Varkaneh, H. K., Gorjipour, M. M., & Esmaillzadeh, A., (2020). The effects of curcumin supplementation on body weight, body mass index and waist circumference: A systematic review and dose-response meta-analysis of randomized trials. *Crit. Rev. Food Sci. Nutr., 60*, 171–180.

Niemann, B., Rohrbach, S., Miller, M. R., Newby, D. E., Fuster, V., & Kovacic, J. C., (2017). Oxidative stress and cardiovascular risk: Obesity, diabetes, smoking, and pollution: Part 3 of a 3-part series. *J. Am. Coll. Cardiol., 70*, 230–251.

Omar, S. H., & Al-Wabel, N. A., (2010). Organosulfur compounds and possible mechanism of garlic in cancer. *Saudi Pharm. J., 18*, 51–58.

Panahi, Y., Badeli, R., Karami, G. R., & Sahebkar, A., (2015a). Investigation of the efficacy of adjunctive therapy with bioavailability-boosted curcuminoids in major depressive disorder. *Phytother. Res., 29*, 17–21.

Panahi, Y., Ghanei, M., Bashiri, S., Hajihashemi, A., & Sahebkar, A., (2015b). Short-term curcuminoid supplementation for chronic pulmonary complications due to sulfur mustard intoxication: Positive results of a randomized double-blind placebo-controlled trial. *Drug Res., 65*, 567–573.

Panahi, Y., Hosseini, M. S., Khalili, N., Naimi, E., Soflaei, S. S., Majeed, M., & Sahebkar, A., (2016). Effects of supplementation with curcumin on serum adipokine concentrations: A randomized controlled trial. *Nutrition, 32*, 1116–1122.

Panahi, Y., Khalili, N., Hosseini, M. S., Abbasinazari, M., & Sahebkar, A., (2014a). Lipid-modifying effects of adjunctive therapy with curcuminoidspiperine combination in patients with metabolic syndrome: Results of a randomized controlled trial. *Complement. Ther. Med., 22*, 851–857.

Panahi, Y., Rahimnia, A. R., Sharafi, M., Alishiri, G., Saburi, A., & Sahebkar, A., (2014b). Curcuminoid treatment for knee osteoarthritis: A randomized double-blind placebo-controlled trial. *Phytother. Res., 28*, 1625–1631.

Parikh, M., Netticadan, T., & Pierce, G. N., (2018). Flaxseed: Its bioactive components and their cardiovascular benefits. *Am. J. Physiol. Heart Circ. Physiol., 314*, H146–H159.

Peter, S., Chopra, S., & Jacob, J. J., (2013). A fish a day keeps the cardiologist away! – A review of the effect of omega-3 fatty acids in the cardiovascular system. *Indian J. Endocrinol. Metab., 17*, 422–429.

Pourmasoumi, M., Hadi, A., Rafie, N., Najafgholizadeh, A., Mohammadi, H., & Rouhani, M. H., (2018). The effect of ginger supplementation on lipid profile: A systematic review and meta-analysis of clinical trials. *Phymed., 43*, 28–36.

Rahmani, S., Asgary, S., Askari, G., Keshvari, M., Hatamipour, M., Feizi, A., & Sahebkar, A., (2016). Treatment of non-alcoholic fatty liver disease with curcumin: A randomized placebo-controlled trial. *Phytother. Res., 30*, 1540–1548.

Ranasinghe, P., Pigera, S., Premakumara, G. A., Galappaththy, P., Constantine, G. R., & Katulanda, P., (2013). Medicinal properties of 'true' cinnamon (*Cinnamomum zeylanicum*): A systematic review. *BMC Complement. Altern. Med., 13*, 275.

Rao, P. V., & Gan, S. H., (2014). Cinnamon: A multifaceted medicinal plant. *Evid. Based Complement. Alternat. Med.*, 642942.

Reaven, G. M., (1988). Banting lecture 1988: Role of insulin resistance in human disease. *Diab., 37*, 1595–1607.

Rochlani, Y., Pothineni, N. V., Kovelamudi, S., & Mehta, J. L., (2017). Metabolic syndrome: Pathophysiology, management, and modulation by natural compounds. *Ther. Adv. Cardiovasc. Dis., 11*, 215–225.

Sahebkar, A., Mohammadi, A., Atabati, A., Rahiman, S., Tavallaie, S., Iranshahi, M., Akhlaghi, S., Ferns, G. A. A., & Ghayour-Mobarhan, M., (2013). Curcuminoids modulate pro-oxidant-antioxidant balance but not the immune response to heat shock protein 27 and oxidized LDL in obese individuals. *Phytother. Res., 27*, 1883–1888.

Saravanan, P., Davidson, N. C., Schmidt, E. B., & Calder, P. C., (2010). Cardiovascular effects of marine omega-3 fatty acids. *Lancet, 376*, 540–550.

Shen, Y., Jia, L. N., Honma, N., Hosono, T., Ariga, T., & Seki, T., (2012). Beneficial effects of cinnamon on the metabolic syndrome, inflammation, and pain, and mechanisms underlying these effects – a review. *J. Tradit. Complement. Med., 2*, 27–32.

Shoaib, M., Shehzad, A., Butt, M. S., Saeed, M., Raza, H., Niazi, S., & Shakeel, A., (2016). An overview: Ginger, a tremendous herb. *J. Global Innov. Agric. Social Sci., 4*, 172–187.

Shureiqi, I., & Baron, J. A., (2011). Curcumin chemoprevention: The long road to clinical translation. *Cancer Prev. Res. (Phila), 4*, 296–298.

Sobenin, I. A., Myasoedova, V. A., Iltchuk, M. I., Zhang, D. W., & Orekhov, A. N., (2019). Therapeutic effects of garlic in cardiovascular atherosclerotic disease. *Chin. J. Nat. Med., 17*, 721–728.

Taniguchi, C. M., Emanuelli, B., & Kahn, C. R., (2006). Critical nodes in signaling pathways: Insights into insulin action. *Nat. Rev. Mol. Cell Biol., 7*, 85–96.

Tomaz, P. J., Beltrame, D. J., & Teles, B. G., (2011). Impact of dietary flaxseed (*Linum usitatissimum*) supplementation on biochemical profile in healthy rats. *Nutr. Hosp., 26*, 798–802.

Turner, T., & Burri, B. J., (2013). Potential nutritional benefits of current citrus consumption. *Agriculture, 3*, 170–187.

Visioli, F., & Davalos, A., (2011). Polyphenols and cardiovascular disease: A critical summary of the evidence. *Mini. Rev. Med. Chem., 11*, 1186–1190.

Wang, H., Nair, M. G., Strasburg, G. M., Chang, Y. C., Booren, A. M., Gray, J. I., & DeWitt, D. L., (1999). Antioxidant and anti-inflammatory activities of anthocyanins and their aglycon, cyanidin, from tart cherries. *J. Nat. Prod., 62*, 294–296.

Wolever, T. M., Tosh, S. M., Gibbs, A. L., Brand-Miller, J., Duncan, A. M., Hart, V., & Wood, P. J., (2010). Physicochemical properties of oat β-glucan influence its ability to reduce serum LDL cholesterol in humans: A randomized clinical trial. *Am. J. Clin. Nutr., 92*, 723–732.

Wu, L. Y., Chen, C. W., Chen, L. K., Chou, H. Y., Chang, C. L., & Juan, C. C., (2019). Curcumin attenuates adipogenesis by inducing preadipocyte apoptosis and inhibiting adipocyte differentiation. *Nutrients, 11*, 2307.

CHAPTER 4

Preventive Measures of Type 2 Diabetes via Nutrition

EKNATH D. AHIRE,[1] VIJAYRAJ N. SONAWANE,[2]
KHEMCHAND R. SURANA,[2] SWATI G. TALELE,[3] GOKUL S. TALELE,[4]
SANJAY J. KSHIRSAGAR,[1] SHUBHAM J. KHAIRNAR,[5]
NILIMA A. THOMBRE,[1] and SUNIL K. MAHAJAN[6]

[1]Department of Pharmaceutics, MET's Institute of Pharmacy, Bhujbal Knowledge City (BKC), Affiliated to Savitribai Phule Pune University, Adgaon, Nashik, Maharashtra – 422003, India, Tel.: +91-9588431965, Fax: (02552)-303-515, E-mail: eknathahire05@gmail.com (E. D. Ahire)

[2]Department of Pharmaceutical Chemistry, Shreeshakti Shaikshanik Sanstha, Divine College of Pharmacy, Satana, Nashik, Maharashtra, India

[3]Department of Pharmaceutics, Sandip Institute of Pharmaceutical Sciences, Nashik, Maharashtra – 422213, India

[4]Department of Pharmaceutics, Matoshri College of Pharmacy, Near Odhagaon, Eklahare, Nashik, Maharashtra – 422105, India

[5]Department of Pharmacology, MET's Institute of Pharmacy, Bhujbal Knowledge City (BKC), Affiliated to Savitribai Phule Pune University, Nashik, Maharashtra – 422003, India

[6]Department of Pharmaceutical Chemistry, MGV's Pharmacy College, Panchvati, Nashik, Maharashtra, India

ABSTRACT

Over a couple of decades, proof from potential observational research and medical trials has converged to help the significance of character nutrients,

The Metabolic Syndrome: Dietary Supplements and Food Ingredients. Raj K. Keservani, MPharm, Durgavati Yadav, PhD, Rajesh K. Kesharwani, PhD, Sippy Singh, DPhil & Kumar Sandeep, PhD (Eds.)

foods, and nutritional styles inside the prevention and control of type 2 diabetes (T2D). The first-rate of nutritional fat and carbohydrates fed on is extra important than the amount of those macronutrients. Diets wealthy in complete grains, fruits, vegetables, legumes, nuts, mild in alcohol consumption, and decrease in subtle grains, red/processed meats, and sugar-sweetened liquids have validated to lessen diabetes hazard and enhance glycemic manage and blood lipids in sufferers with diabetes. Several healthy nutritional styles emphasizing the general food regimen first-rate may be tailored to suitable non-public and cultural meals possibilities and calorie desires for weight management and diabetes prevention and control. Although widespread development has been made in growing and enforcing proof-primarily based totally vitamins tips in advanced countries, concerted international efforts and guidelines are warranted to relieve nearby disparities.

4.1 INTRODUCTION

Diabetes mellitus (DM) was first diagnosed as a sickness around 3,000 years in the past via way of means of the historic Egyptians and Indians, illustrating a few medical capabilities very just like what we now realize as diabetes. DM is an aggregate of words, "diabetes" Greek phrase derivative, and manner siphon – to by skip via and the Latin phrase "Mellitus" manner honeyed or sweet. In 1776, extra sugar in blood and urine become first showed in Great Britain. With the passage of time, a great expertise of diabetes at the side of designated etiology and pathogenesis has been achieved. DM is described as "a metabolic ailment characterized via way of means of hyperglycemia as a result of both the deficiency in insulin secretion or the movement of insulin." The poorly managed DM can result in harm to diverse organs, mainly the eyes, kidneys, nerves, and cardiovascular system. DM may be of the three most important types, primarily based totally on etiology and medical capabilities. These are diabetes mellitus type 1 (T1DM), diabetes mellitus type 2 (T2DM), and gestational diabetes mellitus (GDM). In T1DM, there may be absolute insulin deficiency because of the destruction of β cells within side the pancreas via way of means of a mobile mediated autoimmune process. In T2DM, there may be insulin resistance and relative insulin deficiency. GDM is any diploma of glucose intolerance this is diagnosed at some point of pregnancy. DM can rise up from different illnesses or because of tablets together with genetic syndromes, surgery, malnutrition, infections, and corticosteroids intake. T2DM elements which may be irreversible together with age, genetics, race, and ethnicity or revisable together with diet, bodily pastime, and smoking (Via and Mechanick, 2016; Gilis-Januszewska et al., 2011).

4.2 EPIDEMIOLOGY

Globally, T2DM is at gift one of the maximum not unusual place illnesses and its degrees are step by step at the upward push. It has been evaluated that round 366 million humans international or 8.3% within side the age institution of 20–70 years had T2DM in 2011. This determine is anticipated to upward push to 552 million (9.9%) through 2030. This disorder is related to intense headaches which have an effect on patient's health, productivity, and exceptional of life. More than 50% of humans with diabetes die of cardiovascular disorder (CVD) (mainly coronary heart disorder and stroke) and is a sole motive of cease degree renal disorder which calls for both dialysis or kidney transplantation. It is likewise a chief motive of blindness because of retinal harm in-person age institution known as diabetic retinopathy. People with T2DM have an extended danger of decrease limb amputation that can be 25 instances more than the ones without the disorder. This disorder brought about around 4.6 million deaths within side the age-institution of 20–70 years in 2011. Dietary elements are of paramount significance within side the control and prevention of type 2 diabetes (T2D) (Chen et al., 2012; Ginter and Simko, 2013; Rane and Keservani, 2020). Despite development in formulating proof-primarily based totally nutritional guidance, controversy, and confusion remain (Figure 4.1).

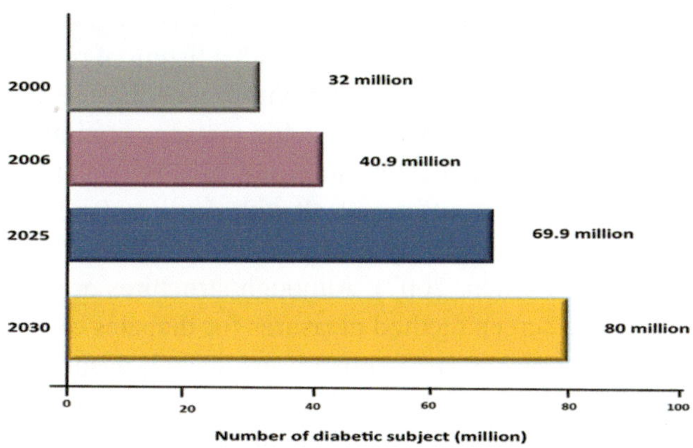

FIGURE 4.1 Estimated number of diabetic subjects in India.

4.3 DIETARY MANAGEMENT AND ITS IMPLEMENT

Diabetes is one in all the largest worldwide public fitness issues: the superiority is envisioned to grow from 425 million humans in 2017 to 629 million via way of means of 2045, with connected fitness, social, and financial costs. Urgent answers for slowing, or maybe reversing, this fashion is needed, particularly from funding in modifiable elements inclusive of eating regimen, bodily interest, and weight. Diet is a main contributor to morbidity and mortality globally, in line with the Global Burden of Disease Study finished in 188 countries. The significance of nutrients within side the control and prevention of T2D thru its impact on weight and metabolic management is clear. However, nutrients are likewise one of the maximum arguable and tough components of the control of T2D. The concept of being on an "eating regimen" for a persistent lifelong circumstance like diabetes is sufficient to position many humans off as understanding what to consume and retaining an ultimate ingesting sample are challenging. Medical nutrients remedy turned into added to manual a scientific and proof-primarily based totally technique to the control of diabetes thru eating regimen, and its effectiveness has been demonstrated, however, problems remain. Although maximum diabetes tips advise beginning pharmacotherapy handiest after first making dietary and bodily interest lifestyle changes, this isn't constantly accompanied in exercise globally (Evert et al., 2014; Rane and Keservani, 2020). Most physicians aren't educated in nutrients interventions, and that is a barrier to counseling sufferers. Moreover, speaker to sufferers approximately nutrients is time consuming. In many settings, out of doors of specialized diabetes centers wherein educated nutritionists/educators are available, recommendation on nutrients for diabetes is, at great, a published menu given to the affected person. In aid terrible settings, while T2D is diagnosed, frequently the affected person leaves the health center with a listing of recent medicines and little else. There is extensive variant within side the use of nutritional amendment on my own to control T2D for instance, estimates of fewer than 5–10% of sufferers with T2D in India and 31% within side the UK are reported, despite the fact that sufferers dealt with via way of means of lifestyle measures can be much less intently controlled than sufferers on remedy for T2D (Schwarz and Peltonen, 2007). Although structures are commonly in the area to report and screen method measures for diabetes care in scientific records, nutritional statistics are frequently neglected, despite the fact that as a minimum modest interest to eating regimen is wanted to obtain good enough glycemic management. Family medical doctors and clinic clinics have to accumulate these statistics mechanically however, a way to do that is a challenge (Teixeira and Budd, 2010). Progress has been made in know-how of the great nutritional recommendation for diabetes however, broader issues exist. For instance, growing vegetable and fruit consumption is usually recommended via way of means of maximum nutritional tips however their fee is prohibitively excessive in lots of settings: the fee of servings of end result and three servings of veggies an afternoon in keeping with individual (to fulfill the "5-a-day" guidance) accounted for 52%, 18%, 16%, and 2% of family profits in low, low to middle, higher to middle, and excessive profits countries, respectively. A highly-priced marketplace of ingredients labeled to be used via way of means of humans with diabetes additionally exists, with merchandise frequently being no healthier, and every so often much less healthy, than everyday ingredients. After new European Union legislation, meals guidelines in a few countries, inclusive of the United Kingdom, have been up to date as lately as July 2016 to prohibit such deceptive labels. This isn't the case elsewhere, however, and what is going to appear to be such law after the United Kingdom leaves the European Union is unclear, which highlights the significance of the political environment (Forouhi et al., 2018a, Rawal et al., 2012). Figure 4.2 indicates that what type of food should be eaten by diabetic patient and which food should be avoided.

FIGURE 4.2 Food to avoid and food to eat in diabetes.

4.4 PHYSICAL ACTIVITY AND LIFESTYLE

A huge quantity of cross-sectional in addition to potential and retrospective research has discovered extensive affiliation among bodily inactiveness and T2DM. A potential look at changed into executed amongst greater than 1,000 non-diabetic people from the high-hazard populace of Pima Indians. During a median follow-up duration of six-year, it changed into discovered that the diabetes prevalence charge remained better in much less lively ladies and men from all BMI groups. The current proof indicates some of viable organic pathways for the protecting impact of bodily pastime at the improvement of T2DM. First, it's been advised that bodily pastime will increase sensitivity to insulin. In a complete document posted through Health and Human Services, USA, 2015 said that bodily pastime notably progressed bizarre glucose tolerance whilst resulting from insulin resistance ordinarily than whilst it changed into resulting from poor quantities of circulating insulin. Second, bodily pastime is in all likelihood to be maximum useful in stopping the development of T2DM at some point of the preliminary stages, earlier than insulin remedy is required (Connelly et al., 2013). The shielding mechanism of bodily pastime seems to have a synergistic impact with insulin. During an unmarried extended consultation of bodily pastime, contracting skeletal muscle complements glucose uptake into the cells. This impact will increase blood waft within side the muscle and complements glucose delivery into the muscle cell (Temelkova-Kurktschiev and Stefanov, 2012). Third, bodily pastime has additionally been discovered to lessen intra-belly fats that are a regarded hazard aspect for insulin resistance. In sure different research, bodily pastime has been inversely related to intra-belly fats distribution and may lessen frame fats stores. Lifestyle and environmental elements are said to be the principal reasons of excessive growth within side the prevalence of T2DM (Sami et al., 2017a; Godino et al., 2016).

4.5 RELATION BETWEEN DIET AND TYPE 2 DIABETES (T2D) MELLITUS

The position of eating regimen within side the etiology of T2DM changed into proposed through Indians, as noted earlier, who discovered that the disorder changed into nearly restricted to wealthy those who fed on oil, flour, and sugar in immoderate quantities. During the First and Second World Wars, declines within side the diabetes mortality costs had been documented because of meals scarcity and famines within side the worried international locations which include Germany and different European international locations. In Berlin, diabetes mortality price declined from 23.1/100,000 in 1914 to 10.9 in 1919. In contrast, there has been no extruding in diabetes mortality price in different international locations without a scarcity of meals on the equal term, which includes Japan and North American international locations. Whereas few research have located robust affiliation of T2DM with excessive consumption of carbohydrates and fat. Many researchers have pronounced a tremendous affiliation among excessive consumption of sugars and improvement of T2DM. In an observer, Ludwig investigated greater than 500 ethnically numerous schoolchildren for 19 months. It changed into located that for every extra serving of carbonated beverages fed on, frequency of weight problems extended, after adjusting for special parameters which includes nutritional, demographic, anthropometric, and lifestyle. A observe changed into carried out which blanketed the diabetic sufferers with differing levels of glycemic control. There had been no variations within side suggest day by day plasma glucose ranges or diurnal glucose profiles. As with carbohydrates, the affiliation among nutritional fat and T2DM changed into additionally inconsistent. Many of potential research have located members of the family among fats consumption and the next chance of growing T2DM. In diabetes observation, carried out at San Louis Valley, a greater than 1,000 topics without a previous prognosis of diabetes had been prospectively investigated for four years. In that observation, the researchers located an affiliation among fats consumption, T2DM and impaired glucose tolerance. Another observes discovered the connection of the numerous eating regimen additives amongst businesses of women, along with fats, fiber plus sucrose, and the chance of T2DM. After adjustment, no institutions had been located among intakes of fats, sucrose, carbohydrate or fiber and the chance of diabetes in each business. Recently, proof counseled a hyperlink among the consumption of gentle beverages with weight problems and diabetes, on account of big quantities of excessive fructose corn syrup used within side the production of gentle beverages, which increases blood glucose ranges and BMI to the harmful ranges. It changed into additionally said through Assay that eating regimen gentle beverages include glycated chemical substances that markedly increase insulin resistance (Sami et al., 2017a). Food consumption has been strongly related with weight problems, now no longer handiest associated with the extent of meals however additionally in phrases of the composition and great of eating regimen. High consumption of purple meat, goodies, and fried foods, make contributions to the extended the chance of insulin resistance and T2DM. In contrast, an inverse correlation changed into discovered among consumption of greens and T2DM. Consumption of end result and greens can also additionally guard the improvement of T2DM, as they're wealthy in nutrients, fiber, and antioxidants which can be taken into consideration as shielding barrier towards the diseases. Recently, in Japanese women, a document found out that multiplied consumption of white rice changed into related to an extended chance of T2DM. This needs a pressing want for converting lifestyle amongst well-known populace and in addition boom the notice of wholesome eating regimen styles in all businesses (Sami et al., 2016; Khazrai et al., 2014; Post et al., 2012).

4.6 EVIDENCE FOR CURRENT DIETARY GUIDELINES

In the main developed countries, nutritional hints for the control of diabetes have advanced from a focal point on a low fats weight-reduction plan to the popularity that greater crucial concerns are macronutrient first-rate (that is, the sort as opposed to the amount of macronutrient), avoidance of processed foods (specially processed starches and sugars), and general nutritional patterns. Many systematic evaluations and countrywide nutritional hints have evaluated the proof for a finest nutritional recommendation, and we can now no longer repeat the proof review. We cognizance alternatively within side the following sections on a few crucial standards in which extensive consensus exists within side the medical and medical network and spotlight regions of uncertainty, however, we start by outlining three underpinning features. Firstly, information of wholesome ingesting for the prevention and control of T2D has in large part been derived from long-time potential research and restricted proof from randomized managed trials in trendy populations, supplemented through proof from humans with T2D. Many posted hints and evaluations have implemented grading standards, and this proof is regularly of slight first-rate within side the hierarchy of proof that locations randomized managed trials on the top. Elsewhere, it's far argued that distinctive types of proof comparing consistency throughout more than one observe designs along with massive populace primarily based totally potential research of medical endpoints, managed trials of intermediate pathways, and in which possible randomized trials of medical endpoints need to be used together for proof-primarily based totally dietary guidance (Godino et al., 2016). Secondly, it's far now acknowledged that nutritional recommendation for each the prevention and control of T2D need to converge, and that they need to know no longer be handled as distinctive entities. However, in people with T2D, the diploma of glycemic manipulate and sort and dose of diabetes remedy need to be coordinated with nutritional intake. With a few nutritional interventions, including very low calorie or low carbohydrate diets, humans with diabetes might commonly forestall or lessen their diabetes remedy and be monitored closely, as reviewed in a later section. Thirdly, at the same time as recognizing the significance of weight-reduction plan for weight control, there's now more information of the more than one pathway thru which nutritional elements exert fitness consequences thru each weight problem-based and weight problems unbiased mechanisms. The impact of weight-reduction plan on weight, glycemia, and glucose-insulin homeostasis is at once applicable to glycemic manipulate in diabetes, at the same time as different results including cardiovascular headaches are similarly stimulated through the impact of weight-reduction plan on blood lipids, apolipoproteins, blood pressure, endothelial function, thrombosis, coagulation, systemic inflammation, and vascular adhesion. The impact of meals and vitamins (Keservani et al., 2010a, b, 2016; Singh et al., 2017) at the intestine microbiome will also be applicable to the pathogenesis of diabetes however, similar studies are needed. Therefore, weight-reduction plans first-rate and amount over the long run are applicable to the prevention and control of diabetes and its headaches thru an extensive variety of metabolic and physiological processes (DeSalvo et al., 2016; Forouhi et al., 2018a).

Second, it is now widely acknowledged that dietary recommendations for T2D prevention and control should converge rather than be viewed as separate entities (Figure 4.3). However, in those with T2D, food intake should be coordinated with glycemic management and the Type and dose of diabetes medication. People with diabetes would normally discontinue or reduce their diabetes medication and be closely managed with some dietary changes, such as very low calorie or low carbohydrate diets, as discussed in a later section (Ley et al., 2014; Micha et al., 2012).

4.6.1 WEIGHT MANAGEMENT

T2D is maximum normally related to obesity or weight problems and insulin resistance. Therefore, decreasing weight and keeping a wholesome weight is middle a part of scientific management. Weight loss is likewise connected to upgrades in glycemia, blood pressure, and lipids and for this reason can put off or save you complications, mainly cardiovascular events (Van Gaal and Scheen, 2015; Wilding, 2014).

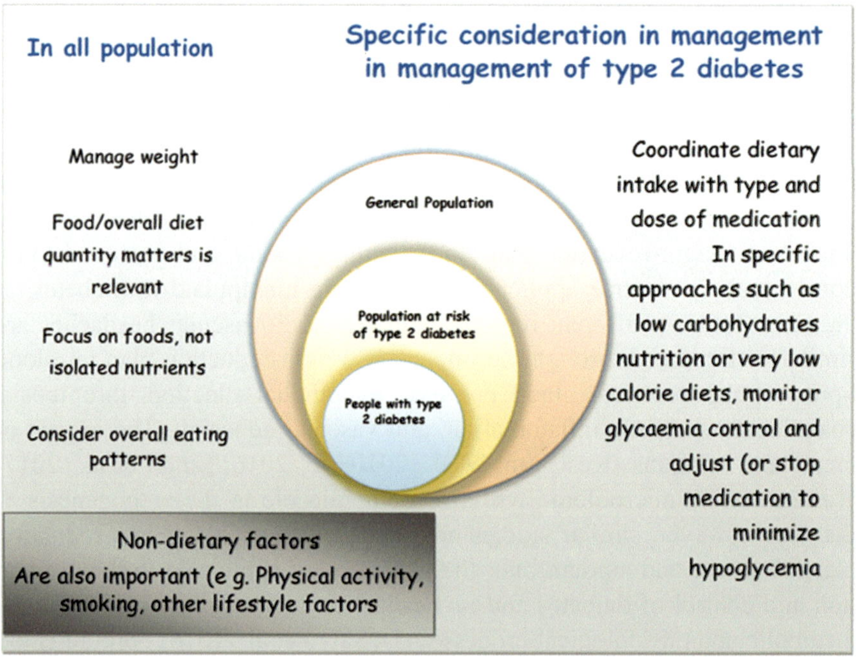

FIGURE 4.3 Dietary recommendations for type 2 diabetes and non-dietary factor for prevention and control.

4.6.2 ENERGY BALANCE

Most pointers suggest selling weight reduction amongst obese or overweight people via way of means of decreasing electricity consumption. Portion manipulate is one method to restriction electricity consumption collectively with a healthful consuming sample that specializes in a weight loss plan composed of entire or unprocessed ingredients mixed with bodily hobby and ongoing support (Romieu et al., 2017; Ferrannini et al., 2015).

4.6.3 DIETARY PATTERNS

The proof factors to selling styles of meals consumption which might be excessive in vegetables, fruit, entire grains, legumes, nuts, and dairy merchandise which include yogurt however with a few cautions. Firstly, a few nutritional approaches (e.g., low carbohydrate diets) advise proscribing the consumption of culmination, entire grains, and legumes due to their sugar or starch content material. For fruit consumption, especially amongst people with diabetes, opinion is split amongst scientists and clinicians. Many tips preserve to advise fruit, however, on the premise that fructose consumption from culmination is top-rated to isocaloric consumption of sucrose or starch due to the extra micronutrient, phytochemical, and fiber content material of fruit. Secondly, regardless of proof from randomized managed trials and potential studies (Esposito, et al., 2013) that nuts can also additionally assist you T2D, a few (probably misplaced) situations exist approximately their excessive power content material. Further studies on human beings with T2D need to assist in making this clear. There is likewise consensus on the blessings of positive named nutritional styles, which include the Mediterranean weight loss plan for prevention and control of T2D. Expert tips additionally assist different wholesome ingesting styles that take account of neighborhood sociocultural elements and private preferences (Tapsell et al., 2016; Alhazmi et al., 2014; Carvajal-Millan et al., 2021; Surana et al., 2021b).

4.6.4 FOODS TO AVOID

Consensus exists on lowering or averting the consumption of processed purple meats, subtle grains, and sugars (especially sugar-sweetened drinks) each for prevention and control of T2D, once more with a few cautions. Firstly, for unprocessed purple meat, the proof of viable damage due to the improvement of T2D is much less steady and of a smaller magnitude. More studies are wanted on unique blessings or harms in humans with T2D. Secondly, the proof is growing at the relevance of carbohydrate quality: this is that complete grains and fiber are higher picks than subtle grains and that fiber consumption ought to be at the least as excessive in humans with T2D as encouraged for the overall population, that diets which have a better glycemic index and cargo are related to an expanded threat of T2D, and that there's a modest glycemic gain in changing ingredients with better glycemic load with ingredients with low glycemic load. However, debate keeps approximately the independence of those consequences from the consumption of nutritional fiber. Some proof exists that intake of potato and white rice might also additionally boom the threat of T2D; however, that is constrained and similarly studies are wanted. Moreover, many tips additionally spotlight the significance of lowering the consumption of in ingredients excessive in sodium and trans fats due to the relevance of those mainly for cardiovascular health (Sami et al., 2017a, Miller et al., 2012).

4.7 AREAS OF UNCERTAINTY IN GUIDELINES

4.7.1 OPTIMAL MACRONUTRIENT COMPOSITION

One of the maximum contentious problems approximately the control of T2D has been at the exceptional macronutrient composition of the weight loss program. Some suggestions retain to suggest macronutrient amount goals, which include the European or Canadian advice of 45–60% of general strength as carbohydrate, 10–20% as protein, and much less than 35% as fats, 10–15% from protein, and much less than 30% from fats. In contrast, the maximum latest dietary guiding principle from the American Diabetes Association (ADA) concluded that there's no perfect blend of macronutrients for absolutely everyone with diabetes and endorsed for my part tailor-made goals. Alternatively, a low carbohydrate weight loss program for weight and glycemic management has received recognition amongst a few experts, clinicians, and the public. Others finish that a low carbohydrate weight loss program mixed with low saturated fats consumption is exceptional. For weight reduction, 3 factors are noteworthy whilst evaluating nutritional macronutrient composition. Firstly, proof from trials factors to doubtlessly more advantages from a low carbohydrate than a low fats weight loss program however the distinction in weight reduction among diets is modest. Secondly, an evaluation of named weight loss program programs with extraordinary macronutrient composition highlighted that the crucial component in effectiveness for weight reduction becomes the extent of adherence to the weight loss program over time. Thirdly, the best of the weight loss program in low carbohydrate or low fats diets is important. Research thus far on weight or metabolic results in diabetes is complex via way of means of the use of various definitions for the extraordinary macronutrient approaches. For instance, the definition of a low carbohydrate weight loss program has ranged from 4% of day-by-day strength consumption from carbohydrates (selling dietary ketosis) to 40%.15 similarly, low fats diets had been described as fats consumption much less than 30% of day-by-day strength consumption or appreciably lower. Given those limitations, the exceptional cutting-edge method can be an emphasis on using person evaluation for nutritional recommendation and a focal point at the sample of ingesting that maximum effectively lets in the person to restriction calorie consumption and enhances macronutrient best (which include fending off delicate carbohydrates) (Esposito et al., 2013; Forouhi et al., 2018a, Copeland et al., 2013).

4.7.1.1 FISH

Regular fish consumption of as a minimum serving a week, such as one serving of oily fish (e.g., salmon, mackerel, and trout) is suggested for cardiovascular threat prevention however fish consumption has exceptional institutions with the threat of growing T2D throughout the world an inverse association, no association, and an effective association. It is the notion that the sort of fish consumed, training or cooking practices, and feasible contaminants (e.g., methyl mercury and polychlorinated biphenyls) range with the aid of using geographical area and contributed to this heterogeneity. More studies are wanted to solve whether or not fish consumption ought to be endorsed for the prevention of diabetes. However, the cutting-edge proof helps a boom in intake of oily fish for people with diabetes because of its useful results on lipoproteins and prevention of coronary heart disease. Most tips agree that omega-three polyunsaturated fatty acid (fish oil) supplementation for cardiovascular prevention in human beings with diabetes ought to know no longer be endorsed, however, greater studies are wanted, and the effects of the ASCEND (a study of cardiovascular events in diabetes) trial ought to assist to make clear this (Comerford and Pasin, 2016).

4.7.1.2 DAIRY

Dairy ingredients are advocated for the prevention of T2D, with greater steady proof of the blessings of fermented dairy merchandise, consisting of yogurt. Similar to populace stage pointers approximately restricting the consumption of ingredients excessive in saturated fat and changing them with ingredients wealthy in polyunsaturated fats, the cutting-edge recommendation for diabetes additionally favors low fats dairy merchandise however that is debated. More studies are wanted to remedy this question (Comerford and Pasin, 2016; Rahati et al., 2014).

4.7.1.3 OILS

Uncertainty maintains approximately positive plant oils and tropical oils consisting of coconut or palm oil as proof from potential research or randomized managed trials on scientific activities is sparse or non-existent. However, olive oil, in particular greater virgin olive oil, has been studied in extra element with proof of capability advantages for the prevention and control of T2D and the prevention of cardiovascular sickness inside the context of a Mediterranean diet (Comerford and Pasin, 2016; Goday et al., 2016).

4.8 DIFFICULTIES IN SETTING GUIDELINES

Where nutritional recommendations exist (in lots of settings there are none, or they're tailored from the ones in evolved international locations and consequently won't be relevant to the neighborhood situation), they range extensively in whether or not they're proof-primarily based totally or opinion pieces, and up to date in step with medical development or outdated. Their accessibility each bodily availability (e.g., thru an internet site or clinic) and comprehensibility for sufferers and healthcare experts varies. They range additionally in scope, content, detail, and emphasis at the significance of individualized nutritional advice, regions of controversy, and in addition studies desires. The first-rate of studies that informs nutritional recommendations additionally desires extra funding from the medical network and funders. Moreover, loss of transparency within side the improvement of recommendations and bias within side the number one dietary research can undermine the improvement of dependable nutritional recommendations; pointers for his or her development ought to be heeded (Conlin et al., 2017).

4.8.1 REVERSING TYPE 2 DIABETES (T2D) THROUGH DIET

T2D becomes as soon as a concept to be irreversible and revolutionary after diagnosis, however, a great deal hobby has arisen approximately the ability for remission. The consensus at the definition of remission is an indication of progress: glucose stages decrease than the diagnostic stage for diabetes within side the absence of medicinal drugs for hyperglycemia for duration of time (regularly proposed to be at the least one year). However, the important position of power deficit as opposed to macronutrient composition of the food regimen in reaching remission remains controversial (Taylor, 2013; McMacken and Shah, 2017).

4.9 MANAGEMENT OF REMISSION

4.9.1 REMISSION THROUGH A LOW-CALORIE ENERGY DEFICIT DIET

Although the medical commentary of the lifelong, regularly revolutionary nature of T2D turned into showed with the aid of using the United Kingdom Prospective Diabetes Study, fast normalization of fasting plasma glucose after bariatric surgical operation recommended that deterioration turned into now no longer inevitable. As the primary extrude turned into one in every of unexpected calorie limit, a low-calorie food plan turned into used as a device to have a look at the mechanisms involved. In one look at sufferers with T2D, fasting plasma glucose normalized inside seven days of following a low-calorie food plan. This normalization thru food plan befell in spite of simultaneous withdrawal of metformin therapy. Gradually over 8 weeks, glucose inspired insulin secretion again to regular. Was this an outcome of calorie limit or composition of the food plan? To attain the diploma of weight reduction obtained (15 kg), approximately 610 kcal an afternoon turned into provided 510 kcal as a liquid components food plan and approximately 100 kcal as non-starchy vegetables. The components food plan consisted of 59 g of carbohydrate (30 g as sugars), 11.4 g of fats, and 41 g of protein, which include required nutrients and minerals. This high "sugar" technique to controlling blood glucose can be unexpected however the crucial element isn't what's eaten however the hole among electricity required and brought in. Because of this deficit, the frames have to use formerly saved electricity. Intrahepatic fats are used (Steven et al., 2016). In addition, pancreatic fats content material fell over 8 weeks and beta molecular feature improved. This is due to the fact insulin secretary feature turned into regained with the aid of using re-differentiation after fat removal. The permanence of those modifications turned into examined with the aid of using a dietary and behavioral technique to attain long time isocaloric ingesting after the extreme weight reduction phase. It turned into a hit in maintaining weight constant over the subsequent six months of the have a look at. Calorie limit turned into related to each hepatic and pancreatic fats content material ultimate on the low degrees accomplished (Goldenberg et al., 2021). The preliminary remission of T2D turned into carefully related to length of diabetes, and the people with T2D of shorter length who accomplished regular degrees of blood glucose maintained regular body structure throughout the six-month follow-up period. Recently, 46% of a UK number one care cohort remained freed from diabetes at twelve months throughout an established low calorie weight reduction program. These consequences are convincing, and four years of follow-up are planned. A not unusual place complaint of the electricity deficit studies has been that very low-calorie diets will not be potential or sustainable. Indeed, adherence to maximum diets withinside the long term is a crucial challenge. However, the biggest randomized have a look at of way of life interventions in T2D (n = 5,145), randomized people to in-depth way of life management, which include the purpose to lessen general calorie consumption to 1,200–1,800 kcal/d thru a low fats food plan assisted with the aid of using liquid meal replacements, and this technique accomplished more weight reduction. A considerable hobby has arisen approximately whether or not low-calorie diets related to diabetes remission also can assist to save you diabetic headaches. Evidence is sparse due to the shortage of long-time follow-up research; however, the current studies are promising. A go back to the non-diabetic nation brings a development in cardiovascular danger (Q danger lowering from 19.8% to 5.4%); case reviews of people going through foot amputation report a go back to a low danger nation over 2–4 years with decision of painful neuropathy; and retinal headaches are not likely to arise or progress. However, different proof highlights that worsening of treatable maculopathy or proliferative retinopathy can also additionally arise following an unexpected fall in plasma glucose degrees, so retinal imaging in four-Six months is usually recommended for people with extra than minimum retinopathy if following a low-calorie remission food plan. Annual evaluation is usually recommended for all the ones withinside the post-diabetic nation, and a "diabetes in remission" code (C10P) is now to be had withinside the UK (Brown and Leeds, 2019).

4.9.2 *MANAGEMENT OR REMISSION THROUGH A LOW CARBOHYDRATE DIET*

Before insulin became advanced as a therapy, lowering carbohydrate consumption became the primary remedy for diabetes. Carbohydrate limit for the remedy of T2D has been a place of extreme hobby because, of all of the macronutrients, carbohydrates have the finest impact on blood glucose and insulin levels. In a evaluate via way of means of the ADA, interventions of low carbohydrate (much less than 40% of calories) diets posted from 2001 to 2010 have been identified. Of 11 trials, 8 have been randomized and approximately 1/2 of mentioned more development in HbA1c at the low carbohydrate weight-reduction plan than the contrast weight-reduction plan (generally a low fats weight-reduction plan), and a more discount withinside the use of drugs to decrease glucose. Notably, calorie discount coincided with carbohydrate limit in most of the research, although it became now no longer regularly exact withinside the nutritional counseling. One of the extra extraordinarily managed research became an inpatient feeding examine, which mentioned a decline in suggest HbA1c from 7.3% to 6.8% (P=0.006) over simply 14 days on a low carbohydrate weight-reduction plan. For glycemia, different critiques of proof from randomized trials on human beings with T2D have various conclusions. Some concluded that low carbohydrate diets have been advanced to different diets for glycemic manipulate, or that a dose-reaction dating existed, with stricter low carbohydrate limit ensuing in more discounts in glycemia. Others counseled approximately brief time period useful outcomes now no longer being sustained withinside the long run, or observed no average benefit over the contrast weight-reduction plan. Narrative critiques have usually been extra emphatic at the advantages of low carbohydrate diets, which include extended satiety, and spotlight the benefits for weight reduction and metabolic parameters. More recently, a 365 days hospital primarily based totally examine of the low carbohydrate weight-reduction plan designed to result in dietary ketosis (generally with carbohydrate consumption much less than 30 g/d) became powerful for weight reduction, and for glycemic manipulate and remedy discount. However, the examination became now no longer randomized, remedy depth differed notably withinside the intervention as opposed to standard care groups, and members have been capable of picking out their group (Wylie-Rosett et al., 2013). Concerns approximately capacity destructive outcomes on cardiovascular fitness were raised as low carbohydrate diets are generally excessive in nutritional fats, which include saturated fats. For lipid markers as predictors of destiny cardiovascular events, numerous researches observed more upgrades in excessive density lipoprotein LDL cholesterol and triglycerides without a relative worsening of low-density lipoprotein LDL cholesterol in sufferers with T2D following carbohydrate limit (Forouhi, et al., 2018b), with comparable conclusions in non-diabetic populations. Low density lipoprotein LDL cholesterol has a tendency to say no extra, however, in a low fats contrast weight-reduction plan and despite the fact that low density lipoprotein LDL cholesterol might not get worse with a low carbohydrate weight-reduction plan withinside the brief time period, the long run outcomes are unclear. Evidence suggests that low carbohydrate consumption can decrease the extra thermogenic small, dense low density lipoprotein particles. Because a few people might also additionally revel in a boom in serum low density lipoprotein LDL cholesterol whilst following a low carbohydrate weight-reduction plan excessive in saturated fats, tracking is important. Another subject is the impact of the doubtlessly better protein content material of low carbohydrate diets on renal characteristics. Evidence from sufferers with T2D with regular baseline renal characteristics and from people without diabetes and with regular or mildly impaired renal characteristics has now no longer proven worsening renal characteristics at one or up to two years of follow-up, respectively. Research in sufferers with extra critically impaired renal characteristics, without or with diabetes, has now no longer been mentioned to our knowledge. Other capacity facet outcomes of a totally low carbohydrate weight-reduction plan consist of headache, fatigue, and muscle cramping however those facet outcomes may be averted via way of means of ok fluid and sodium consumption, specifically withinside the first week or after beginning the weight-reduction plan whilst dieresis is finest. Concern approximately urinary calcium loss and a probable contribution to extended destiny chance of kidney stones or osteoporosis have now no longer been proven however proof is sparse and warrants in addition investigation (Table 4.1) (Hussain et al., 2012; Surana et al., 2021a).

TABLE 4.1 Dietary Patterns in the Diabetes Mellitus Patient

Components and Item Labels	Dietary Patterns				
	Oils, Nuts, Seeds, Sugars, Eggs	**Cereals and Meats**	**Fruits and Potatoes**	**Vegetables and Fish**	**Pulses**
Cereals	0.36	0.59	–0.04	–0.23	0.16
Potatoes and starch	0.07	0.23	0.62	–0.01	0.24
Sugars	0.78	0.15	0.26	0.00	–0.12
Pulses	0.04	0.12	0.09	0.22	0.73
Nuts and seeds	0.81	–0.31	0.09	0.22	–0.06
Green and dark yellow vegetables	–0.21	0.08	0.45	0.48	0.20
Leafy vegetables	–0.08	0.36	–0.02	0.76	0.03
Fruits	0.09	–0.19	0.71	–0.02	–0.15
Mushrooms	–0.09	0.55	–0.04	0.13	0.03

TABLE 4.1 *(Continued)*

Components and Item Labels	Dietary Patterns				
	Oils, Nuts, Seeds, Sugars, Eggs	**Cereals and Meats**	**Fruits and Potatoes**	**Vegetables and Fish**	**Pulses**
Algae	–0.08	0.14	0.51	0.32	–0.31
Fish and seafood	0.21	–0.15	–0.10	0.73	0.18
Meats	0.24	0.56	0.01	–0.24	–0.51
Eggs	0.48	0.18	0.05	–0.26	0.33
Milk and Milk products	–0.10	–0.63	0.44	0.05	0.02
Fats and oils	0.87	0.04	–0.14	–0.01	–0.011
Snacks	0.07	–0.06	0.07	0.09	–0.44
Beverages	–0.03	–0.32	0.19	0.51	–0.26
Seasoning and spices	–0.15	0.56	0.30	0.13	0.28
Cooking and processing	0.14	–0.27	0.58	–0.04	–0.05
Contribution percentages	15	14	12	8	7

Finally, low carbohydrate diets can limit complete grain consumption, and despite the fact that a few low carbohydrate ingredients can offer the fiber and micronutrients contained in grains, it can require more attempts to include such ingredients. This has led a few professionals to emphasize proscribing subtle starches and sugars however preserving complete grains (Stewart et al., 2016).

4.9.3 NUTRITION TRANSITION AND POPULATION SPECIFIC FACTORS

Several nations in sub-Saharan Africa, South America, and Asia (e.g., India and China) have gone through speedy nutrients transition withinside the beyond decades. These modifications have paralleled monetary growth, overseas funding withinside the speedy meals industry, urbanization, direct-to-customer advertising of ingredients excessive in calories, sale of ultra-processed ingredients, and as a result, decrease intake of conventional diets. The impact of those elements on nutrients has brought about weight problems and sort two diabetes on the only hand, and co-current undernutrition and micronutrient deficiencies at different. Dietary shifts in low and center profits nations were stark: in India, those consist of a significant boom in fats consumption withinside the putting of an already excessive carbohydrate consumption, with a moderate boom in general strength and protein, and a lowering consumption of coarse cereals, pulses, fruits, and vegetables; in China, animal protein and fats as a percent of strength has additionally increased, even as cereal consumption has reduced. A nearly prevalent boom withinside the consumption of caloric drinks has additionally occurred, with sugar-sweetened soda liquids being the principal beverage contributing to strength consumption, as an instance amongst adults and kids in Mexico, or the significant upward push in China in income of sugar-sweetened liquids from 10.2 L consistent with capita in 1998 to 55.0 L consistent with capita in 2012. The motion of populations from rural to city regions inside a rustic can also be connected with shifts in diets to extra bad styles, even as acculturation of immigrant populations into their host nations additionally consequences in nutritional shifts. In a few populations, consisting of South Asians, rice and wheat flour bread are staple ingredients, with an associated excessive carbohydrate consumption (60–70% of calories). Although time tendencies display that consumption of carbohydrates has reduced amongst South Asian Indians, the first-class of carbohydrates has shifted in the direction of the use of delicate carbohydrates. The use of oils and conventional cooking practices additionally has precise styles in extraordinary populations. For instance, in India, the import and intake of palm oil, frequently integrated withinside the famous oil Vanaspati (in part hydrogenated vegetable oil, excessive in Trans fats), is excessive. Moreover, the conventional Indian cooking exercise of frying at excessive temperatures and re-heating will increase Tran's fatty acids in oils. Such oils are low price, effortlessly available, and feature a protracted shelf life, and as a consequence are extra appealing to humans from the center and occasional socioeconomic strata however their long-time outcomes on T2D are unknown. Despite the nutrients transition being connected to a growing incidence of T2D, weight problems, and different non-communicable diseases, robust measures to restrict dangerous ingredients aren't in the area in lots of nations. Regulatory frameworks, which include monetary rules consisting of taxation for sugar-sweetened drinks, want to be bolstered to be powerful and different preventive interventions want to be well implemented. Efforts to manipulate Trans fatty acids in ingredients have won momentum however are in large part constrained to advanced nations. To lessen intake in low and center profits nations would require each stringent guidelines and the provision and improvement of opportunity picks of healthful and occasional price oils, equipped made meals products, and customer education. The want for dietary labeling is crucial; however, information nutrients labels is a trouble in populations with low literacy or nutrients awareness, which highlights the want for academic sports and less complicated varieties of labeling. The position of nutritional/ dietary elements withinside the predisposition of a few ethnic organizations to growing T2D at appreciably decrease degrees of weight problems than European populations is poorly researched and desires investigation (Shetty, 2013; Popkin, 2015).

4.10 DIETARY KNOWLEDGE OF TYPE 2 DIABETICS

ADA has described self-nutritional control as an important thing step in imparting the diabetics, the expertise and ability in relation with treatment, dietary aspects, medicinal drugs, and complications. A look at confirmed that the nutritional expertise of the focused organization who had been at excessive chance of growing T2DM turned into poor. Red meat and fried meals had been eating up greater with the aid of using adult males compared to ladies. The percentage of adult males to ladies in day-by-day rice intake turned into drastically excessive. In current instances in Saudi Arabia, meals choices, length of quantities and sedentary way of life have expanded dramatically that led to excessive chance of obesity. Unfortunately, many Saudis are getting greater overweight due to the ease of rapid foods, and this provides to the horrifying diabetes statistics. On the alternative hand, Saudis drink too many excessive-sugar drinks. In addition, Backman suggested nutritional expertise to be a large aspect that impacts nutritional behaviors. In every other look at performed with the aid of using Savoca and Miller said that sufferers' meals choice and nutritional behaviors can be stimulated with the aid of using the robust expertise approximately diabetic weight loss plan recommendations. Significant fine dating turned into found among expertise concerning diabetic weight loss plan and the quantity of calorie needs. The look at concluded that expertise concerning diabetic weight loss plan is vital and is wanted to obtain higher nutritional behaviors. Results of look at performed in Saudi Arabia suggested that greater than 1/2 of the diabetic sufferers denied enhancing their nutritional pattern, discount in weight and carry out exercise. National Centre for Health Statistics suggested that socioeconomic repute performs a vital position withinside the improvement of T2DM; wherein it turned into called a sickness of the rich. On the contrary, the equal reference suggested that T2DM turned into greater familiar in decrease earnings degree and in people with much less education. The variations can be because of the sort of meals eaten up. Nutritionists counseled that nutrients could be very vital in coping with diabetes, now no longer most effective Type however additionally amount of meals which impacts blood sugar. Meals need to be eaten up at everyday instances with low fats and excessive fiber contents, which includes a restricted quantity of carbohydrates. It turned into found that day by day intake of protein, fats, and electricity consumption with the aid of using Saudi citizens had been better than what is usually recommended with the aid of using the International Nutritional Organization (Ley et al., 2014; Sami et al., 2017b).

4.11 ATTITUDE OF TYPE 2 DIABETICS TOWARD FOOD

DM may be managed via development in patient's nutritional knowledge, attitudes, and practices. These elements are taken into consideration as an indispensable a part of complete diabetes care. Although the superiority of diabetes mellitus (DM) is excessive in gulf countries, sufferers are nevertheless poor in information the significance of weight loss program in diabetes management. Studies have proven that assessing sufferers' nutritional mindset might also additionally have a big gain in the direction of remedy compliance and reduce the incidence price of headaches as well. A look at performed in Egypt stated that the mindset of the sufferers in the direction of meals, compliance to remedy, meals manipulate with and without drug use and foot care become inadequate. Another look at supplied that 1–0.33 of the diabetic sufferers had been privy to the significance of weight loss program planning, and restricting LDL cholesterol consumption to save you CVD. Various research has documented improved occurrence of consuming problems and consuming sickness signs in T2DM sufferers. Most of that research has mentioned approximately the binge consuming sickness, because of its sturdy correlation with obesity, a situation that ends in T2DM. Furthermore, a look at discovered that the burden benefit amongst diabetic sufferers become related to the consuming sickness because of mental distress (Keservani et al., 2015a, b). In any other look at that tested consuming sickness-associated signs in T2DM sufferers, cautioned that the dieting-binging series may be implemented to diabetics, in particular overweight diabetic sufferers. Unhealthy consuming conduct and bodily inaction are the main reasons of diabetes. Failure to observe a strict diet regime and workout, alongside prescribed remedies are the main reasons for headaches amongst sufferers of T2DM. Previous research performed in Saudi Arabia have stated that diabetic sufferers do now no longer regard the recommendation given via way of means of their physicians frequently concerning weight loss program planning, weight loss program amendment and exercise (Abolghasemi and Sedaghat, 2015; Okonta et al., 2014).

4.12 DIETARY PRACTICES OF TYPE 2 DIABETICS

Diabetics' nutritional practices are specifically encouraged through cultural backgrounds. Concerning of the scale of nutritional practices, there have been large advantageous relationships among know-how concerning diabetic weight loss plan and nutritional practices. Knowledge turned into a salient component associated with nutritional behaviors control. Moreover, sufferers' know-how on an advocated weight loss plan suggests their expertise of nutritional tips, which encouraged their meals choice and consuming styles. The affiliation among nutritional know-how and nutritional practices amongst T2DM sufferer's withinside the preceding research have been inconsistent. Another observation found out that there has been no courting among nutritional know-how and compliance of nutritional practices. On the opposite hand, the identical observe determined that an excessive nutritional know-how rating turned into related to following nutritional guidelines and informed sufferers carried out self-control sports in a higher way. Dietary know-how drastically impacts nutritional practices. In Indonesia, observe turned into performed to degree nutritional practices amongst diabetic sufferers, which elaborated that the Indonesian people, favored to eat excessive-fats meals which result in an elevated threat of CVD (Keservani and Sharma, 2016). The fashion of skipping breakfast has dramatically elevated during the last 10 years in children, adolescents, and adults. There is growing proof that skipping breakfast is associated with obesity and different fitness issues. In addition, common consuming or snacking may additionally boom the frame weight and threat of metabolic diseases. The prudent nutritional sample turned into characterized through elevated intake of fish, poultry, diverse veggies, and end result whereas; the western nutritional sample turned into characterized through an elevated intake of processed and pink meat, chips, dairy products, delicate grains, and candies and cakes. These styles have been formerly related to T2DM threat. The glycemic index is a trademark of the postprandial blood glucose reaction to meals in step with gram of carbohydrate as compared with reference meals inclusive of white bread or glucose. Hence, the glycemic load represents each the pleasant and amount of the carbohydrates consumed. Another observes performed in Lebanon established direct correlation of the delicate grains and cakes and rapid meals styles with T2DM, however, withinside the identical observe an inverse correlation turned into discovered among the conventional meals sample and T2DM amongst Lebanese adults. Despite the demanding situations of dietary studies, big development has been made in formulating proof-primarily based totally nutritional steerage and a few not unusual place ideas may be agreed that ought to be beneficial to clinicians, sufferers, and the public. Several regions of uncertainty and controversy continue to be and similarly studies are wanted to remedy these. While adherence to nutritional recommendation is a critical challenge, weight control continues to be a cornerstone in diabetes control, supplemented with new developments, inclusive of the capability for the remission of T2D via weight loss plan (Forouhi et al., 2018b; Rajagopalan and Brook, 2012).

4.13 CONCLUSION

Economic increase and environmental transitions have brought about drastic adjustments in meals manufacturing, processing, and distribution structures and feature expanded the accessibility of unhealthful foods. With dietary transitions, women and men around the world have skilled extra frame weight advantage observed through expanded diabetes occurrence and mortality. In the beyond decades, proof from potential cohort research and RCTs has surged to focus on the significance of person nutrients, foods, and nutritional styles in T2D prevention and control. The convergence of nutritional elements for prevention and control of diabetes become observed, and healthy nutritional styles for diabetes prevention and control had been usually wealthy in entire grains, end result and vegetables, nuts, legumes, slight in alcohol consumption, and decrease in subtle grains, red/processed meats, and SSBs. To attain long-time period adherence, people could have flexibility in meal alternatives without compromising standard weight loss program quality. The sizable majority of gift know-how on nutritional prevention and control of diabetes has been derived from Western populations. It is significantly vital to behavior unique investigations in different populations with distinct sickness susceptibility and consuming habits. Evidence-primarily based total vitamins remedy guidelines were advanced and applied in lots of advanced countries (Via, & Mechanick, 2016; Wilding, 2014; Wylie-Rosett, et al., 2013). However, in addition, improvement of area precise hints is wanted to offer sensible academic devices thinking about variants in nutritional styles, accessibility to foods, and agricultural in distinct areas and cultures. Global public fitness regulations are warranted throughout more than one sector to create healthy meal surroundings and sell the company social responsibility. Potential techniques consist of vitamins and agricultural regulations that desire the manufacturing and distribution of healthy meals, for example, taxing incredibly processed meals and instituting agricultural subsidies that boom accessibility and affordability of entire foods. Global efforts, along with homogenizing standardization of front-of-package deal vitamins labels and vitamins records along with public training campaigns; will reshape the trajectory of dietary transition and the worldwide meals delivery supporting to decrease the T2D epidemic.

ACKNOWLEDGMENT

We thank MET, Institute of Pharmacy, BKC, affiliated under Savitribai Phule Pune University, Nashik, Divine College of Pharmacy Satana, and Sandip Institute of Pharmaceutical Science for their constant support and for providing all facilities to complete this work.

KEYWORDS

- **cardiovascular disorder**
- **diabetes mellitus**
- **dietary management**
- **nutrition**
- **preventive measures**
- **type 2 diabetes**

REFERENCES

Abolghasemi, R., & Sedaghat, M., (2015). The patient's attitude toward type 2 diabetes mellitus, a qualitative study. *Journal of Religion and Health, 54*, 1191–1205.

Alhazmi, A., Stojanovski, E., Mcevoy, M., & Garg, M., (2014). The association between dietary patterns and type 2 diabetes: A systematic review and meta-analysis of cohort studies. *Journal of Human Nutrition and Dietetics, 27*, 251–260.

Brown, A., & Leeds, A., (2019). Very low-energy and low-energy formula diets: Effects on weight loss, obesity co-morbidities and type 2 diabetes remission–an update on the evidence for their use in clinical practice. *Nutrition Bulletin, 44*, 7–24.

Carvajal-Millan, E., Yaser, A. Z., & Haghi, A., (2021). *Natural Food Products and Waste Recovery: Healthy Foods, Nutrition Design, and Extraction of Valuable Compounds*, CRC Press.

CHEN, L., Magliano, D. J., & Zimmet, P. Z., (2012). The worldwide epidemiology of type 2 diabetes mellitus—present and future perspectives. *Nature Reviews Endocrinology, 8*, 228–236.

Comerford, K. B., & Pasin, G., (2016). Emerging evidence for the importance of dietary protein source on glucoregulatory markers and type 2 diabetes: Different effects of dairy, meat, fish, egg, and plant protein foods. *Nutrients, 8*, 446.

Conlin, P. R., Colburn, J., Aron, D., Pries, R. M., Tschanz, M. P., & Pogach, L., (2017). Synopsis of the 2017 US department of veterans affairs/US department of defense clinical practice guideline: Management of type 2 diabetes mellitus. *Annals of Internal Medicine, 167*, 655–663.

Connelly, J., Kirk, A., Masthoff, J., & Macrury, S., (2013). The use of technology to promote physical activity in type 2 diabetes management: A systematic review. *Diabetic Medicine, 30*, 1420–1432.

Copeland, K. C., Silverstein, J., Moore, K. R., Prazar, G. E., Raymer, T., Shiffman, R. N., Springer, S. C., Thaker, V. V., Anderson, M., & Spann, S. J., (2013). Management of newly diagnosed type 2 diabetes mellitus (T2DM) in children and adolescents. *Pediatrics, 131*, 364–382.

Desalvo, K. B., Olson, R., & Casavale, K. O., (2016). Dietary guidelines for Americans. *JAMA, 315*, 457–458.

Esposito, K., Ceriello, A., & Giugliano, D., (2013). Does personalized diabetology overcome clinical uncertainty and therapeutic inertia in type 2 diabetes? *Endocrine, 44*, 343–345.

Evert, A. B., Boucher, J. L., Cypress, M., Dunbar, S. A., Franz, M. J., Mayer-Davis, E. J., Neumiller, J. J., et al., (2014). Nutrition therapy recommendations for the management of adults with diabetes. *Diabetes Care, 37*, S120–S143.

Ferrannini, G., Hach, T., Crowe, S., Sanghvi, A., Hall, K. D., & Ferrannini, E., (2015). Energy balance after sodium-glucose cotransporter 2 inhibition. *Diabetes Care, 38*, 1730–1735.

Forouhi, N. G., Misra, A., Mohan, V., Taylor, R., & Yancy, W., (2018a). Dietary and nutritional approaches for prevention and management of type 2 diabetes. *BMJ,* 361.

Forouhi, N. G., Misra, A., Mohan, V., Taylor, R., & Yancy, W., (2018b). Dietary and nutritional approaches for prevention and management of type 2 diabetes. *BMJ, 361*, k2234.

Gilis-Januszewska, A., Szybinski, Z., Kissimova-Skarbek, K., Piwonska-Solska, B., Pach, D., Topor-Madry, R., Tuomilehto, J., et al., (2011). Prevention of type 2 diabetes by lifestyle intervention in primary health care setting in Poland: Diabetes in Europe prevention using lifestyle, physical activity and nutritional intervention (DE-PLAN) project. *The British Journal of Diabetes & Vascular Disease, 11*, 198–203.

Ginter, E., & Simko, V., (2013). Type 2 diabetes mellitus, pandemic in 21st century. *Diabetes,* 42–50.

Goday, A., Bellido, D., Sajoux, I., Crujeiras, A. B., Burguera, B., García-Luna, P. P., Oleaga, A., et al., (2016). Short-term safety, tolerability and efficacy of a very low-calorie-ketogenic diet interventional weight loss program versus hypocaloric diet in patients with type 2 diabetes mellitus. *Nutrition & Diabetes, 6*, 1–7.

Godino, J. G., Van, S. E. M., Marteau, T. M., Sutton, S., Sharp, S. J., & Griffin, S. J., (2016). Lifestyle advice combined with personalized estimates of genetic or phenotypic risk of type 2 diabetes, and objectively measured physical activity: A randomized controlled trial. *PLoS Medicine, 13*, e1002185.

Goldenberg, J. Z., Day, A., Brinkworth, G. D., Sato, J., Yamada, S., Jönsson, T., Beardsley, J., et al., (2021). Efficacy and safety of low and very low carbohydrate diets for type 2 diabetes remission: Systematic review and meta-analysis of published and unpublished randomized trial data. *BMJ,* 372.

Hussain, T. A., Mathew, T. C., Dashti, A. A., Asfar, S., Al-Zaid, N., & Dashti, H. M., (2012). Effect of low-calorie versus low-carbohydrate ketogenic diet in type 2 diabetes. *Nutrition, 28*, 1016–1021.

Keservani, R. K., Kesharwani, R. K., Sharma, A. K., & Ahmad, M. F., (2015a). Stress, Parkinson's, *Alzheimer's* disease: Role of dietary supplements. In: Bagchi, D., Preuss, H. G., & Swaroop, A., (eds.), *Nutraceutical and Functional Foods in Human Life and Disease Prevention* (pp. 241–253). Boca Raton, Fl. Taylor and Francis, Ch 14. ISBN: 9781482237214.

Keservani, R. K., Kesharwani, R. K., Sharma, A. K., Vyas, N., & Chadoker, A., (2010b). Nutritional supplements: An overview. *Int. J. of Curr. Pharm. Rev. and Res., 1*(1), 59–75.

Keservani, R. K., Kesharwani, R. K., Vyas, N., Jain, S., Raghuvanshi, R., & Sharma, A. K., (2010a). Nutraceuticals and functional food as future food. *Der Pharm. Lett., 2*(1), 106–116.

Keservani, R. K., Sharma, A. K., & Kesharwani, R. K., (2016). An overview and therapeutic applications of nutraceutical & functional foods. In: Keservani, R. K., Sharma, A. K., & Kesharwani, R. K., (eds.), *Recent Advances in Drug Delivery Technology* (pp. 160–201). IGI Global Publisher, USA, Ch 6. ISBN:9781522507543.

Keservani, R. K., Sharma, A. K., & Kesharwani, R. K., (2016). Nutraceutical and functional foods for cardiovascular health. In: Meghwal, M., & Goyal, M. R., (eds.), *Food Process Engineering: Emerging Technologies and Innovations in Food Engineering* (pp. 257–278). Apple Academic Press, CRC Press, Taylor and Francis, Ch 10. ISBN: 9781771884020.

Keservani, R. K., Singh, S., Singh, V., Kesharwani, R. K., & Sharma, A. K., (2015b). Nutraceuticals and functional foods in the prevention of mental disorder. In: Bagchi, D., Preuss, H. G., & Swaroop, A., (eds.), *Nutraceutical and Functional Foods in Human Life and Disease Prevention* (pp. 255–268). Boca Raton, Fl. Taylor and Francis, Ch 15. ISBN: 9781482237214.

Khazrai, Y., Defeudis, G., & Pozzilli, P., (2014). Effect of diet on type 2 diabetes mellitus: A review. *Diabetes/Metabolism Research and Reviews,* 30, 24–33.

Ley, S. H., Hamdy, O., Mohan, V., & Hu, F. B., (2014). Prevention and management of type 2 diabetes: Dietary components and nutritional strategies. *The Lancet, 383*, 1999–2007.

Mcmacken, M., & Shah, S., (2017). A plant-based diet for the prevention and treatment of type 2 diabetes. *Journal of Geriatric Cardiology: JGC, 14*, 342.

Micha, R., Michas, G., & Mozaffarian, D., (2012). Unprocessed red and processed meats and risk of coronary artery disease and type 2 diabetes–an updated review of the evidence. *Current Atherosclerosis Reports, 14*, 515–524.

Miller, C. K., Kristeller, J. L., Headings, A., Nagaraja, H., & Miser, W. F., (2012). Comparative effectiveness of a mindful eating intervention to a diabetes self-management intervention among adults with type 2 diabetes: A pilot study. *Journal of the Academy of Nutrition and Dietetics, 112*, 1835–1842.

Okonta, H. I., Ogunbanjo, G. A., & Ikombele, J. B., (2014). Knowledge, attitude and practice regarding lifestyle modification in type 2 diabetic patients: Original research. *African Journal of Primary Health Care and Family Medicine, 6*, 1–6.

Popkin, B. M., (2015). Nutrition transition and the global diabetes epidemic. *Current Diabetes Reports, 15*, 64.

Post, R. E., Mainous, A. G., King, D. E., & Simpson, K. N., (2012). Dietary fiber for the treatment of type 2 diabetes mellitus: A meta-analysis. *The Journal of the American Board of Family Medicine, 25*, 16–23.

Rahati, S., Shahraki, M., Arjomand, G., & Shahraki, T., (2014). Food pattern, lifestyle and diabetes mellitus. *International Journal of High Risk Behaviors & Addiction, 3*.

Rajagopalan, S., & Brook, R. D., (2012). Air pollution and type 2 diabetes. *Mechanistic Insights, 61*, 3037–3045.

Rane, B. R., & Keservani, R. K., (2020). Introduction to nutraceutical & dietary supplements. In: Keservani, R. K., Sharma, A. K., & Kesharwani, R. K., (eds.), *Nutraceutical and Dietary Supplements* (pp. 35–55). Apple Academic Press, Taylor & Francis Group; Chapter 3. ISBN: 9781771888738.

Rawal, L. B., Tapp, R. J., Williams, E. D., Chan, C., Yasin, S., & Oldenburg, B., (2012). Prevention of type 2 diabetes and its complications in developing countries: A review. *International Journal of Behavioral Medicine, 19*, 121–133.

Romieu, I., Dossus, L., Barquera, S., Blottière, H. M., Franks, P. W., Gunter, M., Hwalla, N., et al., (2017). Energy balance and obesity: What are the main drivers? *Cancer Causes & Control, 28*, 247–258.

Sami, W., Ansari, T., Butt, N. S., & Ab Hamid, M. R., (2016). Type 2 diabetes mellitus: Link between diet, HbA1c and complications. *Australasian Medical Journal (Online), 9*, 346.

Sami, W., Ansari, T., Butt, N. S., & Ab Hamid, M. R., (2017a). Effect of diet on type 2 diabetes mellitus: A review. *International Journal of Health Sciences, 11*, 65.

Sami, W., Ansari, T., Butt, N. S., & Hamid, M. R. A., (2017b). Effect of diet on type 2 diabetes mellitus: A review. *International Journal of Health Sciences, 11*, 65–71.

Schwarz, P., & Peltonen, M., (2007). Prevention of type 2 diabetes lessons we have learnt for implementation. *Hormone and Metabolic Research, 39*, 636–641.

Shetty, P., (2013). Nutrition transition and its health outcomes. *The Indian Journal of Pediatrics,* 80, 21–27.

Singh, P., Kesharwani, R. K., & Keservani, R. K., (2017). Antioxidants and vitamins: Roles in cellular functions and metabolism. In: Bagchi, D., (ed.), *Sustained Energy for Enhanced Human Function and Activity*. Elsevier, Ch 24. ISBN: 9780128054130.

Steven, S., Hollingsworth, K. G., Al-Mrabeh, A., Avery, L., Aribisala, B., Caslake, M., & Taylor, R., (2016). Very low-calorie diet and 6 months of weight stability in type 2 diabetes: Pathophysiological changes in responders and nonresponders. *Diabetes Care, 39*, 808–815.

Stewart, R. A. H., Wallentin, L., Benatar, J., Danchin, N., Hagström, E., Held, C., Husted, S., et al., (2016). Dietary patterns and the risk of major adverse cardiovascular events in a global study of high-risk patients with stable coronary heart disease. *European Heart Journal, 37*, 1993–2001.

Surana, K. R., Ahire, E. D., Sonawane, V. N., Talele, S. G., & Talele, G., (2021a). Informatics and methods in nutrition design and development. *Natural Food Products and Waste Recovery: Healthy Foods, Nutrition Design, and Extraction of Valuable Compounds,* 33.

Surana, K. R., Ahire, E. D., Sonawane, V. N., Talele, S. G., & Talele, G., (2021b). Molecular modeling: Novel techniques in food and nutrition development. *Natural Food Products and Waste Recovery: Healthy Foods, Nutrition Design, and Extraction of Valuable Compounds,* 17.

Tapsell, L. C., Neale, E. P., Satija, A., & Hu, F. B., (2016). Foods, nutrients, and dietary patterns: Interconnections and implications for dietary guidelines. *Advances in Nutrition, 7*, 445–454.

Taylor, R., (2013). Type 2 diabetes: Etiology and reversibility. *Diabetes Care, 36*, 1047–1055.

Teixeira, M. E., & Budd, G. M., (2010). Obesity stigma: A newly recognized barrier to comprehensive and effective type 2 diabetes management. *Journal of the American Academy of Nurse Practitioners, 22*, 527–533.

Temelkova-Kurktschiev, T., & Stefanov, T., (2012). Lifestyle and genetics in obesity and type 2 diabetes. *Experimental and Clinical Endocrinology & Diabetes, 120*, 1–6.

Van, G. L., & Scheen, A., (2015). Weight management in type 2 diabetes: Current and emerging approaches to treatment. *Diabetes Care, 38*, 1161–1172.

Via, M. A., & Mechanick, J. I., (2016). Nutrition in type 2 diabetes and the metabolic syndrome. *Medical Clinics, 100*, 1285–1302.

Wilding, J., (2014). The importance of weight management in type 2 diabetes mellitus. *International Journal of Clinical Practice,* 68, 682–691.

Wylie-Rosett, J., Aebersold, K., Conlon, B., Isasi, C. R., & Ostrovsky, N. W., (2013). Health effects of low-carbohydrate diets: Where should new research go? *Current Diabetes Reports, 13*, 271–278.

CHAPTER 5

Potential of Trace Elements and Vitamins in the Management of Oxidative Stress-Linked Diabetes

SHANTI BHUSHAN MISHRA,[1] SHRADHANJALI SINGH,[2]
AMIT KUMAR SINGH,[3] and ANIL KUMAR SINGH[3]

[1]*Department of Pharmacognosy, United Institute of Pharmacy, Prayagraj – 211010, Uttar Pradesh, India, Tel.: +917905848793, E-mail: shantipharma15@gmail.com*

[2]*Department of Pharmaceutical Chemistry, United Institute of Pharmacy, Prayagraj – 211010, Uttar Pradesh, India*

[3]*Department of Pharmaceutics, United Institute of Pharmacy, Prayagraj – 211010, Uttar Pradesh, India*

ABSTRACT

The occurrence of diabetes rises worldwide due to the increasing population, urbanization, aging, and sedentary lifestyle that increase the prevalence of obesity and physical inactivity. Type II diabetes mellitus is the most common and severe metabolic disorder. It is presently supposed to be the fourth cause of death. Type 2 DM is a metabolic disorder with hyperglycemia as the predominant feature, resulting from reduced insulin secretion and enhanced insulin resistance. Globally this metabolic disorder is associated with increased free radical formation leading to oxidative stress, and previous studies have indicated that minor micronutrient deficiencies are more prevalent in diabetic adults rather than in general individuals. However, most of the micronutrients are involved as part of the etiology of this chronic disorder.

The Metabolic Syndrome: Dietary Supplements and Food Ingredients. Raj K. Keservani, MPharm, Durgavati Yadav, PhD, Rajesh K. Kesharwani, PhD, Sippy Singh, DPhil & Kumar Sandeep, PhD (Eds.)

DM complexity increases in the case of discrepancy between free radical production and their regulation by natural antioxidants. Thus, the antioxidant property of micronutrients is of great importance in disease treatment and its complications, whereas other non-antioxidant micronutrients have also shown some relationship.

5.1 INTRODUCTION

Worldwide, diabetes mellitus (DM) is a major cause of death rate and detrimental health. It is also related with life-threatening and potentially deactivating macro- and micro-vascular complications (Levin and Pfeifer, 2009). The critical causes for the occurrence of T2DM consider the interconnected metabolic irregularities, including obesity, hyperlipidemia, and hypertension that brings the metabolic syndrome (Freeman, 2006). About, 150 million people globally have impacted by T2DM and this figure is assumed to double in the next 20 years (Afkhami-Ardekani et al., 2008; Jarouliya and Keservani, 2019; Keservani et al.; 2010a, b). On the basis of a meta-analysis of various studies, 23.7% of people had metabolic syndrome. As per the diabetes atlas of the International Diabetes Federation, the overall occurrence of diabetes is 8.8% as of 2015 and is likely to rise to 10.4% by 2040 (Ogurtsova et al., 2017). Diabetic patients were reported to be over 200 million people globally, 90% of those have Type II diabetes mellitus, characterized by insulin resistance, hyperinsulinemia, cell dysfunction, and subsequent cell failure (Stumvoll et al., 2005). Nutritional therapy will be considered as a factor of quality diabetes care. Although nutritional counseling will improve the risk factors related to diabetes and prediabetes, several micronutrients augment insulin action and encounters consequences of hyperglycemia (Evert et al., 2019). It was also recommended that the augmented rate of type 2 diabetes (T2D) along with the elements of metabolic disorders may be improved by an inclusive lifestyle involvement considering the subsequent quantity of micronutrients and macronutrients in the diet (Ahola et al., 2017). The present chapter is based upon the studies that various micronutrient and macronutrient potentially reduce the onset and improves the conditions associated with metabolic disorders, including DM.

5.2 EFFECT OF MICRONUTRIENTS ON DIABETES

Chronic hyperglycemia causes significant changes in micronutrient status in the body of an individual (Moradian et al., 1994). Disturbances in micronutrient level and increased oxidative damage in DM might affect the occurrence of insulin resistance and diabetic complications (Vincent et al., 2004). The enhancement of DM can also lead to variances in status of micronutrient metabolism (Friedrich et al., 2009). Various researchers have reported an association between DM and alterations in the metabolism of a number of micronutrients. Administration of antioxidants such as magnesium, selenium, zinc, vitamin A, vitamin C, and vitamin E can improve tissue response to insulin and enhance the effectiveness of drugs (Keservani et al., 2010a). The major reason for this pathogenesis of diabetes and its complications are Oxidative stress (Kelly, 1998). Micronutrients as a component of enzymes may be playing an important role in the development and control of T2DM (Figure 5.1). Several possible mechanisms have been postulated by a number of researchers to explain the role of several trace elements in the improvement of oxidative stress-linked diabetes, which are mentioned in subsections (Praveeena et al., 2013).

FIGURE 5.1 Graphic demonstration of the altered levels of trace elements as an indicator of oxidative stress by generating ROS that augmenting the pathways of insulin resistance and development of diabetes.

5.2.1 ZINC

Zinc is an important micronutrient called for appropriate growth and maintenance of systems in the body. Zinc is present in most of the tissues in the human being, primarily in the pancreas. This trace element is important for over 300 enzymatic biochemical reactions (Meunier et al., 2005) and is part of more than 2,000 Zn-dependent transcription factors and other polypeptide molecules (Song et al., 2005). Impairment in the regulation of

Zinc metabolism impacts the biosynthesis, storing, and secretion of insulin and thus glycemic control. A recent study indicated that slight zinc deficiency is more common among Type II DM people, compared to the healthy adult population (Lee et al., 2005). Type 2 diabetic animal models showed decreased amount of Zinc in the pancreas and high urinary excretion of Zn. Irregular zinc and lipid plasma levels present more often in metabolically uncontrolled diabetic patients. Once glucose arrived into the systemic circulation, the preserved insulin from the pancreas which binds with the zinc will get released. It has its anti-inflammatory activities (Keservani et al., 2010b) where its play an important role in eradicating inflammatory markers such as C-reactive proteins (CRPs). Zinc also helps to discard the substances that cause inflammation in cells, assists in preserving insulin sensitivity and cell strength (Ortega et al., 2012). Previous study revealed that a restricted calorie diet (RCD) plays an important factor in the management of obesity and major components related to metabolic disorders (Most et al., 2017). Although zinc is the limiting micronutrient of RCD, the deficiency of zinc increases appetite and worsens insulin resistance (Marreiro et al., 2006). Data from an earlier study showed that daily consumption of 30 mg Zinc gluconate for one month significantly decreased body mass index (BMI) and body weight (BW) (Payahoo et al., 2013). In support with the previous literature, a double-blind randomized clinical trial was also conducted, which states that Zinc levels may play a significant role in appetite and weight management of obese persons (Khorsandi et al., 2019).

5.2.2 COPPER

Oxidative stress plays an important part in the emergence of Type II diabetes mellitus. Imbalance in redox reactions gives rise to the production of reactive oxygen species and induces oxidative damage in islet cells in the pancreas (Savu et al., 2012). Copper (Cu), along with Zn, is essential for a balanced oxidant-antioxidant mechanism. Copper works as a prooxidant and is required for the catalytic activity of SOD, which protects against superoxide radicals (Lyochev and Fridovich, 2010). Imbalance in redox mechanisms may influence the pathogenicity of DM and its related complications (Opara et al., 1992).

Copper takes part as a cofactor for various types of enzymatic reactions. Copper involved in SOD activity. T2DM is associated with the mass production of free radicals, which leads to oxidative damage, and many pathological effects of copper excess are conducive to oxidative damage of the membrane. Once again it has been proved that the imbalance of copper ions between Type II DMs increases their complications (Lefevre et al., 1986). The disequilibrium in Cu and Zn increases the likelihood of oxidative damage to tissues and thus may lead to the pathogenesis of DM or diabetic complications.

5.2.3 CHROMIUM (CR)

Trivalent chromium (Cr) is a vital trace mineral in the metabolism of carbohydrate, protein, and lipids. Thus (Singh et al., 2017), currently it has been prescribed by some physicians in an intention to control Type II DM. Many reports have shown that Chromium could refine insulin receptors also accelerate the hepatic enzyme glucokinase, and enhance the pancreatic islets. Chromium is a vital cofactor mandatory for optimal insulin signaling. It plays a strategic role in the regulation of homeostasis of insulin action. The oligopeptide improves the tyrosine kinase action of the Cr-bound insulin receptor for chromodulin and inhibits phosphotyrosine phosphatase activity, and therefore enhances the intracellular insulin signaling pathway. Acceleration of insulin receptor kinase action by Cr and prohibition of insulin receptor tyrosine phosphatase are accountable for enhanced phosphorylation of the insulin receptor, which is linked with improved insulin sensitivity (Wang et al., 2005). In the state of chromium deficiency, this recognized correlation can be inverted or even stopped, a condition that contributes to insulin resistance over the long term (Figure 5.2). Cr can control hyperglycemia in type 2 DM (Cefalu and Hu, 2004).

5.2.4 MAGNESIUM

Magnesium is a basic mineral involved in maintaining glucose homeostasis. It increases the rate of chemical reaction involved in carbohydrate metabolism. It is involved in the production, binding, and activity of insulin on many levels. Hormones such as parathyroid hormone and calcitonin play important roles in Magnesium metabolism, and the collection of evidence indicates that insulin is an important hormone in regulating Mg metabolism (Barbagallo et al., 2003; Paolisso and Barbagallo, 1997). Magnesium improves the capacity of insulin to activate tyrosine kinase for the metabolism of glucose. Lower level of magnesium has been predictable in T2DM.

Lower level of serum magnesium will result in glucose homeostasis and insulin sensitivity among Type 2 DM patients (Figure 5.3). Lack of Mg in the body's cellular compartments has been strongly associated with T2DM (Vincent, 2018). Hypomagnesemia can result in hyperglycemia and cause insulin resistance, which usually occurs in diabetic persons with poor sugar control (Tosiello, 1996). The level of Magnesium in urine was directly correlated with fasting blood glucose. Large amount of urinary Mg excretion in diabetes was indorsed to be due to poor metabolic control, impaired renal reabsorption, and the use of diuretics. Increased loss of urinary Mg may be a strong indication of diabetes status, especially when there was a strong association between glucose control and urinary Mg loss.

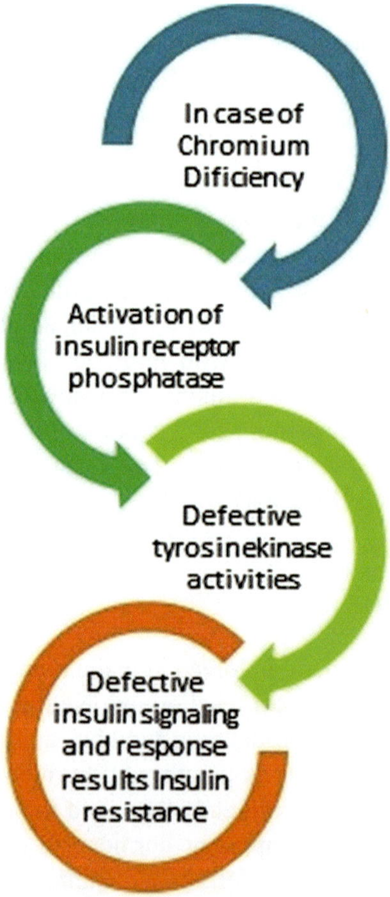

FIGURE 5.2 The relationship between intracellular Cr deficiency and type 2 diabetes.

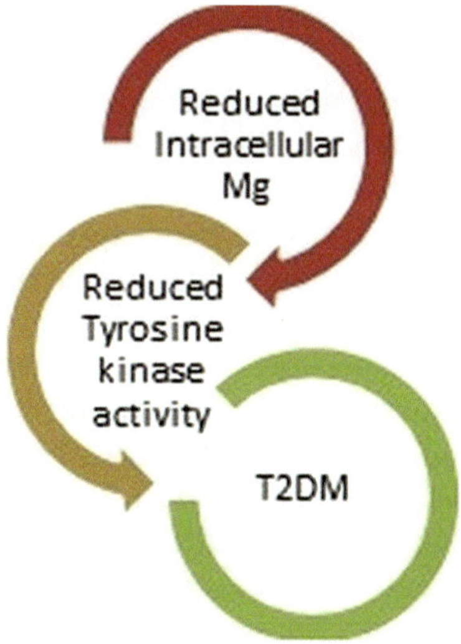

FIGURE 5.3 The relationship between intracellular Mg deficiency and T2DM.

5.2.5 SELENIUM

Selenium (Se) is an important microcomponent and a key ingredient of many functional selenoproteins that defend tissues and membranes from oxidative damage (Rayman, 2000). Selenium has been revealed to enhance insulin sensitivity and establish insulin-like activity in each of the *in vivo* and *in vitro* trials (Mueller and Pallauf, 2006; Stapleton, 2000). Selenium has been mentioned to sustain amelioration of glucose homeostasis via exerting insulin-like action in diabetic rats (Berg et al., 1995; Becker et al., 1996). It regulates specific-cell target genes and promotes recovery of islet function. Though, the role of Se in glucose metabolism has been controversial. *In vitro* studies in the pancreatic islets and mouse insulinoma-cell line have shown an increase in insulin secretion in response to Se treatment, suggesting that Se provides protection to the pancreas (Campbell et al., 2008). Elevated levels of plasma Se were actually found to be associated with elevated serum insulin in mice (Labunskyy et al., 2011). There is limited mechanistic evidence that can explain the association between increased Se exposure and higher risk of T2DM; so, for now, any such discussion is highly anticipated. Selenium is shown to have high inter-individual variability and restricted therapeutic range in terms of metabolic sensitivity (Whanger et al., 1996; Vinceti et al., 2001).

5.2.6 CALCIUM

Calcium, an important element, is the most lavish ion in the human body, mainly preserved in the bones as hydroxyapatite (Laine et al., 1994). To improve individual's metabolic health, a calcium rich diet prevents the detrimental impacts of metabolic disorders by expressing hormonal activities, changes in intracellular calcium levels, intestinal fat absorption, carbohydrate metabolism, renin-angiotensin system, fecal fat excretion, lipid metabolism, oxidative stress, and inflammatory responses. A systemic meta-analysis study was conducted to study dose-response correlation between calcium consumption and chances of metabolic disorder (Rane and Keservani, 2020). The study indicated that dose-response effect showed a 300 mg/day increase in dietetic calcium ingestion is significantly linked with a reduction of metabolic syndrome risk (Das and Chaudhry, 2021). Epidemiologic studies have stated that low calcium consumption may be a lethal for primary hypertension and hence studies suggested a substantially less prevalence of the metabolic disorder among individuals with higher calcium consumption (both diet and supplements). Animal and human studies revealed that high calcium consumption might reduce the levels of parathyroid hormone and 1,25 dihydroxy vitamin D3 [1,25(OH)2D3] and thus influence adipocyte metabolism by preventing lipogenesis and stimulating lipolysis (Simin et al., 2005).

5.2.7 VANADIUM

Vanadium compounds including III, IV, and V chlorodipicolinate (Vdipic-Cl)) or inorganic vanadium salts have been primarily explored as probable therapeutic agents for the therapy of various metabolic syndrome, particularly diabetes. Models of Type 1 diabetes treated with bis (allixinato) oxo vanadium (IV) showed an increase in glycogen in muscle and heart, which recommended an improvement in the insulin signaling pathway associated with the restoration of the GLUT-4 expression (Hiromura et al., 2007).

Studies in Streptozotocin-induced diabetic rats treated with Vdipic-Cl substantially improved hyperglycemia, increased hepatic glycogen synthesis, glucose intolerance, and ameliorate mRNA levels of amylolytic enzymes in the hepatic region such as phosphoenolpyruvate carboxykinase, glucokinase, and L-pyruvate kinase (Xie et al., 2014). In T2D model, the administration of BMOV) bis (3-hydroxy-2-methyl-4-pyronato) oxovanadium (IV), vanadium compound normalizes and improves glycogen synthase stimulation in the skeletal muscle in streptozotocin-diabetic rats (Semiz and McNeill, 2002). In the previous literature, it was suggested that vanadium salts possess protein-tyrosine phosphatase (PTPases) inhibitory properties and thus phosphorylation activity amongst the insulin receptor and its substrate was regulated and inhibition of protein-tyrosine phosphatase 1B activity permits the insulin receptor (IR) to remain activated (Deng et al., 2002). One of the promising results concluded that peroxovanadium complexes could form small amounts of hydroxyl radicals, peroxide, superoxide, and singlet oxygen and improve insulin signaling (Iwakami et al., 2011). With respect to cell apoptosis, vanadate-dependent nicotinamide adenine dinucleotide hydrogen (NADH) oxidation linked to plasma membranes generates hydrogen peroxide (H_2O_2), which facilitates apoptosis via the activation of p53 (Huang et al., 2000).

5.3 EFFECT OF MACRONUTRIENTS ON DIABETES

Over the past decades, macronutrients were known to exert ameliorating effects on complications associated with diabetes. Fatty acids were studied for influencing the transfer of glucose transporters and insulin binding receptor to lessened diabetic complications. Linoleic acid (LA) and Linoleic acid (ALA) are essential fatty acids (FAs), as they vie for the desaturation enzymes. Enzymes including fatty acid desaturases (FADS) and elongases catalyzes fatty acids, such as ALA is converted into the long chain omega 3 fatty acids or docosahexaenoic acid (DHA) (Day et al., 2017) and assures the levels of omega 3 and omega 6 FAs in the plasma and serum. Since ALA is preferred as compare to LA by both FADS 1 and FADS 2 (Mathias et al., 2014), maintaining an optimal balance between omega 6 and omega 3 FAs (1:1 to 2:1) in the diet declines the prevalence of diabetes, obesity, and atherosclerosis (Wu et al., 2014). Various sources of micronutrients and macronutrients from plant and animal origin are mentioned in Table 5.1.

TABLE 5.1 Sources of Micronutrients and Macronutrients

SL. No.	Micronutrients/Macronutrients	Source
1.	Omega-3-fatty acids	Nuts, seeds, vegetables, some fruit, egg yolk, white and red meats, as well as marine organisms (Behl et al., 2019)

TABLE 5.1 *(Continued)*

SL. No.	Micronutrients/Macronutrients	Source
2.	Alpha-linolenic acid (ALA)	Plant oils (rapeseed oil, soybean, and walnuts), flax, and perilla (Racine et al., 2007)
3.	Zinc	Pumpkin seeds, yogurt, cashews, chickpeas, almonds, kidney beans, peas, cheese, chicken, and lobster (McKillop et al., 2021)
4.	Copper	Tomatoes, apples, sesame seeds, asparagus, avocado, salmon, millet, sunflower seed kernels lobster (McKillop et al., 2021)
5.	Chromium	Grape juice, orange juice, Brewer's yeast, lettuce, turkey breast, green beans, banana, whole wheat bread, peanut butter, chicken breast, spaghetti, carrot, egg, and orange (Anderson et al., 1992)
6.	Magnesium	Black beans, spinach, pumpkin seeds, almonds, oatmeal, cashews, peanuts, soymilk, broccoli, brown rice, yogurt, and raisins (McKillop et al., 2021)
7.	Selenium	Brazil nuts, tuna, yellowfin, shrimp, macaroni, turkey, cottage cheese, boiled lentils, white bread, spinach, corn flakes, and green peas (McKillop et al., 2021)
8.	Calcium	Milk, cheese, kale, turnip green, chia seeds, broccoli, whole wheat bread, yogurt, and orange juice (McKillop et al., 2021)
9.	Vanadium	Gelatin, skim milk powder, dried lentils, lobster, pea green, radishes, buckwheat, cabbage, oats, potato, rye seed, lettuce, vegetable oils (peanut, safflower, soybean, olive, corn, cedar, and cottonseed), wheat, millet, meats, fish, butter, and cheese (Badmaev et al., 1999; Schroeder et al., 1963)

5.3.1 ALPHA LINOLEIC ACID

Abundant *in-vivo* and *in-vitro* literature has considered the potential of ALA in regulating glucose homeostasis and ameliorating insulin sensitivity by fat metabolism, gene regulation, and adipocyte formation (Kelley et al., 2009; Ghafoorunissa and Natarajan, 2005; Adachi et al., 2006, 2008; Opara et al., 1992; Xiong et al., 2013; Poudyal et al., 2013). Although, reports suggest that reduction in glycated hemoglobin (HbA1c %) and fasting blood glucose (FBG mmol/L), the effect of Alpha linoleic acid on the biological markers of diabetic control exclusively in persons with T2DM can be assessed effectively (Jovanovski et al., 2017).

5.3.2 OMEGA-3-FATTY ACIDS

Preclinical, lipidomics, and anti-inflammatory studies on omega 3 fatty acids suggested a preventable measure for insulin resistance, obesity, and other complications associated with diabetes (Flachs et al., 2014). In previous studies, it was suggested that omega 3 fatty acids decrease the glucose level by alleviating glucose uptake in differentiated adipocytes through the induction of glucose transporter-1 in association with fibroblast growth factor 21 which is produced in consequences of peroxisome proliferator-activated receptor (Li et al., 2012). Various advantages of omega 3 fatty acids in diabetes related complications exist. Some of the uses are listed in Table 5.2.

TABLE 5.2 Uses of Omega 3 Fatty Acids in T2DM

Parameters	Benefits
Cardiovascular disease	Treatment in atrial fibrillation and hypertension (Nestel et al., 2015)
Diabetic nephropathy	Ameliorates urine albumin to creatinine ratio (ACR) and preserving glomerular filtration rate (GFR) (Han et al., 2016)
Diabetic neuropathy	Reduced neuro-inflammation and oxidative stress and activation of cell survival pathways (Michael-Titus et al., 2014)
Enhance insulin sensitivity/ insulin signaling pathway	Increasing diacylglycerol, reactive oxygen species (ROS) and protein kinase C (PKC), insulin receptor substrate-1 (IRS1) serine phosphorylation and decreases IRS1 tyrosine phosphorylation, thus inhibiting the action of PI3K (Shetty et al., 2021). Attenuate ER stress, enhance FA oxidation in mitochondria and uncouple mitochondria, causing reduction in lipid buildup and ROS accumulation (Lopez et al., 2011)
Modulate pancreatic β-cell insulin secretion	Enhance the adipokine production from adipose tissue, and directly affect cell activity by binding to GPR120, PPARs, and GPR40 thereby promoting insulin secretion (Vessby et al., 2012).

5.4 ROLE OF VITAMINS AS TRACE ELEMENTS IN T2DM

5.4.1 VITAMIN A

It is stated that T2DM affected population will suffer from lower levels of plasma vitamin A and carotenoids. One study showed that a high level of carotene is associated with a lower risk of diabetes (Hussain et al., 2009).

5.4.2 VITAMIN B1

Thiamine (B1) plays many roles in the body, such as acting as a cofactor in the active displacement of aldehyde groups and glycations. It also plays as a neuro-transmission and neuronal conductivity, lower level of thiamine may cause various diabetic complications. Reduced levels of thiamine and rise in the renal excretion have been found in both types of diabetic patients (Gavrilov et al., 2012).

5.4.3 VITAMIN B6

Pyridoxal-5'-phosphate is a functional form of vitamin B6. This vitamin B6 will It is an aminotransferase, also acting as a coenzyme for glucose-phosphorylase for the utilization of glycogen in muscle and liver, thus actively participating in the metabolism of glucose (Al-Attas et al., 2012).

5.4.4 VITAMIN B3

NAD and NADH have nicotic acid as a vital compound, and this plays a major role in the production of adenosine triphosphate and energy proficiency at the cellular level. Supplementing niacin towards DM patients will reduce the monocyte adhesion to endothelial cells (Gonzalez-Ortiz et al., 2011).

5.4.5 VITAMIN B7

A study stated that supplementation of biotin in diabetic rats done with anti-diabetic effects, clearly inhibiting insulin resistance in skeletal muscle by enhancing the activity of the glucose transporter protein GLUT4 (Hellmann et al., 2010).

5.4.6 *VITAMIN B12*

Inadequacy of vitamin B12 and folic acid exists in Type II diabetic patients, which is interrelated with oxidative stress, resulting in hyperhomocysteinemia (Tavintharan et al., 2011). As a result of this correlation, it is reasonable that lack of vitamin B12 level should be viewed as an important risk factor for diabetes-related complications.

5.4.7 *VITAMIN B9*

The glycosylated hemoglobin fasting blood glucose got reduced with the supplementation of Folate and also improved the insulin resistance, glycemic control, and serum insulin level as well as homocysteinemia in T2DM patients. The vitamin B12, pyridoxine, and folate supplementation exhibit promising effects on manifestations of diabetic retinopathy (Sahin et al., 2013).

5.4.8 *VITAMIN C*

Studies have observed that Type II diabetes mellitus patients have lower circulating levels of vitamin C and enhanced levels of oxidative stress biomarkers compared to nondiabetic patients. Present evidences suggest that the supplementation of Vitamin-C showed the diverging results on plasma concentration of antioxidants (Paolisso et al., 1995).

5.4.9 *VITAMIN D*

Vitamin D deficiency in individuals with T2D is hazardous for cardiovascular disease. Elderly patients with Type II DM have a higher risk of vitamin D deficiency, which is very frequent in all over European countries, due to inadequate intake, little sun exposure, and alterations in biosynthesis, absorption, and metabolism of the vitamin D (Keservani et al., 2016; Hirani et al., 2014).

5.4.10 VITAMIN E

The antioxidant activity of vitamin E alleviate the risk of glucotoxicity and its related problems; decreased plasma alpha-tocopherol levels have been observed in diabetic patients with longer period of illness. Few studies proved that supplementation with vitamin E have shown positive effects on blood glucose on T2DM (Polidori et al., 2000; Keservani et al., 2016).

5.5 CONCLUSION

Although micronutrients have an impact on the risk of Type II Diabetes development and its related complications, there is inadequate evidence to recommend micronutrients supplementation in the overall diabetic population. Hence viewed in this way, it is essential to conduct a dietary assessment to identify specific intake deficits. The use of supplements brings a risk of toxicity, however, there is sufficient scientific evidence to recommend micronutrient intake in persons with Type II diabetes who are treated with oral antihyperglycemic drugs to diminish the risk of neuropathy and its consequences.

KEYWORDS

- **diabetes mellitus**
- **fatty acid desaturases**
- **hydrogen peroxide**
- **insulin receptor**
- **micronutrients**
- **oxidative stress**
- **trace elements**

REFERENCES

Adachi, T., Tanaka, T., Takemoto, K., et al., (2006). Free fatty acids administered into the colon promote the secretion of glucagon-like peptide-1 and insulin. *Biochem. Biophys. Res. Commun., 340*, 332–337.

Adachi, T., Yanaka, H., Kanai, H., et al., (2008). Administration of perilla oil coated with calshell increases glucagon-like peptide secretion. *Biol. Pharm. Bull., 31*, 1021–1023.

Afkhami-Ardekani, M., Modarresi, M., & Amirchaghmaghi, E., (2008). Prevalence of microalbuminuria and its risk factors in type 2 diabetic patients. *Indian J. Nephrol., 18*, 112–117.

Ahola, A., Harjutsalo, V., Thorn, L., Freese, R., Forsblom, C., Mäkimattila, S., & Groop, P., (2017). The association between macronutrient intake and the metabolic syndrome and its components in type 1 diabetes. *Br. J. Nutr., 117*, 450–456.

Al-Attas, O. S., Al-Daghri, N. M., Alfadda, A. A., Abd-Alrahman, S. H., & Sabico, S., (2012). Blood thiamine and its phosphate esters as measured by high-performance liquid chromatography: Levels and associations in diabetes mellitus patients with varying degrees of microalbuminuria. *J. Endocrinol. Invest., 35*, 951–956.

Anderson, R. A., Bryden, N. A., & Polansky, M. M., (1992). Dietary chromium intake. Freely chosen diets, institutional diet, and individual foods. *Biol. Trace. Elem. Res., 32*, 117–121.

Badmaev, V., Prakash, S., & Majeed, M., (1999). Vanadium: A review of its potential role in the fight against diabetes. *J. Altern. Complement. Med., 5*, 273–291.

Barbagallo, M., Dominguez, L. J., Galioto, A., Ferlisi, A., Cani, C., Malfa, L., et al., (2003). Role of magnesium in insulin action, diabetes and cardio-metabolic syndrome X. *Mol. Aspects Med., 24*, 39–52.

Becker, D., Reul, B., Ozcelikay, A., Buchet, J., Henquin, J., & Brichard, S., (1996). Oral selenate improves glucose homeostasis and partly reverses abnormal expression of liver glycolytic and gluconeogenic enzymes in diabetic rats. *Diabetologia., 39*, 3–11.

Behl, T., Grover, M., Shah, K., Makkar, R., Kaur, L., Sharma, S., & Gupta, J., (2019). Role of omega-3-fatty acids in the management of diabetes and associated complications. In: Watson, R. R., & Preedy, V. R., (ed.), *Bioactive Food as Dietary Interventions for Diabetes* (pp. 185–192). Academic Press: London.

Berg, E. A., Wu, J. Y., Campbell, L., & Stapleton, S. R., (1995). Insulin-like effects of vanadate and selenate on the expression of glucose-6-phosphate dehydrogenase and fatty acid synthase in diabetic rats. *Biochimie., 77*, 919–924.

Campbell, S. C., Aldibbiat, A., Marriott, C. E., Landy, C., Ali, T., Ferris, W. F., et al., (2008). Selenium stimulates pancreatic beta-cell gene expression and enhances islet function. *FEBS Lett., 582*(15), 2333–2337.

Cefalu, W. T., & Hu, F. B., (2004). Role of chromium in human health and in diabetes. *Diab. Care., 27*, 2741–2751.

Das, S., & Choudhuri, D., (2021). Role of dietary calcium and its possible mechanism against metabolic disorders: A concise review. *J. Food. Biochem., 45*, e13697.

Day, E. A., Ford, R., & Steinberg, G. R., (2017). AMPK as a therapeutic target for treating metabolic diseases. *Trends. Endocrinol. Metab.,* 28, 545-560.

Deng, H., Callender, R., Huang, Z. H., & Zhang, Z. Y., (2002). Is the PTPase–vanadate complex a true transition state analogue? *Biochem., 41*, 5865–5872.

Evert, A. B., Dennison, M., Gardner, C. D., Garvey, W. T., Lau, K. H. K., MacLeod, J., Mitri, J., et al., (2019). Nutrition therapy for adults with diabetes or prediabetes: A consensus report. *Diab. Care, 42*, 731–754.

Flachs, P., Rossmeisl, M., & Kopecky, J., (2014). The effect of n–3 fatty acids on glucose homeostasis and insulin sensitivity. *Physiol. Res., 63*, S93–118.

Freeman, H. R. D., (2006). Cox. type-2 diabetes: A cocktail of genetic discovery. *Hum. Mol. Genet., 15*, 202–209.

Friedrich, M., Hansell, P., & Palm, F., (2009). Diabetes, oxidative stress, nitric oxide and mitochondria function. *Curr. Diab. Rev., 5*, 120–144.

Gavrilov, V., Harman-Boehm, I., Amichay, D., Tessler, G., Shuster, T., Friger, M., & Gorodischer, R., (2012). Kidney function and retinol status in type 2 diabetes mellitus patients. *Acta Diabetol., 49*, 137–143.

Ghafoorunissa, I. A., & Natarajan, S., (2005). Substituting dietary linoleic acid with alpha-linolenic acid improves insulin sensitivity in sucrose fed rats. *Biochim. Biophys. Acta, 1733*, 67–75.

Gonzalez-Ortiz, M., Martínez-Abundis, E., Robles-Cervantes, J. A., Ramirez-Ramirez, V., & Ramos-Zavala, M. G., (2011). Effect of thiamine administration on metabolic profile, cytokines and inflammatory markers in drug-naïve patients with type 2 diabetes. *Eur. J. Nutr., 50*, 145–149.

Han, E., Yun, Y., Kim, G., Lee, Y. H., Wang, H. J., Lee, B. W., et al., (2016). Effects of omega-3 fatty acid supplementation on diabetic nephropathy progression in patients with diabetes and hypertriglyceridemia. *PLoS One, 11*, e0154683.

Hellmann, H., & Mooney, S., (2010). Vitamin B6: A molecule for human health. *Molecules, 15*, 442–459.

Hirani, V., Cumming, R. G., Le Couteur, D. G., Naganathan, V., Blyth, F., Handelsman, D. J., Waite, L. M., & Seibel, M. J., (2014). Low levels of 25-hydroxy vitamin D and active 1,25-dihydroxyvitamin D independently associated with type 2 diabetes mellitus in older Australian men: The concord health and ageing in men project. *J. Am. Geriatr. Soc., 62*, 1741–1747.

Hiromura, M., Nakayama, A., Adachi, Y., Doi, M., & Sakurai, H., (2007). Action mechanism of bis (allixinato) oxovanadium (IV) as a novel potent insulin-mimetic complex: Regulation of GLUT4 translocation and FoxO1 transcription factor. *J. Biol. Inorg. Chem., 12*, 1275–1287.

Huang, C., Zhang, Z., Ding, M., Li, J., Ye, J., Leonard, S. S., Shen, H. M., et al., (2000). Vanadate induces p53 transactivation through hydrogen peroxide and causes apoptosis. *J. Biol. Chem., 275*, 32516–32522.

Hussain, F., Maan, M. A., Sheikh, M. A., Nawaz, H., & Jamil, A., (2009). Trace elements status in type 2 diabetes. *Bangladesh. J. Med. Sci., 8*, 52–56.

Iwakami, S., Misu, H., Takeda, T., Sugimori, M., Matsugo, S., Kaneko, S., & Takamura, T., (2011). Concentration-dependent dual effects of hydrogen peroxide on insulin signal transduction in H4IIEC hepatocytes. *PLoS One, 6*, e27401.

Jarouliya, U., & Keservani, R. K., (2019). Pathways leading to child obesity. In: Bagchi, D., (ed.), *Global Perspectives on Childhood Obesity: Current Status, Consequences and Prevention* (2nd edn., pp. 137–146). Academic Press, Elsevier, Chapter 12. ISBN: 9780128128404.

Jovanovski, E., Li, D., Thanh Ho, H. V., et al., (2017). The effect of alpha-linolenic acid on glycemic control in individuals with type 2 diabetes: A systematic review and meta-analysis of randomized controlled clinical trials. *Medicine (Baltimore), 96*, e6531.

Kelley, D. S., Vemuri, M., Adkins, Y., et al., (2009). Flaxseed oil prevents trans-10, cis12-conjugated linoleic acid-induced insulin resistance in mice. *Brit. J. Nutr., 101*, 701–708.

Kelly, F. J., (1998). Use of antioxidants in the prevention and treatment of disease. *J. Int. Fed. Clin. Chem., 10*, 21–23.

Keservani, R. K., Kesharwani, R. K., Sharma, A. K., Vyas, N., & Chadoker, A., (2010b). Nutritional supplements: An overview. *Int. J. of Curr. Pharm. Rev. and Res., 1*(1), 59–75.

Keservani, R. K., Kesharwani, R. K., Vyas, N., Jain, S., Raghuvanshi, R., & Sharma, A. K., (2010a). Nutraceuticals and functional food as future food. *Der Pharm. Lett., 2*(1), 106–116.

Keservani, R. K., Sharma, A. K., & Kesharwani, R. K., (2016). An overview and therapeutic applications of nutraceutical & functional foods. In: Keservani, R. K., Sharma, A. K., & Kesharwani, R. K., (eds.), *Recent Advances in Drug Delivery Technology* (pp. 160–201). IGI Global Publisher, USA, Ch 6. ISBN:9781522507543.

Khorsandi, H., Nikpayam, O., Yousefi, R., et al., (2019). Zinc supplementation improves body weight management, inflammatory biomarkers and insulin resistance in individuals with obesity: A randomized, placebo-controlled, double-blind trial. *Diabetol. Metab. Syndr., 11*, 101.

Labunskyy, V. M., Lee, B. C., Handy, D. E., Loscalzo, J., Hatfield, D. L., & Gladyshev, V. N., (2011). Both maximal expression of selenoproteins and selenoprotein deficiency can promote the development of type 2 diabetes-like phenotype in mice. *Antioxid. Redox. Signal., 14*, 2327–2336.

Laine, M., Weckstrom, M., Vuolteenaho, O., & Arjamaa, O., (1994). Effect of ryanodine on atrial natriuretic peptide secretion by contracting and quiescent rat atrium. *Pflugers. Arch., 426*, 276–283.

Lee, J. H., Lee, H. J., Lee, I. K., & Yoon, J. S., (2005). Zinc and copper status of middle- and old-aged woman in type 2 diabetes. *Korean. J. Nutr., 38*, 56–66.

Lefevre, M., Keen, C. L., Lonnerdal, B., Hurley, L. S., & Schneeman, B. O., (1986). Copper deficiency-induced hypercholesterolemia: Effects on HDL subfractions and hepatic lipoprotein receptor activity in the rat. *J. Nutr., 116*, 1735–1746.

Levin, M., & Pfeifer, M., (2009). *Uncomplicated Guide to Diabetes Complications* (3rd edn.). American Diabetes Association, Alexandria, Virginia.

Li, K., Li, L., Yang, M., Liu, H., Boden, G., & Yang, G., (2012). The effects of fibroblast growth factor-21 knockdown and over-expression on its signaling pathway and glucose & lipid metabolism *in vitro. Mol. Cell. Endocrinol., 348*, 21–26.

Lopez, S., Bermudez, B., Ortega, A., Varela, L. M., Pacheco, Y. M., Villar, J., Abia, R., & Muriana, F. J., (2011). Effects of meals rich in either monounsaturated or saturated fat on lipid concentrations and on insulin secretion and action in subjects with high fasting triglyceride concentrations. *Am. J. Clin. Nutr., 93*, 494–499.

Lyochev, S. I., & Fridovich, I., (2010). Mechanism of the peroxidase activity of Cu, Zn superoxide dismutase. *Free Radic. Biol. Med., 48*, 1565–1569.

Marreiro, D. N., Geloneze, B., Tambascia, M. A., Lerário, A. C., Halpern, A., & Cozzolino, S. M., (2006). Effect of zinc supplementation on serum leptin levels and insulin resistance of obese women. *Biol. Trace. Elem. Res., 112*, 109–118.

Mathias, R. A., Pani, V., & Chilton, F. H., (2014). Genetic variants in the FADS gene: Implications for dietary recommendations for fatty acid intake. *Curr. Nutr. Rep., 3*, 139–148.

McKillop, K., Harnly, J., Pehrsson, P., Fukagawa, N., & Finley, J., (2021). Food data central, USDA's updated approach to food composition data systems. *Curr. Dev. Nutr., 5*, 596.

Meunier, N., O'Connor, J. M., Maiani, G., Cashman, K. D., Secker, D. L., Ferry, M., et al., (2005). Importance of zinc in the elderly: The ZENITH study. *Eur. J. Clin. Nutri., 59*, S1–S4.

Michael-Titus, A. T., & Priestley, J. V., (2014). Omega-3 fatty acids and traumatic neurological injury: From neuroprotection to neuroplasticity? *Trends. Neurosci., 37*, 30–38.

Mooradian, A. D., Failla, M., Hoogwerf, B., Maryniuk, M., & Wylie-Rosett, J., (1994). Selected vitamins and minerals in diabetes. *Diab. Care, 17*, 464–479.

Most, J., Tosti, V., Redman, L. M., & Fontana, L., (2017). Calorie restriction in humans: An update. *Ageing Res. Rev., 39*, 36–45.

Mueller, A. S., & Pallauf, J., (2006). Compendium of the anti-diabetic effects of supranutritional selenate doses. *In vivo* and *in vitro* investigations with type II diabetic db/db mice. *J. Nutri. Biochem., 17*, 548–560.

Nestel, P., Clifton, P., Colquhoun, D., et al., (2015). Indications for omega-3 long-chain polyunsaturated fatty acid in the prevention and treatment of cardiovascular disease. *Heart. Lung. Circ., 24*, 769–779.

Ogurtsova, K., Fernandes, J. D., Huang, Y., et al., (2017). IDF diabetes atlas: Global estimates for the prevalence of diabetes for 2015 and 2040 *Diab. Res. Clin. Prac., 128*, 40–50.

Opara, E. C., Hubbard, V. S., Burch, W. M., et al., (1992). Characterization of the insulinotropic potency of polyunsaturated fatty acids. *Endocrinology, 130*, 657–662.

Ortega, R. M., Rodríguez-Rodríguez, E., Aparicio, A., Jiménez, A. I., López-Sobaler, A. M., Gonzalez-Rodriguez, L. G., & Andres, P., (2012). Poor zinc status is associated with increased risk of insulin resistance in Spanish children. *Br. J. Nutr., 107*, 398–404.

Paolisso, G., & Barbagallo, M., (1997). Hypertension, diabetes mellitus, and insulin resistance: The role of intracellular magnesium. *Am. J. Hyperten., 10*, 346–355.

Paolisso, G., Balbi, V., Volpe, C., Varricchio, G., Gambardella, A., Saccomanno, F., Ammendola, S., et al., (1995). Metabolic benefits deriving from chronic vitamin C supplementation in aged non-insulin dependent diabetics. *J. Am. Coll. Nutr., 14*, 387–392.

Payahoo, L., Ostadrahimi, A., Mobasseri, M., Bishak, Y. K., Farrin, N., Jafarabadi, M. A., et al., (2013). Effects of zinc supplementation on the anthropometric measurements, lipid profiles and fasting blood glucose in the healthy obese adults. *Adv. Pharm. Bull., 3*, 161–165.

Polidori, M. C., Mecocci, P., Stahl, W., Parente, B., Cecchetti, R., Cherubini, A., Cao, P., et al., (2000). Plasma levels of lipophilic antioxidants in very old patients with type 2 diabetes. *Diabetes. Metab. Res. Rev., 16*, 15–19.

Poudyal, H., Panchal, S. K., Ward, L. C., et al., (2013). Effects of ALA, EPA and DHA in high-carbohydrate, high-fat diet-induced metabolic syndrome in rats. *J. Nutr. Biochem., 24*, 1041–1052.

Praveeena, S., Pasula, S., & Sameera, K., (2013). Trace elements in diabetes mellitus. *J. Clin. Diagn. Res., 7*, 1863–1865.

Racine, R. A., & Deckelbaum, R. J., (2007). Sources of the very-long-chain unsaturated omega-3 fatty acids: Eicosapentaenoic acid and docosahexaenoic acid. *Curr. Opin. Clin. Nutri. Metab. Care, 10*, 123–128.

Rane, B. R., & Keservani, R. K., (2020). Introduction to nutraceutical & dietary supplements. In: Keservani, R. K., Sharma, A. K., & Kesharwani, R. K., (eds.), *Nutraceutical and Dietary Supplements* (pp. 35–55.). Apple Academic Press, Taylor & Francis Group; Chapter 3. ISBN: 9781771888738.

Rayman, M. P., (2000). The importance of selenium to human health. *The Lancet, 356*, 233–241.

Sahin, K., Tuzcu, M., Orhan, C., Sahin, N., Kucuk, O., Ozercan, I. H., Juturu, V., & Komorowski, J. R., (2013). Anti-diabetic activity of chromium picolinate and biotin in rats with type 2 diabetes induced by high-fat diet and streptozotocin. *Br. J. Nutr., 110*, 197–205.

Savu, O., Ionescu-Tirgoviste, C., Atanasiu, V., Gaman, L., Papacocea, R., & Stoian, I., (2012). Increase in total antioxidant capacity of plasma despite high levels of oxidative stress in uncomplicated type 2 diabetes mellitus. *Int. J. Med. Res., 40*, 709–716.

Schroeder, H. A., Balassa, J. J., & Tipton, I. H., (1963). Abnormal trace metals in man-vanadium. *J. Chronic. Dis., 16*, 1047–1071.

Semiz, S., & McNeill, J. H., (2002). Oral treatment with vanadium of Zucker fatty rats activates muscle glycogen synthesis and insulin-stimulated protein phosphatase-1 activity. *Mol. Cell. Biochem., 236*, 123–131.

Shetty, S. S., & Kumari, S., (2021). Fatty acids and their role in type-2 diabetes (Review). *Exp. Ther. Med., 22*, 706.

Simin, L., Yiqing, S., Earl, S., Ford, J., Manson, A. E., Buring, J. E., Paul, M., & Ridker, P. M., (2005). Dietary calcium, vitamin D, and the prevalence of metabolic syndrome in middle-aged and older U.S. Women *Diab. Care, 28*, 2926–2932.

Singh, P., Kesharwani, R. K., & Keservani, R. K., (2017). Protein, carbohydrate and fats: Energy metabolism. In: Bagchi, D., (ed.), *Sustained Energy for Enhanced Human Function and Activity*. Ch 6. ISBN: 9780128054130.

Song, Y., Wang, J., Li, X. K., & Cai, L., (2005). Zinc and the diabetic heart. *Biometals., 18*, 325–332.

Stapleton, S. R., (2000). Selenium: An insulin mimetic. *Cell. Mole. Life Sci., 57*, 1874–1879.

Stumvoll, M., Goldstein, B. J., & Van, H. T. W., (2005). Type 2 diabetes: Principles of pathogenesis and therapy. *Lancet, 365*, 1333–1346.

Tavintharan, S., Woon, K., Pek, L. T., Jauhar, N., Dong, X., Lim, S. C., & Sum, C. F., (2011). Niacin results in reduced monocyte adhesion in patients with type 2 diabetes mellitus. *Atherosclerosis, 215*, 176–179.

Tosiello, L., (1996). Hypomagnesemia and diabetes mellitus: A review of clinical implications. *Arch. Int. Med., 156*, 1143–1148.

Vessby, B., Ahrén, B., Warensjo, E., & Lindgärde, F., (2012). Plasma lipid fatty acid composition, desaturase activities and insulin sensitivity in Amerindian women. *Nutr. Metab. Cardiovasc. Dis., 22*, 176–181.

Vincent, A. M., Russell, J. W., Low, P., & Feldman, E. L., (2004). Oxidative stress in the pathogenesis of diabetic neuropathy. *Endocrine. Rev., 25*, 612–628.

Vincent, J. B., (2018). Beneficial effects of chromium (III) and vanadium supplements in diabetes. In: Bagchi, D., & Nair, S., (ed.), *Nutritional and Therapeutic Interventions for Diabetes and Metabolic Syndrome* (p 365). Academic Press: New York.

Vinceti, M., Wei, E. T., Malagoli, C., Bergomi, M., & Vivoli, G., (2001). Adverse health effects of selenium in humans. *Rev. Environ. Health, 16*, 233–251.

Wang, H., Kruszewski, A., & Brautigan, D. L., (2005). Cellular chromium enhances activation of insulin receptor kinase. *Biochem., 44*, 8167–8175.

Whanger, P., Vendeland, S., Park, Y. C., & Xia, Y., (1996). Metabolism of subtoxic levels of selenium in animals and humans. *Ann. Clin. Lab. Sci., 26*, 99–113.

Wu, Y., Ding, Y., Tanaka, Y., & Zhang, W., (2014). Risk factors contributing to type 2 diabetes and recent advances in the treatment and prevention. *Int. J. Med. Sci., 11*, 1185–1200.

Xie, M., Chen, D., Zhang, F., Willsky, G. R., Crans, D. C., & Ding, W., (2014). Effects of vanadium (III, IV, V)-chlorodipicolinate on glycolysis and antioxidant status in the liver of STZ-induced diabetic rats. *J. Inorg. Biochem., 136*, 47–56.

Xiong, Y., Swaminath, G., Cao, Q., et al., (2013). Activation of FFA1 mediates GLP1 secretion in mice. Evidence for allosterism at FFA1. *Mol. Cell. Endocrinol., 369*, 119–129.

CHAPTER 6

Gut Microbiome and the Metabolic Syndrome

SHIVANI SRIVASTAVA,[1] PRERANA ADITI,[2] and DURGAVATI YADAV[3]

[1]*Department of Physiology and Cell Biology, Davis Heart and Lung Research Institute, The Ohio State University Wexner Medical Center, Columbus, OH, USA*

[2]*Department of Medicinal Chemistry, Banaras Hindu University, Varanasi, Uttar Pradesh, India*

[3]*Department of Urology, AIIMS, New Delhi, India, E-mail: durgavati45yadav@gmail.com*

ABSTRACT

Microbiota (as a microbial endocrine organ), symbiotically living inside our gut, performs a vital role in the protection of physiology of human body, metabolism, and function, thus mutually benefited. The alteration in microbial composition is known to be associated with various complications leading to generation of diseases like obesity, chronic kidney disease (CKD), cardiovascular diseases (CVDs), inflammatory bowel diseases (IBD), and fatty liver diseases like non-alcoholic fatty liver disease (NAFLD) and non-alcoholic steatohepatitis (NASH). Epigenetic is a heritable sequence-independent DNA change. These changes enforced by the environmental factors like chemicals, nutrition, stress, aging, microbiome, exercise, circadian rhythm, smoking, and early environment. The epigenetic modifications of genes expression include (i) DNA methylation; (ii) histone modifications such as acetylation, methylation, phosphorylation, SUMOylation, and ubiquitination; (iii)

The Metabolic Syndrome: Dietary Supplements and Food Ingredients. Raj K. Keservani, MPharm, Durgavati Yadav, PhD, Rajesh K. Kesharwani, PhD, Sippy Singh, DPhil & Kumar Sandeep, PhD (Eds.)

histone tail modifications; (iv) chromatin remodeling; and (v) RNA-based mechanisms like microRNAs (miRNAs), piwi-interacting RNA (piRNAs), circular RNA (circRNAs), and long non-coding RNAs (lncRNAs). Recently, there have been wide studies in linking the role of the gut microbiome in inflammation and insulin resistance. Methodologies should be developed to identify the importance of gut microbial composition and function. It needs to be standardized to enhance the presentation of meta-analyzes, and aids the understanding of the function of mechanistic pathways concerning to short-chain fatty acids, propionate, butyrate, bile acids, lipopolysaccharide, TLRs (toll-like receptors), and NLRP (NOD-LRR containing protein family members) inflammasomes in the pathogenesis of difficulties of obesity, metabolic syndrome, and diabetes.

6.1 INTRODUCTION

During 460–377 B.C., Hippocrates (father of modern medicine) gave a famous quote, "All diseases begin in the gut." The collective genomes of the microorganisms at any specific environment are referred to as microbiome and community of microorganisms is known as microbiota. The microbiome is believed to be as a virtual organ of our body since it is not involved in structural parts. There are around 100 trillion microorganisms (principally bacteria) survive in the human gut. The whole genome of humans consists of over 2,300 genes in which microbiome encoded 3 million genes and releasing thousands of metabolites, thus affecting many vital functions of host phenotype and health (Valdes, Walter, Segal, and Spector, 2018).

Since the last decade, various evidences have supported the role of the gut microbiota (through analysis of fecal samples) in type 2 diabetes (T2D) and glycemic control. The gut microbiota study with respect to health and disease revolutionized followed by the development of PCR-based techniques leading to quantification and characterization of bacterial composition through sequencing to bacterial genes. Shotgun metagenomics sequencing and 16S ribosomal RNA gene amplicon sequencing are the two main methods for the microbiome next-generation sequencing (Byndloss et al., 2017). Gut microbiota has been studied to have a significant role in metabolic disorders. It is reported that gut microbiota differs significantly in the Caucasian population with obesity and T2D, but limited reports are available for developing countries. The integrity of the gut barrier is influenced by metabolites produced by gut microorganisms, leads to inflammatory mediators' leakage into systemic circulation results in insulin resistance (Upadhyaya and Banerjee, 2015). It is recently estimated that our total number of germ and somatic cells equals our microbiota (Sender, Fuchs, and Milo, 2016).

6.2 MICROBIOME AND METABOLIC SYNDROME

Metabolic syndrome is believed to predict or raise the likelihood of heart failure and diabetes, constitute the risk factors mainly low HDL cholesterol level, high blood pressure and high fasting blood sugar (Tzika, Dreker, and Imhof, 2018). In insulin resistance, body cells are unable to respond to the action of insulin.

In the case of metabolic syndrome (MetS), obesity, and insulin resistance etiopathogenesis are involved and its related metabolic defects. Inflammation-induced metabolic abnormalities are exacerbated by intestinal permeability in chronic low-grade inflammation, making the microbiome a key role. Inflammation is a set of reactions by vascularized tissue in response to damage or infection (Chen et al., 2018). These responses are required as a defensive response for endurance, but long-lasting contact to stimuli and mobilization of immune cells can become unfavorable. Hotamisligl et al. in 1993 reported the first study linking chronic low-grade inflammation to obesity and insulin resistance; over expression of TNF-α in the adipose tissue of rodents is responsible for insulin resistance (Hotamisligil, Shargill, and Spiegelman, 1993). Again, the study was extended to obese humans, and it was found that there was increased expression of TNF-α mRNA in obese adipose tissue than normal controls (Hotamisligil, Shargill, and Spiegelman, 1993). Adipocytes and hepatocytes, among other metabolic cells, are in intimate relation with immune cells and blood arteries, according to the scientists. This demonstrates that metabolism and the immune system are inextricably linked (Chait and den Hartigh, 2020). Several studies have found increased proinflammatory markers in obese people's adipose tissue (Lindhorst et al., 2021), as well as enlarged adipocytes in mice and humans with macrophage infiltration (Russo and Lumeng, 2018) and dying adipocytes surrounded by crown-like structures made up of phagocytosis macrophages (Russo and Lumeng, 2018). Proinflammatory cytokines and chemokines are produced by inflamed adipocytes in collaboration with macrophages. Cai et al. discovered that HFD (high fat diet)-induced obesity leads to the chronic activation of NF-κB in the liver. This kind of subacute-chronic inflammation led to hyperglycemia and hepatic insulin resistance (Cai et al., 2005). These studies have recognized chronic low-grade inflammation as a causal factor to the expansion of diabetes. HFDs are being investigated to observe if they can cause inflammatory responses that lead to changes in metabolism. Intestinal pathobionts (commensal bacteria that turn pathogenic) are proving to be a potential option. Vaziri et al. found that the microbiome may contribute to oxidative stress conditions in chronic kidney disease (CKD) by generating uremic toxins like indoxyl sulfate and p-cresylsulfate, which increase oxidative stress and inflammation in the kidneys (Lau, Kalantar-Zadeh, and Vaziri, 2015).

Thus, it can be inferred from the above that special types of microbiomes may be flourish and creating diseased state. According to Byndloss et al. Study the normal intestinal anaerobic dynamic homeostasis was experimentally disrupted, indicating that an oxygenated gut increases pathobionts which leads to inflammation (Byndloss et al., 2017).

The intestinal PPARγ limits the oxygen bioavailability via β-oxidation in colonocytes. The natural process which works as a stimulant is short-chain butyrate produced bacterially. When bacteria producing butyrate were removed by antibiotics in the gut, oxygen levels increased in the lumen, and pathogenic forms like *E. coli* and *Salmonella* flourished (Molinaro et al., 2020; Rivera-Chávez et al., 2016). Obese people have higher levels of ECB (endocannabinoid system) in their adipose tissue and plasma, whereas bacterial LPS is a powerful activator of ECB production, producing dysmetabolism through low-grade inflammation (Geurts et al., 2011) cannabinoid receptor (CB1) inhibition protects against obesity and low-grade inflammation. Various researchers concluded that the gut microbiome commands ECB activity in the adipose tissue and colon by testing dietary factors and altered gut microbiome composition (Muccioli et al., 2010) the connection between inflammation, altered glycemia, and intestinal permeability was studied by Thaiss et al. By meticulously studying the functions of leptin, the gut microbiome, and obesity in obstruction in many animal models and people, then they conclude dysmetabolism may be both a driver and an outcome of impaired barrier function. They measured 27 human serum markers and correlated them with intestinal permeability surrogate markers, i.e., serum pathogen recognition receptors (PRR). Hemoglobin A1c was the only marker that was associated positively with circulating PRR ligands (Dorcely et al., 2017).

Collectively, these findings suggest that chronic hyperglycemia, intestinal permeability, and an increased risk of systemic infections all have a link.

6.3 GUT MICROBIOME – HEART AXIS

There is a close association between CVD and type 2 DM, which makes the common soil hypothesis, suggesting that both pathological conditions influencing each other by sharing common environmental and genetic risk factors, such as inflammation, high blood pressure, oxidative stress, dyslipidemia, obesity, hypercoagulability, thrombophilia, and insulin resistance (Dorcely et al., 2017). Approximately 451 million people are affected by diabetes mellitus (DM) around the world and according to epidemiological studies; these diabetic patients are at more risk towards the condition of heart failure. Various pathological conditions related to diabetes such as hyperglycemia, inflammation, oxidative stress, additional development of advanced glycation end products (AGEs), disturbed calcium homeostasis, endothelial dysfunction (ED), autonomic dysfunction, abnormalities in glucose and fatty acid consumption, small vessel disease and myocardial fibrosis work in coordination at molecular level to damage myocardial function (WHO report; https://www.who.int/health-topics/diabetes#tab=tab_1).

6.4 MICROBIOME AND ALZHEIMER'S DISEASE (AD)

Alzheimer's disease (AD), one of the neurodegenerative disorders is the most frequent cause of dementia. It is characterized by the impaired cerebral and cognition accumulation of amyloid-β plaques (Aβ) and neurofibrillary tangles (Kowalski and Mulak, 2019; Jiang, Li, Huang, Liu, and Zhao, 2017). Additionally, the large amounts of lipopolysaccharides and amyloids can be released by the community of bacteria present in the gut, which might result in the AD pathogenesis-associated modulation of signaling pathways and cytokines production through proinflammation (Jiang et al., 2017).

6.5 GUT MICROBIOME – KIDNEY AXIS

The altered intestinal epithelial barrier and enhanced gut-derived toxins production can be caused by the change in intestinal flora due to increasing concentration of urea during CKD. Because of this, the process of kidney injury can be accelerated (Hobby et al., 2019).

The potent endotoxin called lipopolysaccharide (LPS), exist in the outer membrane of gram-negative bacteria is known to be accountable for chronic immune responses linked with inflammation.

6.6 GUT MICROBIOME – LUNG AXIS

Gut microbiota dysbiosis is implicated with several respiratory infections, lung disorders and diseases, involving cystic fibrosis, asthma, and allergy (Anand and Mande, 2018).

6.7 GUT MICROBIOME – LIVER AXIS

6.7.1 MICROBIOME AND FATTY LIVER DISEASES (NAFLD/NASH)

NAFLD is the most common liver disease in the world, characterized by abnormal fat accumulation in the liver, with an estimated prevalence of approximately 25–30%. The interactions of microbiome with the host play a vital role in the pathogenesis of NAFLD and also promotes its severity to NASH, that may develop towards liver cirrhosis, fibrosis, hepatocellular carcinoma, and the requirement for liver transplantation (Bashiardes, Shapiro, Rozin, Shibolet, and Elinav, 2016).

6.7.2 GUT MICROBIOME AND OBESITY

Obesity is defined as an accumulation of excess fat which impairs the state of resulting from changes in gut microbiota, irregular energy intake and energy balance, and inappropriate diet with the influence of environmental factors and genetic makeup (Bhagavathi et al., 2019).

Through exacerbation of cardiac insulin-stimulated glucose oxidation, obesity impairs cardiac energy metabolism and worsens the severity of heart failure. For every 1 kg/m^2 rise in the BMI (body mass index), the possibility of heart failure in men increases by 5% and in women by 7%. Loss of weight through dietary interventions blunts cardiac hypertrophy and enhances cardiac insulin sensitivity and cardiac function in obesity-related heart failure (Keservani et al., 2016, 2010a, b). Improving the sensitivity of cardiac insulin increases the rate of glucose oxidation and also results in the ATP production in the failing heart (Karwi et al., 2019).

Obesity and obesity-associated metabolic problems diabetes (Jarouliya and Keservani, 2019) are characterized via way of means of particular changes within the composition and characteristics of the human intestine microbiome. Mechanistic research has indicated that the gastrointestinal microbiota can impact each aspect of the electricity stability equation; namely, as an element influencing electricity usage from the food regimen and as an element that impacts host genes that adjust electricity expenditure and storage (Davis, 2016).

Gut microbes play a chief function in power extraction from meals through lots of mechanisms. Many plant polysaccharides and complicated carbohydrates cannot be digested with the aid of using the host; however, the intestine microbes can metabolize those to short-chain fatty acids (SCFA), along with butyrate, propionate, and acetate. Butyrate is used because the number one power supply for colonic epithelial cells, even as propionate and acetate are vital for lipogenesis and gluconeogenesis withinside the liver (Turnbaugh et al., 2006). Differences in SCFA tiers were located in overweight and lean mice. For example, in a generic version of obesity, ob/ob mice have multiplied butyrate and acetate concentrations of their ceca and much less power, decided with the aid of using bomb calorimetric, of their feces as compared to their lean counterparts (Turnbaugh et al., 2006).

6.8 MICROBIOME AND AGING

In comparison to SAMR1 (senescence-accelerated mouse resistance 1), vascular aging phenotype is found to be increased in SAMP8 mice with activation of p53 and p21, inhibition of cyclin-dependent kinases CDK2 and CDK4, activates retinoblastoma (Rb), increase of MT (media thickness), MT/LD (the ratio of media thickness to lumen diameter), MA (media area), and collagen/MA (ratio of collagen content to media area), decrease of eNOS expression, an enhanced SAβ-gal (senescence-associated β-galactosidase) activity, decrease of EDD (endothelium-dependent dilation), increases the expression of liver FMO3, ED. These gets further aggravated and worsened after TMAO treatment. During the aging process, the SIRT1 expression may be downregulated by the enhanced circulating gut flora-dependent metabolite TMAO levels leading to increased oxidative stress and excessive ROS accumulation in endothelial cells. This causes ED by impairing NO production and induces EC senescence through activation of p53-p21-Rb pathway, ultimately leads to vascular aging. The inhibitor of TMA formation, 3,3-dimethyl-1-butanol (DMB) could improve the eNOS expression and EDD throughout the aging process (Ke et al., 2018).

6.9 INTERCONNECTION BETWEEN MICROBIOME AND EPIGENETIC ALTERATION

6.9.1 MICROBIOTA REGULATES DNA METHYLATION AND HISTONE MODIFICATION

The changes in feature and composition of the intestine microbiota were recognized to be concerned within the pathogenesis of metabolic sicknesses through initiation of epigenetic adjustments like DNA methylation, histone adjustments and regulation through noncoding RNAs. The metabolites produced through the intestine microbiota along with short-chain fatty acids, folates, biotin, and trimethylamine-N-oxide can also regulates epigenetic modification. Metabolites inclusive of folate, diet B12, betaine, and choline are probably concerned within the synthesis of 6-methyl tetrahydrofolate, that is a major methyl donor, for the era of S-adenosylmethionine (SAM) that play a major role in DNA methylation approaches (Kowalski and Mulak, 2019). These methyl donor vitamins are determined to be regulated through particular intestine microbial groups inclusive of Lactobacillus and Bifidobacteria which might be recognized for folate manufacturing (Rossi, Amaretti, and Raimondi, 2011). In order to recognize the function performed through Lactobacillus and Bifidobacterium in the regulation of diabetes, a study through Murri et al. determined their degrees had been decreased in kind 1 diabetic healthful Caucasian children (Murri et al., 2013). The histone acetylation system may be regulated through diverse intestine-microbial derived metabolites inclusive of SCFAs (Fellows et al., 2018). In a rat model of LPS-precipitated neuro-inflammation, supplementation with acetate increased the acetylation levels of brain histones H3 at lysine nine and H4 at lysine 8 and 16, resulting in neuroglial activation and decrease inside the cholinergic cell (a nerve cell) (Soliman, Smith, and Rosenberger, 2012). The microbiome of the intestine can influence HDAC activity by producing epigenetic metabolites like SCFAs. Butyrate and propionate were diagnosed as ability individuals to HDAC inhibition (Woo and Alenghat, 2022). Butyrate is vital for retaining homeostasis within the intestine and is likewise essential in the regulation of many approaches inclusive of epigenetic mechanisms, gluconeogenesis, lipogenesis, and inflammatory circumstances. Among diverse epigenetic adjustments, butyrate is in particular referred to as a category I and sophistication II HDAC inhibitor (Kelly et al., 2018).

6.9.2 MICROBIOTA REGULATES MIRNA, CIRCRNAS, LONG NONCODING RNAS (LNCRNAS)

Unlike linear mRNAs, formed through canonical splicing and removing pre-mRNAs introns with the help of small nuclear ribonucleoproteins (snRNPs) and involves polyadenylation, circular RNAs (circRNAs) are derived by the back-splicing of precursor mRNAs and the cutted segment circularize by joining 5' end covalently to the 3' end without polyadenylated tails (Bayoumi et al., 2018; Workman et al., 2019). At the junction site, circRNAs have 3,'5'-phosphodiester bond. Exon regions of the encoded protein are the major source of circRNAs, but some are also generated from the antisense transcripts, intron regions or long non-coding RNAs (Wang et al., 2017). Many CVDs like cardiac fibrosis, myocardial infarction, cardiomyopathy, and atherosclerosis are predominantly regulated by CircRNAs. Based on the genomic locations, circRNAs are grouped into intergenic, exon-intronic, multiple exonic, one exonic, intronic, and antisense circRNAs (Bayoumi et al., 2018).

Microbiome remotely governs the expression miR-204 while the latter controls the expression of SIRT1 (Vikram et al., 2016). Reports suggested that the use of broad-spectrum antibiotics to suppress the gut microbiota rescue high fat diet-induced vascular dysfunction by repression of miR-204 via endothelial transcription factor Stat3 signaling activation, leading to the upregulation of SIRT1 (Vikram et al., 2016; Lou et al., 2014). SIRT1 activation plays a beneficial role against cardiovascular aging through the promotion of autophagy and suppression of oxidative stress and inflammation (Luo et al., 2014; Kida and Goligorsky, 2016; Cencioni et al., 2015).

The intestine microbiota has been suggested to have an effect on host fitness with the aid of using regulating host miRNAs. For example, *Fusobacterium nucleatum* could exaggerate chemoresistance of colorectal cancer by diminishing the expression of miR-18a* and miR-4802. MiR-515-5p and miR-1226-5p should sell the boom of *Fusobacterium nucleatum* (Fn) and *Escherichia coli (E. coli),* that have been suggested to power colorectal cancer (Li, Chen, and Wang, 2020). In the human gastrointestinal tract, approximately 1,014 microorganisms are present, accounting for approximately 95% of the human microbiome (Thursby and Juge 2017).

6.10 CERAMIDE, GUT MICROBIOME, AND AUTOPHAGY

For a few decades, autophagic pathways have been widely studied with respect to many diseases, like diabetes, cancer, etc. Autophagy gets stimulated in response to external stimuli, like stress, starvation, etc., leading to the degradation of non-functional and harmful sub-cellular components within the autolysosomes to become the supplier of energy and needful components/molecules to other parts of the cells. Because of autophagic flux diminished due to chronic cellular damage, autophagy becomes insufficient with age. The phenomenon of autophagy is essential in order to maintain homeostasis under stressful conditions.

One of the sphingolipids intermediate, ceramide has been studied to play a crucial role in autophagy induced cell death through both mTOR and Beclin:Bcl 2 dissociation pathways. Ceramide has been shown to negatively regulate cell survival and to disturb the homeostatic autophagic phenomenon in T2D.

Ceramides and sphingomyelin belong to the family of sphingolipids, known to act as mediators in the crosstalk between intestinal cells and microbiota. Ceramides also perform as anti-cancerous representative in the body. Thus, Ceramides have paradoxical behavior. In contrast to this, sphingomyelin also induces starvation induced autophagy. Probably, the various intermediates of sphingolipid metabolism have different approaches to perform either constructive or destructive messenger among microbiome and host cells.

In mammalian cells, sphingolipids play an important role to balance between cell death and survival via ceramide (pro-apoptotic)/sphingolipid-1-phosphate (anti-apoptotic) rheostat. Both metabolic products have the potential to induce autophagy, sphingolipid-1-phosphate regulates autophagy during starvation while ceramide induced autophagy is starvation independent (Pattingre et al., 2009). Also, Sphingolipid act as a communication mediator between host intestinal cells and microbiota (Bryan et al., 2016).

6.11 MICROBIOME AND AUTISM

Autism, or autism spectrum disorder (ASD) is a heterogeneous category of neurodevelopmental disorders exhibited by social contact and communication impairments and limited or stereotyped interests and behaviors generally occurring prior to the fourth year of life (Tonacci et al., 2019).

Blood-brain barrier permeability is known to get affected by microbial constitution and early childhood development. Allergies, metabolic problems, yeast overgrowth, inflammation, diabetes, gastrointestinal complications, and immune impairment can be caused by an unevenness between pathogenic and beneficial microbes due to over and early use of antibiotics (Eshraghi et al., 2018b).

Autism is currently diagnosed in one out of every 59 children in the United States. Although initial research focused mainly on the genome to identify the possible causes of autism, further recent studies suggested a considerable role for epigenetic regulation of microbiome and gene expression. Differences in the intestinal microbiome may lead to altered genetic expression and potentially contribute to the cause of autism (Eshraghi et al., 2018a).

6.12 ROLE OF PREBIOTICS AND PROBIOTICS

It is reported that the protein intake via red or white meat may be overall advantageous over non-meat protein. The V4–V5 region sequencing examination of 16S ribosomal RNA gene from rat's caecum showed that the favorable genus *Lactobacillus* was found to be enhanced in white meat (chicken and fish) fed rat group compared to red (beef and pork) or non-meat (casein and soy) group. Rats fed with red meat have decreased levels of visceral fat as compared to soya protein intake. Additionally, the lipopolysaccharide-binding proteins were significantly reduced in rats fed with casein and meat proteins, leading to downregulation of host inflammation. If taken at recommended level, meat, and dairy protein may be healthier over soy protein through maintenance of gut bacterial composition (Zhu et al., 2015).

In the scientific community, the concept of probiotics has been introduced by the Nobel laureate Elie Metchnikoff. Lactobacillus species are recognized as probiotics in a range of metabolic diseases such as diabetes and obesity because of their health-promoting and defensive properties (Sivamaruthi, et al., 2019).

The reduction in heart performance and atrial and ventricular remodeling represents the complex process called cardiac aging. Concomitantly, liver flavin monooxygenase 3 (FMO3) converts TMA into trimethylamine-*N*-oxide (TMAO) (Warrier et al., 2016). The evidence of links between CVD to gut microbiota formation of TMAO (trimethylamine-N-oxide) is considered as novel opportunities for therapeutic intervention. This choline derived metabolite, TMAO, has also been implicated in massive production of ROS, which in turn can impact epigenetic programming because it can lead to nucleic acid depurination or deamination, which can trigger mechanisms for DNA repair and substitution with a non-methylated cytosine (Sharma, Li, Stoll, and Tollefsbol, 2020) (Figure 6.1).

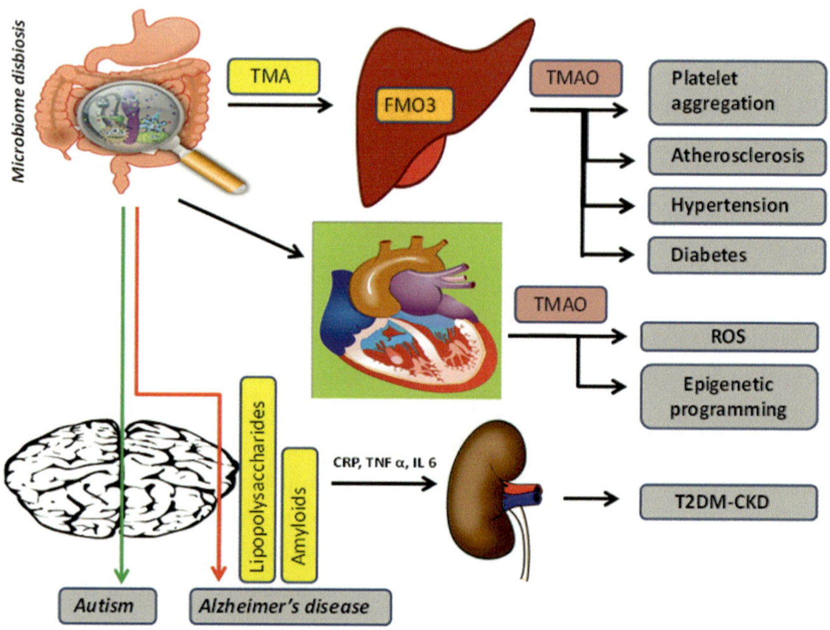

FIGURE 6.1 Pathways connecting microbial dysbiosis with diseases.

6.13 GUT MICROBIOME AND ANTI-DIABETIC THERAPY

There are numerous cures that can be used to deal with insulin resistance and diabetes, one of the most famous cures which is used as a first-line remedy is metformin (McCreight, Bailey, and Pearson, 2016). Some of the useful consequences of metformin may be attributed to alteration in intestine microbiota (Rodriguez, 2018). Metformin remedy, similarly to enhancing the glycemic profile of mice, additionally will increase large quantity of Akkermansia, a mucin-degrading bacterium, whilst as compared to controls without metformin. Human research from Danish, Swedish, and Chinese members with T2DM on metformin remedy have corroborated those findings (Hao et al., 2013; Shin et al., 2014; Jiang et al., 2017). Multivariate evaluation has proven that there are full-size variations in intestine composition among metformin-untreated members with T2DM vs controls and full-size will increase in Escherichia species and reduce in Intestinibacter after metformin remedy. Diet-brought on obese (DIO) C57BL/6 J mice that had been fed a high-fat food regimen underwent Roux-en-Y gastric bypass (RYGB) surgical operation, sham surgical operation, or sham surgical operation alongside caloric restriction (Hao, Zhao, Berthoud, and Ye, 2013). RYGB altered intestine microbial composition as soon as 1-week post-surgical operation and stabilized after five weeks. Mostly, RYGB produced enrichment of Bacteroidetes, Verrucomicrobia, and Proteobacteria (Crommen, Mattes, and Simon, 2020).

6.14 FECAL TRANSPLANTS AND METABOLIC CORRECTION

In this system, feces from a donor are transferred to a recipient through a nasogastric tube, colonoscope, enema, capsule, or any aggregate of these. Therapeutically, this technique has verified fulfillment in curing colitis because of recurrent Clostridium difficile infection (86). It restores the diversity of the recipient's microbiome, which had been lost due to long-term use of broad-spectrum antibiotics, and it competes with *C. difficile* for key intestinal niches. This increases the enquiry of whether or not FMT may be used for different indications, for example, to switch a metabolically healthful microbiome right into a recipient with MetS within the desire that the healthful phenotype can be transferred. This has been verified in lots of rodent studies. Among the primary changed into a take a look at from Bäckhed et al. wherein microbiota transferred from traditional mice into germ-unfastened mice ended in a 60% growth in frame fats content material and insulin resistance inside 14 days no matter decreased meals intake (Bäckhed et al., 2004). In humans, the switch of microbiota from donors that are lean to overweight people with MetS brought about a growth in insulin sensitivity inside 6 weeks, in addition to a growth in butyrate-generating bacteria (Vrieze et al., 2012). After FMT, the intestine microbiota composition and insulin sensitivity again to baseline levels (Vrieze et al., 2012) in a follow-up take a look at. These statistics show that a metabolically healthful microbiome exists, and the associated phenotype may be transferred, by the help of microbiome, to people with MetS. Metformin, for example, can sell blooms of *Akkermansia muciniphila*, generally seemed as a useful bacterium (Shin et al., 2014; Jiang et al., 2017). When the metformin-dealt with human microbiome changed into transplanted into germ-unfastened mice with glucose intolerance, the discovered glucose defects had been corrected (Wu et al., 2017).

6.15 CONCLUSION

The elements contributing to MetS are the end result of complicated host intrinsic elements consisting of genetics and the intestine microbiome, and extrinsic elements consisting of weight loss programs and lifestyle. The interactions of certain participants of the intestine microbiota network procedure outside cues that contributes to persistent low-grade inflammation, obesity, hyperglycemia, and dyslipidemia. Fascinatingly, nutritional elements were maximum broadly implicated in inflicting metabolic defects, and are also time-honored because they are the number one driving force of our intestine microbiota composition. The intestine microbiome sensitivity debts now no longer handiest to the quantity of meals, however to nutritional composition, appearing as nutritional biosensors for the host, makes it almost impossible to split the location of the intestine microbiome from concerns of the way meals alternatives have an effect on host metabolism. Environmental elements consisting of shifts in light/darkish cycles, i.e., circadian disruption, can disrupt the herbal diurnal version within the intestine microbiota, which additionally has an effect on host metabolism. The microbiome subject remains in its infancy as compared with the long-status take a look at endocrinology and metabolic diseases.

KEYWORDS

- **cardiovascular diseases**
- **chronic kidney disease**
- **inflammatory bowel diseases**
- **metabolic syndrome**
- **microbiome**
- **prebiotics**

REFERENCES

Anand, S., & Mande, S. S., (2018). Diet, microbiota, and gut-lung connection. *Frontiers in Microbiology*. https://doi.org/10.3389/fmicb.2018.02147.

AS, B., T, A., JP, T., YL, T., & IM, K., (2018). Circular noncoding RNAs as potential therapies and circulating biomarkers for cardiovascular diseases. *Acta Pharmacologica Sinica, 39*(7), 1100–1109. https://doi.org/10.1038/APS.2017.196.

Bashiardes, S., Shapiro, H., Rozin, S., Shibolet, O., & Elinav, E., (2016). Non-alcoholic fatty liver and the gut microbiota. *Molecular Metabolism*. https://doi.org/10.1016/j.molmet.2016.06.003.

Bryan, P. F., Karla, C., Edgar, A. M. T., Sara, E. E. P., Gemma, F., & Luz, C., (2016). Sphingolipids as mediators in the crosstalk between microbiota and intestinal cells: Implications for inflammatory bowel disease. *Mediators of Inflammation, 2016*(Article ID 9890141), 1–11. https://doi.org/10.1155/2016/9890141.

BS, S., P, K., N, S., & C, C., (2019). A review on role of microbiome in obesity and antiobesity properties of probiotic supplements. *BioMed Research International, 2019*(Article ID 3291367), 1–20. https://doi.org/10.1155/2019/3291367.

Byndloss, M. X., Olsan, E. E., Rivera-Chávez, F., Tiffany, C. R., Cevallos, S. A., Lokken, K. L., & Bäumler, A. J., (2017). Microbiota-activated PPAR-γ signaling inhibits dysbiotic Enterobacteriaceae expansion. *Science, 357*(6351), 570–575. https://doi.org/10.1126/SCIENCE.AAM9949.

Cai, D., Yuan, M., Frantz, D. F., Melendez, P. A., Hansen, L., Lee, J., & Shoelson, S. E., (2005). Local and systemic insulin resistance resulting from hepatic activation of IKK-β and NF-κB. *Nature Medicine, 11*(2), 183–190. https://doi.org/10.1038/nm1166.

Cencioni, C., Spallotta, F., Mai, A., Martelli, F., Farsetti, A., Zeiher, A. M., & Gaetano, C., (2015). Sirtuin function in aging heart and vessels. *Journal of Molecular and Cellular Cardiology, 83*, 55–61. https://doi.org/10.1016/J.YJMCC.2014.12.023.

Chait, A., & Den, H. L. J., (2020). Adipose tissue distribution, inflammation and its metabolic consequences, including diabetes and cardiovascular disease. *Frontiers in Cardiovascular Medicine, 5*, 87–100. https://doi.org/10.3389/FCVM.2020.00022.

Chen, L., Deng, H., Cui, H., Fang, J., Zuo, Z., Deng, J., & Zhao, L., (2018). Inflammatory responses and inflammation-associated diseases in organs. *Oncotarget, 9*(6), 7204–7218. https://doi.org/10.18632/ONCOTARGET.23208.

Crommen, S., Mattes, A., & Simon, M. C., (2020). Microbial adaptation due to gastric bypass surgery: The nutritional impact. *Nutrients, 12*(4), 1–21. https://doi.org/10.3390/NU12041199.

Dorcely, B., Katz, K., Jagannathan, R., Chiang, S. S., Oluwadare, B., Goldberg, I. J., & Bergman, M., (2017). Novel biomarkers for prediabetes, diabetes, and associated complications. *Diabetes, Metabolic Syndrome and Obesity: Targets and Therapy, 10*, 345. https://doi.org/10.2147/DMSO.S100074.

Eshraghi, R. S., Deth, R. C., Mittal, R., Aranke, M., Kay, S. I. S., Moshiree, B., & Eshraghi, A. A., (2018a). Early disruption of the microbiome leading to decreased antioxidant capacity and epigenetic changes: Implications for the rise in autism. *Frontiers in Cellular Neuroscience, 12*(Article 256), 1–19. https://doi.org/10.3389/FNCEL.2018.00256.

Eshraghi, R. S., Deth, R. C., Mittal, R., Aranke, M., Kay, S. I. S., Moshiree, B., & Eshraghi, A. A., (2018b). Early disruption of the microbiome leading to decreased antioxidant capacity and epigenetic changes: Implications for the rise in autism. *Frontiers in Cellular Neuroscience*. https://doi.org/10.3389/fncel.2018.00256.

Geurts, L., Lazarevic, V., Derrien, M., Everard, A., Van, R. M., Knauf, C., & Cani, P. D., (2011). Altered gut microbiota and endocannabinoid system tone in obese and diabetic leptin-resistant mice: Impact on apelin regulation in adipose tissue. *Frontiers in Microbiology, 2*(149), 1–17. https://doi.org/10.3389/FMICB.2011.00149.

Hao, Z., Zhao, Z., Berthoud, H. R., & Ye, J., (2013). Development and verification of a mouse model for Roux-En-Y gastric bypass surgery with a small gastric pouch. *PLoS One, 8*(1), 1–9. https://doi.org/10.1371/JOURNAL.PONE.0052922.

Hobby, G. P., Karaduta, O., Dusio, G. F., Singh, M., Zybailov, B. L., & Arthur, J. M., (2019). Chronic kidney disease and the gut microbiome. *American Journal of Physiology –Renal Physiology*. American physiological society. https://doi.org/10.1152/ajprenal.00298.2018.

Hotamisligil, G.S., Shargill, N.S., &Spiegelman, B.M., (1993). Adipose expression of tumor necrosis factor-alpha: Direct role in obesity-linked insulin resistance. *Science (New York, N.Y.), 259*(5091), 87–91. https://doi.org/10.1126/SCIENCE.7678183.

Jarouliya, U., & Keservani, R. K., (2019). Pathways leading to child obesity. In: Bagchi, D., (ed.), *Global Perspectives on Childhood Obesity: Current Status, Consequences and Prevention* (2nd edn., pp. 137–146). Academic Press, Elsevier. Chapter 12. ISBN: 9780128128404.

Jiang, C., Li, G., Huang, P., Liu, Z., & Zhao, B., (2017). The gut microbiota and Alzheimer's disease. *Journal of Alzheimer's Disease*. https://doi.org/10.3233/JAD-161141.

Ke, Y., Li, D., Zhao, M., Liu, C., Liu, J., Zeng, A., Shi, X., Cheng, S., Pan, B., Zheng, L., & Hong, H., (2018). Gut flora-dependent metabolite trimethylamine: N-oxide accelerates endothelial cell senescence and vascular aging through oxidative stress.

Keservani, R. K., Kesharwani, R. K., Sharma, A. K., Vyas, N., & Chadoker, A., (2010b). Nutritional supplements: An overview. *Int. J. of Curr. Pharm. Rev. and Res., 1*(1), 59–75.

Keservani, R. K., Kesharwani, R. K., Vyas, N., Jain, S., Raghuvanshi, R., & Sharma, A. K., (2010a). Nutraceuticals and functional food as future food. *Der Pharm. Lett, 2*(1).; p106–116.

Keservani, R. K., Sharma, A. K., & Kesharwani, R. K., (2016). Nutraceutical and functional foods for cardiovascular health. In: Meghwal, M., Goyal, M. R., *Food Process Engineering: Emerging Technologies and Innovations in Food Engineering* (pp. 257–278). Apple Academic Press, CRC Press, Taylor and Francis, Ch 10. ISBN: 9781771884020.

Kida, Y., & Goligorsky, M. S., (2016). Sirtuins, cell senescence, and vascular aging. *Canadian Journal of Cardiology, 32*(5), 634–641. https://doi.org/10.1016/J.CJCA.2015.11.022.

Kowalski, K., & Mulak, A., (2019). Brain-gut-microbiota axis in Alzheimer's disease. *Journal of Neuro Gastroenterology and Motility, 25*(1), 48–60. https://doi.org/10.5056/JNM18087.

Kowalski, K., & Mulak, A., (2019). Brain-gut-microbiota axis in Alzheimer's disease. *Journal of Neurogastroenterology and Motility*. https://doi.org/10.5056/jnm18087.

Lau, W. L., Kalantar-Zadeh, K., & Vaziri, N. D., (2015). The gut as a source of inflammation in chronic kidney disease. *Nephron, 130*(2), 92–98. https://doi.org/10.1159/000381990.

Lindhorst, A., Raulien, N., Wieghofer, P., Eilers, J., Rossi, F. M. V., Bechmann, I., & Gericke, M., (2021). Adipocyte death triggers a pro-inflammatory response and induces metabolic activation of resident macrophages. *Cell Death & Disease, 12*(6), 1–15. https://doi.org/10.1038/s41419-021-03872-9.

McCreight, L. J., Bailey, C. J., & Pearson, E. R., (2016). Metformin and the gastrointestinal tract. *Diabetologia, 59*(3), 426. https://doi.org/10.1007/S00125-015-3844-9.

Molinaro, A., Bel, L. P., Henricsson, M., Wu, H., Adriouch, S., Belda, E., & Bäckhed, F., (2020). Imidazole propionate is increased in diabetes and associated with dietary patterns and altered microbial ecology. *Nature Communications, 11*(1), 1–10. https://doi.org/10.1038/s41467-020-19589-w.

Muccioli, G. G., Naslain, D., Bäckhed, F., Reigstad, C. S., Lambert, D. M., Delzenne, N. M., & Cani, P. D., (2010). The endocannabinoid system links gut microbiota to adipogenesis. *Molecular Systems Biology, 6*(392), 1–15. https://doi.org/10.1038/MSB.2010.46.

Rivera-Chávez, F., Zhang, L. F., Faber, F., Lopez, C. A., Byndloss, M. X., Olsan, E. E., & Bäumler, A. J., (2016). Depletion of butyrate-producing clostridia from the gut microbiota drives an aerobic luminal expansion of Salmonella. *Cell Host & Microbe, 19*(4), 443–454. https://doi.org/10.1016/J.CHOM.2016.03.004.

Russo, L., & Lumeng, C. N., (2018). Properties and functions of adipose tissue macrophages in obesity. *Immunology, 155*(4), 407–417. https://doi.org/10.1111/IMM.13002.

Sender, R., Fuchs, S., & Milo, R., (2016). Are we really vastly outnumbered? Revisiting the ratio of bacterial to host cells in humans. *Cell, 164*(3), 337–340. https://doi.org/10.1016/J.CELL.2016.01.013.

Sharma, M., Li, Y., Stoll, M. L., & Tollefsbol, T. O., (2020). The epigenetic connection between the gut microbiome in obesity and diabetes. *Frontiers in Genetics, 10*(Article-1329), 1–15. https://doi.org/10.3389/FGENE.2019.01329.

Tonacci, A., Bagnato, G., Pandolfo, G., Billeci, L., Sansone, F., Conte, R., & Gangemi, S., (2019). MicroRNA cross-involvement in autism spectrum disorders and atopic dermatitis: A literature review. *Journal of Clinical Medicine, 8*(1), 88. https://doi.org/10.3390/jcm8010088.

Tzika, E., Dreker, T., & Imhof, A., (2018). Epigenetics and metabolism in health and disease. *Frontiers in Genetics, 9*(361), 1–8. https://doi.org/10.3389/FGENE.2018.00361.

Upadhyaya, S., & Banerjee, G., (2015). Type 2 diabetes and gut microbiome: At the intersection of known and unknown. *Gut Microbes, 6*(2), 85. https://doi.org/10.1080/19490976.2015.1024918.

Valdes, A. M., Walter, J., Segal, E., & Spector, T. D., (2018). Role of the gut microbiota in nutrition and health. *BMJ, 361*, 36–44. https://doi.org/10.1136/BMJ.K2179.

Vikram, A., Kim, Y. R., Kumar, S., Li, Q., Kassan, M., Jacobs, J. S., & Irani, K., (2016). Vascular microRNA-204 is remotely governed by the microbiome and impairs endothelium-dependent vasorelaxation by downregulating Sirtuin1. *Nature Communications, 7*(1), 1–9. https://doi.org/10.1038/ncomms12565.

Wang, L., Meng, X., Li, G., Zhou, Q., & Xiao, J., (2018). Circular RNAs in cardiovascular diseases. *Advances in Experimental Medicine and Biology, 1087*, 191–204. https://doi.org/10.1007/978-981-13- 1426-1_15.

Zhu, Y., Lin, X., Zhao, F., Shi, X., Li, H., Li, Y., & Zhou, G., (2015). Meat, dairy and plant proteins alter the bacterial composition of rat gut bacteria. *Scientific Reports, 5*(1), 1–15. https://doi.org/10.1038/srep15220.

CHAPTER 7

Role of Dietary in the Prevention and Management of Obesity

EKNATH D. AHIRE,[1] AFSAR S. PATHAN,[2] PRATIKSHA M. PANDAGALE,[3] SHUBHAM J. KHAIRNAR,[4] KHEMCHAND R. SURANA,[5] SANJAY J. KSHIRSAGAR,[1] SWATI G. TALELE,[6] and GOKUL S. TALELE7

[1]Department of Pharmaceutics, MET's Institute of Pharmacy, Bhujbal Knowledge City (BKC), Affiliated to Savitribai Phule Pune University, Adgaon, Nashik, Maharashtra – 422003, India, Tel.: +91-9588431965, Fax: (0253)-2303203, E-mail: eknathahire05@gmail.com (E. D. Ahire)

[2]Department of Pharmacology, RC Patel Institute of Pharmaceutical Education and Research, Dhule, Maharashtra, India

[3]Department of Pharmacology, Oriental College of Pharmacy, Mumbai, Maharashtra, India

[4]Department of Pharmacology, MET's Institute of Pharmacy, Bhujbal Knowledge City (BKC), Affiliated to Savitribai Phule Pune University, Nashik, Maharashtra – 422003, India

[5]Department of Pharmaceutical Chemistry, Shreeshakti Shaikshanik Sanstha, Divine College of Pharmacy, Satana, Nashik, Maharashtra, India

[6]Department of Pharmaceutics, Sandip Institute of Pharmaceutical Sciences, Nashik, Maharashtra – 422213, India

[7]Department of Pharmaceutics, Matoshri College of Pharmacy, Near Odhagaon, Eklahare, Nashik, Maharashtra – 422105, India

ABSTRACT

Obesity is an ever-growing ailment globally this is frequently underestimated. Identification of the ailment and its remedy stays low. Overweight

The Metabolic Syndrome: Dietary Supplements and Food Ingredients. Raj K. Keservani, MPharm, Durgavati Yadav, PhD, Rajesh K. Kesharwani, PhD, Sippy Singh, DPhil & Kumar Sandeep, PhD (Eds.)

and weight-associated troubles need to be addressed in a multidisciplinary and multimodal manner. Obesity may be averted now no longer best through healing way, however, additionally weight reduction techniques which include nutritional therapy, bodily interest, and way of life and behavioral adjustments, help to a tremendous extent. Food control is, in particular, an exercise used to offer dietary alternatives to a specific age institution with recognize to an ailment or person and better bodily interest performs an essential position in stopping and treating weight problems. Obesity is a situation that develops steadily in particular because of terrible weight loss program alternatives, which include, Eating large quantities of processed/ rapid foods, Excess consumption of alcohol, Irregular consuming habits, and Soft beverages containing sugar.

7.1 INTRODUCTION

Obesity is an ever-increasing disease worldwide that is often underestimated. Identification of the disease and its treatment remains low. It has become a common issue often leading to other serious diseases such as coronary heart diseases, hypertension, type 2 diabetes (T2D), dyslipidemia, and the list goes on. It adversely influences almost all physiological features of the body. According to the World Health Organization (WHO), greater than 1.9 billion humans have been obese in 2016, with 650 million of them being obese. Obesity is defined as adipose tissue fat buildup that is abnormal or excessive to the point of compromising health. It can also be defined as a BMI of 30 kg/m^2 or above, as well as a Waist Circumference (WC) of 102 cm (male) or 88 cm (female), but this doesn't account for the morbidity and mortality related to milder ranges of obesity. Obesity (Jarouliya and Keservani, 2019) is measured by BMI and waist circumference (WC). BMI is related to normal weight problems; however, WC is related to stomach weight problems and metabolic disorders. The disorder is unexpectedly progressing to end up the main reason of infection and a key barrier to persistent disorder prevention. It is likewise related to improved morbidity and loss of life rates. Obesity can be prevented not only by therapeutic means, but also weight-loss strategies such as dietary therapy, physical activity, and lifestyle and behavioral modifications help to a significant extent (Rane and Keservani, 2020; Keservani et al., 2015). Obesity and overweight should be addressed in a multidisciplinary and multimodal manner. Drug remedy and surgical remedy is reserved for overweight or obese sufferers who've concomitant obesity-associated chance elements or diseases (Astrup, 2008). A healthy diet can to a greater extent prevent the condition. Healthy dietary fats, such As polyunsaturated fats, also can assist lower cholesterol and reduce obesity risks. Diet composition desires to be controlled to save you the disease. Carbohydrates, proteins, lipids, and fats all play a role in the prevention and control of obesity (Singh et al., 2017a). Epidemiology refers to the analysis of distribution, pattern, and determinants of fitness and sickness situations in a described population. Obesity has steadily increased over the last 40–50 years all throughout the world. In 2015, 1.9 billion and 609 million adults worldwide were predicted to be overweight or obese, representing approximately 39% of the world's population. Obesity prevalence grew in every country between 1975 and 2016, according to the most recent study that provided BMI developments for all nations within side the global primarily based totally on measured frame weight and peak facts from 128.9 million children, adolescents, and adults. Also, a significant change in BMI was observed over the time. South Asia (Bangladesh, Bhutan, Nepal, Pakistan, and India), South East Asia (Sri Lanka, Malaysia, Vietnam, Philippines, Thailand, and Indonesia), the Caribbean (Belize, Cuba, Dominican Republic, Jamaica, and Puerto Rico), and southern Latin America (Belize, Cuba, Dominican Republic, Jamaica, and Puerto Rico) all saw an increase in BMI (Brazil, Argentina, Paraguay, Uruguay, and Chile). Overweight and weight problems are anticipated to be popular in Saudi Arabia, and 28.7%, respectively. Worldwide incidence of weight problems extended at an alarming charge in kids and adolescents from 0.7% to 5.6% in boys and 0.9% to 7.8% in ladies among 1975 and 2016. Between 1975 and 2014, the prevalence of obesity raised from 3.2% to 10.8% in grownup guys and from 6.4% to 14.9% in grownup women. The above observation shows heterogeneous data with an increase in the occurrence of the disease. Obesity and related co-morbidities are the main motive of about 2.8-million-persons deaths international each year. The standards used for classifying weight problems are BMI. The criteria used for classifying obesity are BMI. It is a value produced from a Person's mass and height and is described because the frame mass divided through the rectangular root of the frame peak, expressed in kilograms consistent with meter squared, expressed in kilograms consistent with meter squared. It can range from being underweight (<18.5 kg/m^2) to being morbidly obese (>40 kg/m^2) (Table 7.1) (Dyson, 2010; Butterwick and Hawthorne, 1998; Riccardi et al., 2005; Hwalla and Jaafar, 2021; Surana et al., 2021).

TABLE 7.1 Classification of Weight Status by BMI

Classification	BMI (kg/m^2)
Underweight	<18.5
Normal weight	18.5–24.9
Overweight	25–29.9
Obesity Class 1	30–34.9
Obesity Class 2	35–39.9
Extreme obesity Class 3	>40

7.2 DIETARY MANAGEMENT MATTERS BUT IS DIFFICULT TO IMPLEMENT

Food management is mainly a practice used to provide nutritional options to a particular age group with respect to a disease or person. While adhering to a weight-reduction plan with numerous regulations is challenging, hunger, and taste are also altered when obesity is linked to other issues. The best results are achieved by using a low-calorie diet, followed by a low-fat, high-carb diet. Low-calorie diet refers to eating approximately 800 calories per day or less. Excessive dietary fat intake has led to an increase in fat, so that weight loss diets tend to manage obesity. Therefore, over the last 20 years, it has been emphasized on reducing fat consumption. Weight loss takes place in each the fast and longtime whilst the full fats content material of the food regimen is reduced. In post-hoc analysis, reduction in dietary fats and increase in fiber have been the most powerful predictors of weight loss. Complex carbs found in various vegetables, fruits, and whole grains, which are more satiating for less calories than fatty foods and are an awesome supply of vitamins, minerals, hint elements, and fiber (Singh et al., 2017b), have to make up the bulk of the carbohydrate content of low-fat diets. Recent research also suggests that a high sugar intake from soft drinks may have a unique fattening effect. Hence, sugars from soft drinks should be replaced by proteins from low-fat milk, high-fat meat, and dairy products without reducing the intake of fruits, vegetables, and whole-grain products (Keservani et al., 2010a, b, 2016). One of the motives for the partial success in causing weight loss without decreasing total calories is the high protein content. Observational research displays that people with an excessive consumption of dairy products are less likely to develop obesity, but no mechanism has yet been put forth to support this observation (Nanney et al., 2017; Organization, 2000; Gidding et al., 2009; Ahire et al., 2020).

7.3 PHYSICAL ACTIVITY AND LIFESTYLE

A low stage of Physical activity and lifestyle modifications has turned out to be a cause of obesity however this has now no longer but been proven. Increased physical activity, on the other hand, is critical in preventing and treating obesity. Hence, physical activity is inversely associated with obesity. Physical workout is described through the WHO as "anybody action requiring energy expenditure produced by skeletal muscles." Physical interest is a vital lifestyle conduct that contributes to frame weight and regulation. An important aspect of physical activity is exercising but this is often crucial to follow. Enhanced physical activity (EPA) is the handiest one detail within side the control of obesity. Other factors contain nutritional management, behavioral modifications, and drug therapy. However, no surgical or pharmacological intervention is found to be safe and efficacious in lowering body fatness. The fundamental purpose for any such distinction is that the terrible power stability finished thru EPA is notably smaller than that prompted through dieting. Other consequences of EPA consist of discount in blood pressure, extended sensitivity to insulin, advanced lipoprotein profile, and a boom in bodily fitness. In addition to approaches to prevent weight gain, there is also a considerable need to implement effective strategies for the same (Singh and Purohit, 2012; Kurdaningsih et al., 2016; Al-Nakeeb et al., 2012).

7.3.1 PHYSICAL ACTIVITY COMBINED WITH DIETARY MODIFICATION

Clinical suggestions for the remedy of obesity and weight problems propose an approach that includes behavioral modifications along with physical activity combined with dietary modifications. These aggregate consequences in extra weight reduction than nutritional adjustments alone. It also leads to shorter intervention periods that range from 12–24 weeks in duration. However, there are also considerable improvements in cardiovascular and musculoskeletal fitness (Strasser, 2013; Wadden et al., 2012).

7.3.2 PHYSICAL ACTIVITY IN THE ABSENCE OF PRESCRIBED DIETARY RESTRICTION

As per the recommendation of the above guideline, not all interventions need to include all the aspects. Some patients may rely solely on physical activity while eliminating others. Several systematic critiques have concluded that the imply weight reduction with bodily pastime by myself generally does now no longer exceed 3 kg, while the mixture of nutritional restrict and bodily pastime elicits weight reduction of about 8–10 kg. Also, withinside the absence of weight reduction, bodily pastime has been proven to lessen stomach adiposity. This is also of foremost significance for the removal of aerobic metabolic dangers impartial of frame weight. Lifestyle truly way the manner you live. This is also referred to as non-pharmacological treatment. It is a mixture of diet, exercise, and behavioral interventions. If this undergoes severe modifications, life-threatening diseases are at risk. Concerning obesity, a major component of lifestyle modification is behavioral changes (Strasser, 2013). Behavioral adjustments closer to a superb method can result in higher weight manipulation than regimented workout prescriptions. Neither weight loss program nor EPA, even if combined, induces a long-lasting impact past the length of this system. This trouble is of huge realistic and theoretical significance due to the fact maximum overweight sufferers who reply nicely at the same time as on a weight-manipulate software lose floor unexpectedly as soon as this system ends. Comprehensive way of life intervention is for that reason had to create poor electricity stability via amendment of meals and bodily pastime behaviors. A method to that is to comprise behavioral amendment as a quintessential part of the intervention. Hence, increasing the activity of the patients when combined with nutritional habits and behavioral modifications can help manage obesity. The best management of obesity is indeed, prevention (Jakicic et al., 2018; Chin et al., 2016; Wadden et al., 2012).

7.4 RELATION BETWEEN DIET AND OBESITY

Obesity is a complex, multifactorial, and in large part preventable disease. Dietary pattern of an individual provides useful insights into the influence of diet on overall obesity. Over the beyond few decades, this has been keen research of interest, and many studies have been performed concerning it. Diet additionally performs a critical function withinside the genesis of obesity. Hence, the etiology needs to be understood for the prevention and management of the disease. The position of eating regimen withinside the etiology of obesity remains a matter of discussion. Therefore, measurement of overall dietary intake pattern has been recommended, which reflects the dietary intake complexity as an approach to investigate the links between diet and disease. High consumption of energy-dense food might be responsible for the multiplied deposition of energy in adipose tissues. Behavioral modification concerning food intake can be effective in the treatment of obesity. The relation between diet and weight problems is attributed to bad consuming behavior and a lack of information approximately a healthful diet. Obesity is a condition that develops gradually mainly because of negative food regimen selections, such as eating massive quantities of processed/rapid foods, excess consumption of alcohol, irregular eating habits, and soft drinks containing sugar (Clarke et al., 2012).

7.4.1 PROTEIN AND WEIGHT

A higher protein diet offers a much great advantage for weight loss. Also, higher protein and lower carbohydrate diet improves blood lipid profiles which in turn helps prevent heart diseases and diabetes (Asghari et al., 2017).

7.4.2 CARBOHYDRATES AND WEIGHT

Carbohydrates have an excessive glycemic index and glycemic load. Such meals cause an increase in blood sugar and insulin which leads to over-eating and an increase risk of weight gain and diabetes (Clarke et al., 2012).

7.5 CURRENT DIETARY GUIDELINES

The 2010 dietary guidelines for Americans assists in maintaining and achieving a healthy weight as well as reducing the risk of chronic, diet-related diseases through promoting health and a healthy eating pattern. There are many such guidelines approved for patients. The Dietary Advisory Committee suggested that overweight and obese individuals should adopt a healthy eating pattern in order to achieve weight loss. The dietary approaches to stop hypertension (DASH) nutritional tips have additionally published a chart for daily intake of food. The Physical Activity Guidelines for Americans offers with unique key factors and are mainly categorized based on population. In India, the National Institute of Nutrition has proposed dietary guidelines for Indians which include guidelines 1–15 and also dietary suggestions for special group populations (DeSalvo et al., 2016; Millen et al., 2016).

7.5.1 DIETARY PATTERNS

Weight gain in adulthood is a gradual process and hence, the dietary pattern needs to be followed for long periods. Several nutritional patterns, each macronutrient, and food-primarily based totally can result in weight loss. There is increasing evidence that dietary choices for cardiovascular, diabetes, and other conditions associated with obesity often seem to help reduce weight gain. The key points of the dietary pattern can be summarized as minimally processed and whole foods to be chosen. This consists of entire grains, vegetables, fruits, nuts, plant oils, fish, poultry, beans, etc. Sugared beverages, delicate grains, potatoes, purple, and processed meats, highly processed foods (fast foods) need to be limited (Figure 7.1) (Tapsell et al., 2016; Ambrosini, 2014; Schulze et al., 2018).

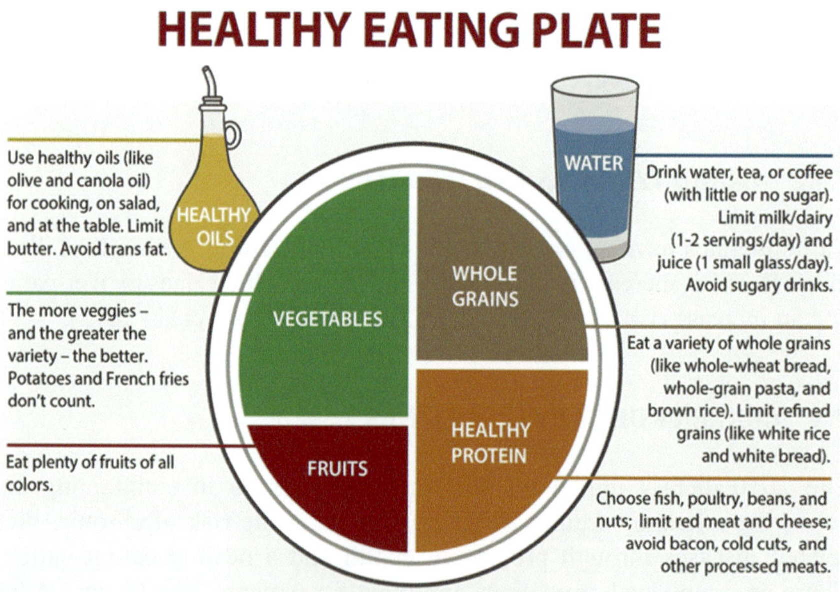

FIGURE 7.1 Complete healthy eating plate.
Copyright © 2011, Harvard University. For more information about The Healthy Eating Plate, please see The Nutrition Source, Department of Nutrition, Harvard T.H. Chan School of Public Health, www.thenutritionsource.org, and Harvard Health Publications, www.health.harvard.edu.

7.5.2 MACRONUTRIENT PATTERNS FOR WEIGHT LOSS

The composition of macronutrients in a diet has been the foundation of many weight-loss diets. An evidence-primarily based totally document from the American College of Cardiology/American Heart Association (AHA) Task Force on Practice Guidelines and The Obesity Society helps numerous energy-restrained nutritional techniques for weight reduction that concentrate on the macronutrients, which include low-fat, lower-carbohydrate, moderate-and higher-protein, and macronutrient-targeted diets. A large clinical trial revealed that altering the composition of macro-nutrients across various diets did not differ much in achieving weight loss. It emphasized that reduction in dietary energy density and boom in fiber consumption had been sturdy predictors for six-month weight reduction in all food regimen groups (Truby et al., 2016; Fleming and Kris-Etherton, 2016).

7.5.3 FOOD-BASED PATTERNS FOR WEIGHT LOSS

DASH refers to a dietary approach to stop hypertension (The DASH diet). The important components of this diet are:

- Low-fat dairy (less than 2% fat);
- Fish, chicken, and lean meat;
- Fruits and vegetables;
- Nuts, whole grains, and legumes.

Such a weight-reduction plan enables lessen saturated fats consumption and boom calcium, potassium, magnesium, and nutritional fiber consumption. Research counseled that the DASH weight-reduction plan caused a continuously decrease BMI among a long time of 9 and 19 years (Yannakoulia et al., 2019; Soeliman and Azadbakht, 2014).

7.5.4 LOW-CARBOHYDRATE DIETS

These diets consist of low carbohydrate/excessive protein diets and occasional glycemic index diets. Major additives consist of, High protein intake; Low carbohydrate intake, a few limitation fats, and others do not. Because of the shortage of consensus amongst fitness experts' approximately low carbohydrate diets, the use of those diets to control weight is controversial. Food-primarily based totally styles additionally consist of the Mediterranean weight-reduction plan with is elaborated in the next section (Hu et al., 2012).

7.5.5 FOODS TO AVOID

Healthy consuming conduct are key to keeping a healthful weight and hence the following foods need to be avoided or limited to control and manage obesity (Allman-Farinelli, 2015):

- Sugar-sweetened beverages;
- Fruit juice;
- Refined grains and sweets;
- Potatoes;
- Trans fat;
- Food high in sodium;
- Alcohol;
- High-calorie coffee drinks;
- Baked foods;
- Artificial sweeteners;
- Saturated fats;
- Red meat and processed meats;
- Other highly processed foods.

7.6 AREAS OF UNCERTAINTY

Obesity occurs when the homeostatic mechanisms underlying the disease become distorted. Also, while energy consumption exceeds energy expenditure, weight problems are probably to occur. The pathogenesis of obesity is complex as a lot of factors are involved in it. Ongoing research has identified intricate humoral and neural mechanisms that respond to genetic, nutritional, environmental, and physiologic signals and trigger a metabolic response through stimulation of centers located in the hypothalamus (Stevens et al., 2012).

7.7 REVERSING OBESITY THROUGH DIET

The first and foremost principle is to set a goal whilst discussing a nutritional eating regimen with an individual. The set goal should include steps to prevent further weight gain. If this purpose is achieved, it is able to lessen threat elements of cardiovascular sicknesses, including hypertension, diabetes mellitus (DM) and dyslipidemia (Figure 7.2) (Zhao et al., 2017).

FIGURE 7.2 A healthy diet.

7.8 DIETARY THERAPY

The goal of dietary therapy is primarily to reduce the intake of calories. With the increasing prevalence of concomitant diseases, it has become important to manage obesity. Dietary pointers at the side of an ordinary plan of exercise are the key elements in the management of obesity. A vast spectrum of options has evolved. Some plans decrease carbohydrate consumption without fats limit, e.g., Atkin's eating regimen and lots of modulate macronutrient stability and glycemic load, e.g., quarter eating regimen and others limit of fats (Goutzourelas et al., 2018). A proper diet must provide all necessary nutrients (other than calories) in sufficient amounts and should be palatable and easily available. A balanced diet containing no less than 12% to 14% of protein, no more than 30% of fat (with saturated fats cut down), and the rest carbohydrates (with sucrose cut down to very low levels), supplied through meals of enough range is infinitely premiere to the fat diets that are promoted to the public (Salwa et al., 2012). Traditionally, the method advocated via way of means of most for weight reduction is an intake of a "low-calorie low-fat diet." This is a technique used to induce negative energy balance. The other conventional strategies are as in subsections.

7.8.1 BALANCED LOW-CALORIE DIET

It consists of an aggregate of good enough quantity of macronutrients (i.e., carbohydrates, proteins, and fat) (Fock and Khoo, 2013).

7.8.2 LOW-FAT DIET

All nutritional pointers suggest decreasing the fat intake by 30% of the total intake. Such a diet decreases weight more than a moderate-fat diet (Blaisdell et al., 2014).

7.8.3 MEDITERRANEAN DIET

This diet varies with respect to components. Some of the common components include (D'Innocenzo et al., 2019):

- High stage of monosaturated fats relative to saturated;
- High intake of vegetables, fruits, legumes, and grains;
- Moderate intake of milk and dairy products, in the main withinside the shape of cheese;
- Relatively low consumption of meat and meat products.

7.8.4 HIGH-FAT CONTENT, LOW-CARBOHYDRATE, AND HIGH-PROTEIN DIETS

A low carbohydrate diet is an example of Atkins's diet, Stillman diet, protein energy lifestyle plan, and zone diet. Such diets are high in fats (55%–65%) and low carbohydrate (<100 gm of carbohydrate per day). Restricting carbohydrates to low degrees results in a circumstance referred to as ketosis. Ketosis is an indicator of fats mobilization. The key advantages are that blood glucose and blood insulin degrees are decreased and urge for food is suppressed. This results in weight loss, frame fats loss, upkeep of lean frame mass, and correction of great scientific complications. Common examples are the Scarsdale Diet and the Stillman Quick Weight Loss diet (Floegel and Pischon, 2012).

7.8.5 LOW-FAT AND VERY LOW-FAT DIETS

Low-fats weight-reduction plan refers to fats among 11% and 19% while a very low-fat diet represents <10% fat. In such diets, there is a very high amount of carbohydrate intake and a moderate amount of protein intake. Ornish diet and Pitkin's diet are examples of very low-fat diets. The main aim is the reduction of calorie intake and increases in energy expenditure to achieve weight loss. Such diets lower total cholesterol levels, especially LDL and hence, decrease the danger of coronary heart disease (Sackner-Bernstein et al., 2015).

7.8.6 VERY LOW-CALORIE DIETS

They are defined as diets that provide lesser than 800 kcals/day. Such diet produces rapid weight Loss at the same time as keeping lean body mass. Large amounts of dietary proteins (70–100 gm/day) are consumed. The above-mentioned diets are not permanent solutions for weight loss. However, they can help to a particular level. Overall, weight problems control needs to be taken into consideration in different phases, achieving weight loss and maintaining weight loss. It needs to be stated right here that weight reduction diets need to continually be individualized. The crux of a weight loss program is the incorporation of a healthier lifestyle and the right food choices. Therefore, patient counseling for lifestyle modification is an important part of a weight loss program and is the right dietary prescription (Haywood et al., 2018).

7.9 MANAGEMENT OF REMISSION

This section deals with weight maintenance. Despite the hit weight loss, many people regain weight over a whiled. Patients that lost weight through lifestyle modification for 20–30 weeks regain almost 35% of the lost weight in one year after treatment, and via way of means of five years more than 50% of patients will return to their baseline weight. There has been a vast increase in the mechanism and hormones underlying the increase in weight. The mechanism includes decreased energy expenditure, increased hunger, and dietary non-adherence. After successful weight loss, there is a plateau phase which is the critical period for maintaining the lost weight. Although this phase can be a bit distressing and disappointing because there is no further loss in weight. This phase needs to be carefully examined and the necessary caloric intake should be adjusted to promote further weight loss or at least to prevent weight gain. This phase, if combined with pharmacotherapy proves beneficial for the long run. Also, portion-controlled servings and meal replacement has been shown to improve and maintain weight loss. Consumption of low calorie-dense foods, particularly by increasing consumption of fiber and replacing fat with carbohydrates and proteins, has shown a short-term decrease in caloric intake. The National Weight Control Registry has identified persons who lost at least 13.6 kg and feature succeeded in keeping weight reduction for at least year (Ruban et al., 2019).

7.9.1 NUTRITION TRANSITION

Over the last few decades, the world has undergone a rapid nutritional transition and has mostly shifted towards a higher intake of fat and sugar Most nations in Asia, Latin America, North Africa, the Middle East, and the city regions of Sub-Saharan Africa have all skilled a shift withinside the universal shape of its nutritional sample with associated ailment styles over a previous couple of decades. The growing urbanization, westernization, and mechanization happening in maximum nations round the arena are related to modifications withinside the weight loss plan in the direction of certainly considered one among excessive fat, excessive energy-dense foods, and a sedentary lifestyle. This shift has highly impacted childhood and adult cases of obesity. Even in lots of low-profits countries, weight problems are now swiftly increasing because of a marked decline in the cost of sugar and vegetable oils. According to Dr. Walter Willett, Chairmen, Department of Nutrition, Harvard University, a main supply of fit to be eaten oil in India, Dalda, a vegetable ghee, has Tran's fatty acid ranges of approximately 50%. Dietary shifts are accelerating from conventional to western weight loss plan consuming over 30% energy from fat (Popkin et al., 2012).

7.9.1.1 EDIBLE OIL

There is evidence of an increase in edible oil use on the countrywide and family degree in the course of the world. In many growing countries, there was a growth withinside the import of vegetable oils compared to others. Principle vegetable oils encompass soybean, sunflower, rapeseed, palm, and groundnut oil (Belahsen, 2014).

7.9.1.2 CALORIC SWEETENERS

Sugar is the most predominant sweetener used worldwide. One teaspoon of sugar consists of approximately 20 calories and five grams of carbohydrates. Every gram of carbohydrate has approximately 4 calories. Many non-sugars merchandise also are been derived from animal reassets which include insects, sugar crops, and fruits. Sweetening will increase the palatability of meals and its miles believed that sweeteners in meals result in over-consumption. In the remaining numerous decades, an increasing number of big portions of cereals had been used to supply sweeteners derived from starch. The overall trend shows a large increase in caloric sweeteners consumed (Belahsen, 2014).

7.9.1.3 ANIMAL SOURCE FOODS

This refers to the increase in demand for meat, fish, and milk in developing countries. China gives an instance of the shift withinside the use of animal supply food. Several surveys confirmed that annual intake of animal ingredients greater than tripled beginning from a completely low stage of intake (Figure 7.3) (Belahsen, 2014).

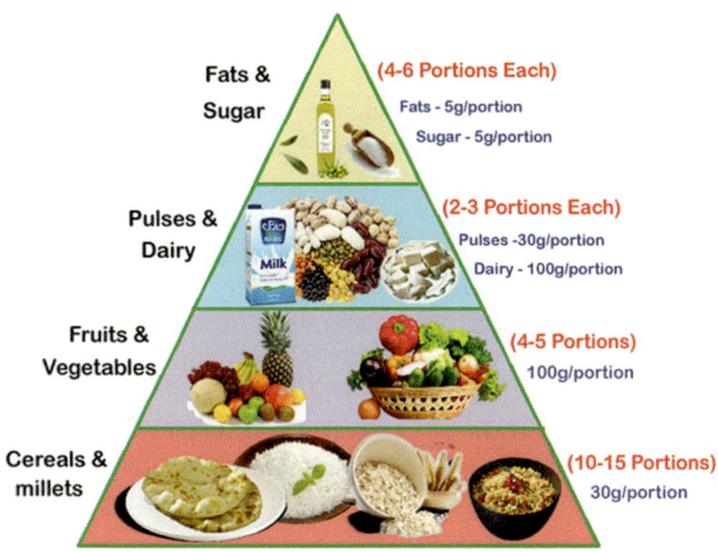

Courtesy: National Institute of Nutrition INDIA

FIGURE 7.3 Indian dietary guideline.

7.10 THE ATTITUDE OF OBESE PATIENTS TOWARDS FOOD

This segment discusses the implicit and express attitudes closer to food between patients with obesity. Also studying behavior and feeling of eating habits, the amount of food intake and attitude towards food is of utmost importance. Recent work has demonstrated that inactivity could also perpetuate obesity, if not cause it. Over-eating has traditionally been considered a cause of obesity. A hypothesis states that patients with weight problems might have a greater fine implicit mindset toward dangerous meals and sedentary activities, which is a contributing factor to obesity. Also, behavioral factors are responsible, which focuses mainly on unhealthy consumption and physical inactivity. Research has proven that children with weight problems eat greater fat, sugar, and proteins than friends without weight problems. One of the feasible factors of this behavioral samples perhaps their attitude and knowledge towards food. This may not always be the case. On the contrary, studies have also revealed that some patients with obesity reveal a positive attitude towards unhealthy food. One of the reasons for this inconsistency may be different attributes (which include the flavor of food) are extra salient than health. Another purpose can be the styles imply sufferers do not in reality pick dangerous food, however surely like ingesting. Assuming that attitudes are crucial precursors of behavior, it appears that evidently weight problems isn't especially associated with ingesting dangerous food, however instead to devour a lot (Sikorski et al., 2013).

To recognize the inconsistency among attitudes and behaviors, Fazio (1990) proposed one of kind modes with the aid of using which attitudes can manually conduct. According to him, express attitudes' might also additionally manual one's conduct with the aid of using a planned and aware evaluation of the expenses and the blessings of that conduct. Alternatively, implicit attitudes might also additionally manually conduct in an extra spontaneous and effective manner, without actively thinking about the professionals and the cons. It is feasible that attitudes are multidimensional and they'll collectively engage with every other. On the other hand, whether, and which attitude gets activated and suppressed is also a point of consideration. It is also possible that behavioral patterns of eating too much in obesity are less related to "liking" than to "wanting" (Beeken and Wardle, 2013).

7.11 DIETARY PRACTICES OF OBESITY

One method for weight reduction is to apply diets that reduce the efficiency of utilization of the energy ingested. It has also been suggested that a high-fat diet is more promising than a high carbohydrate diet in depositing the fats into adipose tissue. When excess amounts of calories are consumed, fat can be used with 79–85% efficiency whereas carbohydrate will be utilized with 62 to 77% (Wojcicki et al., 2012).

7.11.1 IMPAIRED ABSORPTION

This strategy can be employed with foods that impair absorption of calories ingested, so that they remain in the gut and thus, excreted rather than utilized. An example of such foods is high-fiber diets (Wojcicki et al., 2012).

7.11.2 EXCRETORY LOSS OF CALORIES

Diets that produce loss of calories via urine are being introduced. Ketogenic substances can help in the process as excess generation of ketones results in a caloric loss. But the highest limit that is lost does not exceed 2–3% of the total caloric intake (Damms-Machado et al., 2015).

7.11.3 UNBALANCED LOW-CALORIE DIETS

The maximum not unusual place manner of decreasing weight is the usage of unbalance weight loss program comprising carbohydrates, proteins, fats, and a low caloric intake. Such diets are easier to follow and hence are popular. They have a marked imbalance in macronutrients which thus leads to an imbalance in micronutrients. Such diets are defined below (Fock and Khoo, 2013):

- High-fat low carbohydrate high protein diets;
- Low-fat and very low-fat diets;
- Very low-calorie diets.

7.11.4 PROTEIN-SUPPLEMENTED FASTING

Total fasting if tried may be hazardous and require hospitalization, which makes it an impractical method. Hence, fasting along with protein supplements have proven to be useful for fast weight reduction whilst keeping lean body mass. Example of such includes "The Last Chance Diet," in which a protein powder is the sole dietary item (Piccolo et al., 2015).

7.11.5 BALANCED DIETS

There is the need to follow a restricted diet for long periods to maintain the condition; the nutritional adequacy of the diet becomes crucial. The elements which want to be considered are acceptable meal patterns, proper food selection, and allowance for individual preferences. A mixed, balanced diet is a much more sensible and healthy approach to weight reduction (Gueugnon et al., 2012).

7.12 CONCLUSION

Obesity is a multi-factorial condition, with each person and environmental factors affecting dietary obedience. According to the American Paradox, the conclusion is that obesity pervasiveness is increasing in spite of a slight decrease in citizens consumption of dietary fat, is easily described by the accordingly decreasing physical activity, which decreases the fat necessity and prevents the beneficial effect of a slight decrease in dietary fat. Dietary techniques with a lower in electricity consumption which have caused fulfillment at weight reduction, have targeted on macronutrient composition and meals styles. A unifying precept for weight reduction throughout ingesting styles is nutritional electricity density. There are distinctive techniques and gear that people can use to reap a customized wholesome ingesting pattern, and physicians or vitamins experts provide resources and particular dietary advice on modifications to beautify an individual's ingesting behaviors. Low fats diets with an excessive content material of complicated carbohydrates and protein do now no longer produce any unfavorable impact on cardiovascular threat elements while weight reduction is permitted to occur, and they were proven to lower mortality amongst excessive threat subjects.

ACKNOWLEDGMENT

We thank MET, Institute of Pharmacy, BKC, affiliated under Savitribai Phule Pune University, Nashik, Divine College of Pharmacy Satana, and Sandip Institute of Pharmaceutical Science for their constant support and for providing all facilities to complete this work.

KEYWORDS

- **diet**
- **dietary supplements**
- **enhanced physical activity**
- **hypertension**
- **obesity**
- **waist circumference**
- **World Health Organization**

REFERENCES

Ahire, E. D., Surana, K. R., Patil, C. D., Shah, H. S., Sonwane, G. B., & Talele, S. G., (2020). Role of omega-3 fatty acids in different neurodegenerative disorders. *Applied Pharmaceutical Science and Microbiology: Novel Green Chemistry Methods and Natural Products, p.* 173.

Allman-Farinelli, M. A., (2015). *Nutrition Promotion to Prevent Obesity in Young Adults* (pp. 809–821). Healthcare Multidisciplinary Digital Publishing Institute.

Al-Nakeeb, Y., Lyons, M., Collins, P., Al-Nuaim, A., Al-Hazzaa, H., Duncan, M. J., & Nevill, A., (2012). Obesity, physical activity and sedentary behavior amongst British and Saudi youth: A cross-cultural study. *International Journal of Environmental Research and Public Health, 9*, 1490–1506.

Ambrosini, G. L., (2014). Childhood dietary patterns and later obesity: A review of the evidence. *Proceedings of the Nutrition Society, 73*, 137–146.

Asghari, G., Mirmiran, P., Yuzbashian, E., & Azizi, F., (2017). A systematic review of diet quality indices in relation to obesity. *British Journal of Nutrition, 117*, 1055–1065.

Astrup, A., (2008). Dietary management of obesity. *Journal of Parenteral and Enteral Nutrition, 32*, 575–577.

Beeken, R. J., & Wardle, J., (2013). Public beliefs about the causes of obesity and attitudes towards policy initiatives in Great Britain. *Public Health Nutrition, 16*, 2132–2137.

Belahsen, R., (2014). Nutrition transition and food sustainability. *Proceedings of the Nutrition Society, 73*, 385–388.

Blaisdell, A. P., Lau, Y. L. M., Telminova, E., Lim, H. C., Fan, B., Fast, C. D., Garlick, D., & Pendergrass, D. C., (2014). Food quality and motivation: A refined low-fat diet induces obesity and impairs performance on a progressive ratio schedule of instrumental lever pressing in rats. *Physiology & Behavior, 128*, 220–225.

Butterwick, R. F., & Hawthorne, A. J., (1998). Advances in dietary management of obesity in dogs and cats. *The Journal of Nutrition, 128*, 2771S–2775S.

Chin, S. H., Kahathuduwa, C., & Binks, M., (2016). Physical activity and obesity: What we know and what we need to know. *Obesity Reviews, 17*, 1226–1244.

Clarke, S. F., Murphy, E. F., Nilaweera, K., Ross, P. R., Shanahan, F., O'toole, P. W., & Cotter, P. D., (2012). The gut microbiota and its relationship to diet and obesity: New insights. *Gut Microbes, 3*, 186–202.

D'innocenzo, S., Biagi, C., & Lanari, M., (2019). Obesity and the Mediterranean diet: A review of evidence of the role and sustainability of the Mediterranean diet. *Nutrients,* 11, 1306.

Damms-Machado, A., Mitra, S., Schollenberger, A. E., Kramer, K. M., Meile, T., Königsrainer, A., Huson, D. H., & Bischoff, S. C., (2015). Effects of surgical and dietary weight loss therapy for obesity on gut microbiota composition and nutrient absorption. *BioMed Research International,* 2015.

Desalvo, K. B., Olson, R., & Casavale, K. O., (2016). Dietary guidelines for Americans. *JAMA, 315*, 457, 458.

Dyson, P. A., (2010). The therapeutics of lifestyle management on obesity. *Diabetes, Obesity and Metabolism, 12*, 941–946.

Fleming, J. A., & Kris-Etherton, P. M., (2016). Macronutrient content of the diet: What do we know about energy balance and weight maintenance? *Current Obesity Reports, 5*, 208–213.

Floegel, A., & Pischon, T., (2012). Low carbohydrate-high protein diets. *British Medical Journal.* Publishing Group.

Fock, K. M., & Khoo, J., (2013). Diet and exercise in management of obesity and overweight. *Journal of Gastroenterology and Hepatology, 28*, 59–63.

Gidding, S. S., Lichtenstein, A. H., Faith, M. S., Karpyn, A., Mennella, J. A., Popkin, B., Rowe, J., et al., (2009). Implementing American Heart Association pediatric and adult nutrition guidelines: A scientific statement from the American heart association nutrition committee of the council on nutrition, physical activity and metabolism, council on cardiovascular disease in the young, council on arteriosclerosis, thrombosis and vascular biology, council on cardiovascular nursing, council on epidemiology and prevention, and council for high blood pressure research. *Circulation, 119*, 1161–1175.

Goutzourelas, N., Orfanou, M., Charizanis, I., Leon, G., Spandidos, D. A., & Kouretas, D., (2018). GSH levels affect weight loss in individuals with metabolic syndrome and obesity following dietary therapy. *Experimental and Therapeutic Medicine, 16*, 635–642.

Gueugnon, C., Mougin, F., Simon-Rigaud, M. L., Regnard, J., Nègre, V., & Dumoulin, G., (2012). Effects of an in-patient treatment program based on regular exercise and a balanced diet on high molecular weight adiponectin, resistin levels, and insulin resistance in adolescents with severe obesity. *Applied Physiology, Nutrition, and Metabolism, 37*, 672–679.

Haywood, C. J., Prendergast, L. A., Purcell, K., Le Fevre, L., Lim, W. K., Galea, M., & Proietto, J., (2018). Very low calorie diets for weight loss in obese older adults—a randomized trial. *The Journals of Gerontology: Series A, 73*, 59–65.

Hu, T., Mills, K. T., Yao, L., Demanelis, K., Eloustaz, M., Yancy, Jr. W. S., Kelly, T. N., et al., (2012). Effects of low-carbohydrate diets versus low-fat diets on metabolic risk factors: A meta-analysis of randomized controlled clinical trials. *American Journal of Epidemiology, 176*, S44–S54.

Hwalla, N., & Jaafar, Z., (2021). Dietary management of obesity: A review of the evidence. *Diagnostics, 11*, 24.

Jakicic, J. M., Rogers, R. J., Davis, K. K., & Collins, K. A., (2018). Role of physical activity and exercise in treating patients with overweight and obesity. *Clinical Chemistry, 64*, 99–107.

Jarouliya, U., & Keservani, R. K., (2019). Pathways leading to child obesity. In: Bagchi, D., (ed.), *Global Perspectives on Childhood Obesity: Current Status, Consequences and Prevention* (2nd edn., pp. 137–146). Academic Press, Elsevier; Chapter 12. ISBN: 9780128128404.

Keservani, R. K., Kesharwani, R. K., Sharma, A. K., & Ahmad, M. F., (2015a). Stress, Parkinson's, alzemer's disease: Role of dietary supplements. In: Bagchi, D., Preuss, H. G., & Swaroop, A., (eds.), *Nutraceutical and Functional Foods in Human Life and Disease Prevention* (pp. 241–253). Boca Raton, Fl. Taylor and Francis, Ch 14. ISBN: 9781482237214.

Keservani, R. K., Kesharwani, R. K., Sharma, A. K., Vyas, N., & Chadoker, A., (2010b). Nutritional supplements: An overview. *Int. J. of Curr. Pharm. Rev. and Res., 1*(1), 59–75.

Keservani, R. K., Kesharwani, R. K., Vyas, N., Jain, S., Raghuvanshi, R., & Sharma, A. K., (2010a). Nutraceuticals and functional food as future food. *Der Pharm. Lett., 2*(1), 106–116.

Keservani, R. K., Sharma, A. K., & Kesharwani, R. K., (2016). An overview and therapeutic applications of nutraceutical & functional foods. In: Keservani, R. K., Sharma, A. K., & Kesharwani, R. K., (eds.), *Recent Advances in Drug Delivery Technology* (pp. 160–201). IGI Global Publisher, USA, Ch 6. ISBN:9781522507543.

Kurdaningsih, S. V., Sudargo, T., & Lusmilasari, L., (2016). Physical activity and sedentary lifestyle towards teenagers' overweight/obesity status. *Int. J. Community Med. Public Health, 3*, 630–635.

Millen, B. E., Abrams, S., Adams-Campbell, L., Anderson, C. A., Brenna, J. T., Campbell, W. W., Clinton, S., et al., (2016). The 2015 dietary guidelines advisory committee scientific report: Development and major conclusions. *Advances in Nutrition, 7*, 438–444.

Nanney, M. S., Larowe, T. L., Davey, C., Frost, N., Arcan, C., & O'meara, J., (2017). Obesity prevention in early child care settings: A bi-state (Minnesota and Wisconsin) assessment of best practices, implementation difficulty, and barriers. *Health Education & Behavior, 44*, 23–31.

Organization, W. H., (2000). *Obesity: Preventing and Managing the Global Epidemic, 1*, 1–268.

Piccolo, B. D., Comerford, K. B., Karakas, S. E., Knotts, T. A., Fiehn, O., & Adams, S. H., (2015). Whey protein supplementation does not alter plasma branched-chained amino acid profiles but results in unique metabolomics patterns in obese women enrolled in an 8-week weight-loss trial. *The Journal of Nutrition, 145*, 691–700.

Popkin, B. M., Adair, L. S., & Ng, S. W., (2012). Global nutrition transition and the pandemic of obesity in developing countries. *Nutrition Reviews, 70*, 3–21.

rane, B. R., & Keservani, R. K., (2020). Introduction to nutraceutical & dietary supplements. In: Keservani, R. K., Sharma, A. K., & Kesharwani, R. K., (eds.), *Nutraceutical and*

Dietary Supplements (pp. 35–55). Apple Academic Press, Taylor & Francis Group; Chapter 3. ISBN: 9781771888738.

Riccardi, G., Capaldo, B., & Vaccaro, O., (2005). Functional foods in the management of obesity and type 2 diabetes. *Current Opinion in Clinical Nutrition & Metabolic Care, 8*, 630–635.

Ruban, A., Stoenchev, K., Ashrafian, H., & Teare, J., (2019). Current treatments for obesity. *Clinical Medicine, 19*, 205.

Sackner-Bernstein, J., Kanter, D., & Kaul, S., (2015). Dietary intervention for overweight and obese adults: Comparison of low-carbohydrate and low-fat diets. A meta-analysis. *PloS One, 10*, e0139817.

Salwa, M., Maha, I., Hamed, I., Salwa, T., Laila, M., & Lobna, A., (2012). Dietary therapy of obesity: Effect on some hormonal and biochemical blood indices. *African Journal of Food, Agriculture, Nutrition and Development,* 12.

Schulze, M. B., Martínez-González, M. A., Fung, T. T., Lichtenstein, A. H., & Forouhi, N. G., (2018). Food based dietary patterns and chronic disease prevention. *BMJ,* 361.

Sikorski, C., Luppa, M., Glaesmer, H., Brähler, E., König, H. H., & Riedel-Heller, S. G., (2013). Attitudes of health care professionals towards female obese patients. *Obesity Facts, 6*, 512–522.

Singh, A., & Purohit, B., (2012). Physical activity, sedentary lifestyle, and obesity among Indian dental professionals. *Journal of Physical Activity and Health, 9*, 563–570.

Singh, P., Kesharwani, R. K., & Keservani, R. K., (2017a). Protein, carbohydrate and fats: Energy metabolism. In: Bagchi, D., (ed.), *Sustained Energy for Enhanced Human Function and Activity*. Elsevier. Ch 6. ISBN: 9780128054130.

Singh, P., Kesharwani, R. K., & Keservani, R. K., (2017b). Antioxidants and vitamins: Roles in cellular functions and metabolism. In: Bagchi, D., (ed.), *Sustained Energy for Enhanced Human Function and Activity*. Elsevier, Ch 24. ISBN: 9780128054130.

Soeliman, F. A., & Azadbakht, L., (2014). Weight loss maintenance: A review on dietary related strategies. *Journal of Research in Medical Sciences: The Official Journal of Isfahan University of Medical Sciences, 19*, 268.

Stevens, G. A., Singh, G. M., Lu, Y., Danaei, G., Lin, J. K., Finucane, M. M., Bahalim, A. N., et al., (2012). National, regional, and global trends in adult overweight and obesity prevalences. *Population Health Metrics, 10*, 1–16.

Strasser, B., (2013). Physical activity in obesity and metabolic syndrome. *Annals of the New York Academy of Sciences, 1281*, 141.

Surana, K. R., Ahire, E. D., Sonawane, V. N., Talele, S. G., & Talele, G., (2021). Informatics and methods in nutrition design and development. *Natural Food Products and Waste Recovery: Healthy Foods, Nutrition Design, and Extraction of Valuable Compounds,* 33.

Tapsell, L. C., Neale, E. P., Satija, A., & Hu, F. B., (2016). Foods, nutrients, and dietary patterns: Interconnections and implications for dietary guidelines. *Advances in Nutrition, 7*, 445–454.

Truby, H., Baxter, K., Ware, R. S., Jensen, D. E., Cardinal, J. W., Warren, J. M., Daniels, L., et al., (2016). A randomized controlled trial of two different macronutrient profiles on weight, body composition and metabolic parameters in obese adolescents seeking weight loss. *PloS One, 11*, e0151787.

Wadden, T. A., Webb, V. L., Moran, C. H., & Bailer, B. A., (2012). Lifestyle modification for obesity: New developments in diet, physical activity, and behavior therapy. *Circulation, 125*, 1157–1170.

Wojcicki, J. M., Schwartz, N., Jiménez-Cruz, A., Bacardi-Gascon, M., & Heyman, M. B., (2012). Acculturation, dietary practices and risk for childhood obesity in an ethnically heterogeneous population of Latino school children in the San Francisco bay area. *Journal of Immigrant and Minority Health, 14*, 533–539.

Yannakoulia, M., Poulimeneas, D., Mamalaki, E., & Anastasiou, C. A., (2019). Dietary modifications for weight loss and weight loss maintenance. *Metabolism, 92*, 153–162.

Zhao, L., Zhang, Q., Ma, W., Tian, F., Shen, H., & Zhou, M., (2017). A combination of quercetin and resveratrol reduces obesity in high-fat diet-fed rats by modulation of gut microbiota. *Food & Function, 8*, 4644–4656.

CHAPTER 8

Effects of Phenolic Compounds in Plants for Treatment of Type 2 Diabetes

BUI THANH TUNG and NGUYEN THI YEN

Department of Pharmacology and Clinical Pharmacy, VNU University of Medicine and Pharmacy – Vietnam National University, Vietnam

ABSTRACT

Diabetes mellitus (DM) is one of the dead causes in the world. Globally, about 1 in 11 adults have DM, and 90 of them have type 2 diabetes (T2D). Studies indicated that genetic predisposition is the main reason, but a sedentary lifestyle and an unhealthy diet are related to that disease too. Therefore, controlling postprandial hyperglycemia in type 2 diabetics is of great interest content of worldwide health concerns. However, lifestyle changes such as doing exercise, consuming a healthy diet, not drinking alcohol, and not smoking can prevent people from getting T2D. Many plant foods are beneficial to control postprandial hyperglycemia, such as *Punica granatum* (Pomegranate), *Cynara scolymus* L. (artichoke), The common bean (*Phaseolus vulgaris* L.), Finger Millet (*Eleusine coracana L.*), They have several phenolic compounds (PCs) that can preventively effect on insulin signal and both produce and receive glucose process in T2D patients. This review gives updated information about T2D and phenolic compounds effective against T2D.

8.1 DIABETES IS A COMPLEX DISEASE

Diabetes is a chronic disease characterized by defects in insulin secretion, insulin action, or both leading to hyperglycemia. It is requiring special caring

The Metabolic Syndrome: Dietary Supplements and Food Ingredients. Raj K. Keservani, MPharm, Durgavati Yadav, PhD, Rajesh K. Kesharwani, PhD, Sippy Singh, DPhil & Kumar Sandeep, PhD (Eds.)

medicine with multifactorial risk reduction strategies and glycemic control (Care, 2020). There are two popular types of diabetes, including:

- **Type 1 Diabetes:** β-cell is destroyed, often because of deficient absolute insulin.
- **Type 2 Diabetes:** β-cell frequently is continuing to lose insulin on the background of insulin resistance.

Besides, gestational diabetes mellitus (GDM) is known as a common disease in pregnancy. It is found out in the second or third trimester of pregnancy that was not overt diabetes prior to gestation. Other types of diabetes due to abundant causes, exempli gratia diseases of the exocrine pancreas, drug or chemical included diabetes (Association, 2019).

Overall, diabetes is a popular disease in the world. There are 415 million people worldwide who get diabetes and more than 90% of patients get type 2 diabetes (T2D). Most of the patients can suffer from macrovascular complications that make the pain of psychological and physical distress to both patients and their families and put a giant burden on the community (Chatterjee et al., 2017). Diabetes is on the rise and steadily increasing everywhere (World Health Organization, 2016). Therefore, it is necessary to organize regular screening programs for early detection and effective treatment of T2D to reduce morbidity and mortality by preventing or delaying its complications.

8.1.1 TYPE 2 DIABETES (T2D)

Type 2 diabetes is adult-onset diabetes. It used to start at the age of 40. It is more common than type 1 diabetes, accounts for 90% of diabetes patients, and is a different disease. Feature of T2D is high blood sugar, anti-insulin, and lack of insulin (Yang et al., 2018).

8.1.1.1 BASIC MECHANISM

Insulin is a hormone secreted by the pancreas. It is a signal which announces for cells absorb glucose. In T2D, insulin is secreted normally but cell receptors do not allow insulin to bind to them. Therefore, sugar levels in the blood remain too high. Besides, a reduced number of beta cells in the pancreas leading to reduce insulin is another cause of T2D (Figure 8.1) (Roberts et al., 2010).

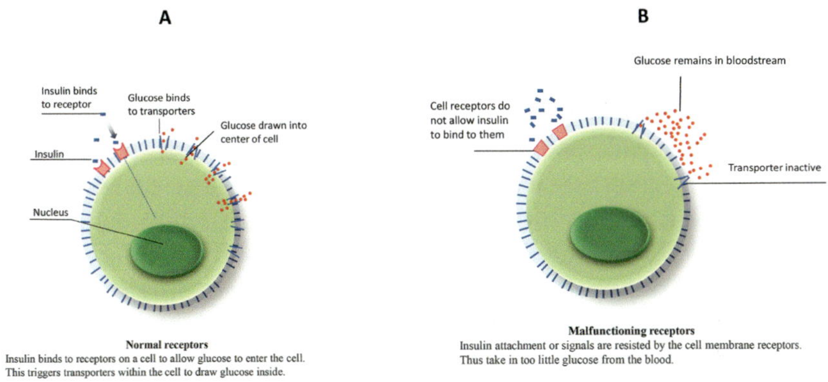

FIGURE 8.1 Normal receptors (A); and malfunctioning receptors (B) in type 2 diabetes.

It is said that T2D is linked to not only genetics but also obesity. The significant increase in patients who got this disease is thought to relate to a weight problem, less exercise, and an unhealthy diet.

8.2 CURRENT TREATMENT

Currently, T2D treatment is included in insulin and oral hypoglycemic. However, they cause a lot of undesirable effects.

The hypoglycemic effect of insulin can be caused by brain damage. Normal expressions are swelling, erythema, and stinging in the beginning. Besides, using insulin in the treatment of T2D can cause allergies to humans. Some patients can be temporarily dependent on edema due to Na^+ retention.

Another common drug is used to therapy T2D is metformin. The most popular unexpected effects of metformin are gastrointestinal disturbances, abdominal pain, and metallic taste. Rarely, patients get lactic acidosis, which has a potentially toxic effect. Therefore, patients with hepatic or renal disease, hypoxic pulmonary disease, heart failure could not treat T2D with metformin. A high dose of metformin intercedes absorption process vitamin B12 in the body. It leads to vitamin B12 deficiency.

Both sulfonylureas and thiazolidinediones are responsible for other serious adverse effects. Sulfonylureas's unwanted effects are hypoglycemia severely and lengthen. Patients using sulfonylureas can get allergic skin rashes and damage bone marrow. The adverse effect of thiazolidinediones is hepatotoxicity, weight gain, gastrointestinal disturbances (Patil et al., 2011).

Thus, current drugs for T2D make lots of negative problems. It is important to explore new compounds that can be used to replace the current treatment. Not only in different ethnic societies of Asia, Africa, and South America but also in developed countries of Europe, North America, and Japan, people use several plant products to treat T2D (Risérus et al., 2009). Scientists have shown that those plants contain compounds that are useful in T2D therapy, one of which is phenolic compounds. PCs can inhibitive α-glucosidase and α-amylase, which leads to decrease hyperglycemia after means by retarding the absorption of glucose. Because phenolic compounds are essential in defense responses, it is positive to eat plant foods including a high concentration of phenol compounds (Lin et al., 2016).

8.3 PHENOLIC COMPOUNDS

8.3.1 WHAT ARE PHENOLIC COMPOUNDS?

Phenolic compounds (PCs) are phytochemicals distributed in most plant tissues, including vegetables and fruits. PCs are secondary metabolites, which are produced through the shikimic acid and phenylpropanoid pathways of plants (Yahia et al., 2018). Chemical structures of PCs consist of benzenes rings combined with one or more hydroxyl substituents. They are ranged from simple phenolic molecules to highly polyphenol compounds (Figure 8.2) (Keservani and Sharma, 2014).

8.3.2 CLASSIFICATION OF PHENOLIC COMPOUNDS

Depending on structure, PCs can be classified into two main groups, contain flavonoids and nonflavonoids.

8.3.2.1 FLAVONOIDS

Flavonoids account for approximately two-thirds of PCs in vegetables and fruits. Till now more than 8,000 phenolic structures are identified, and among them, more than 4,000 flavonoids have been known. Many foods have flavonoids such as fruits, vegetables, flowers, wine, tea, dark chocolate, … and absorbed by humans.

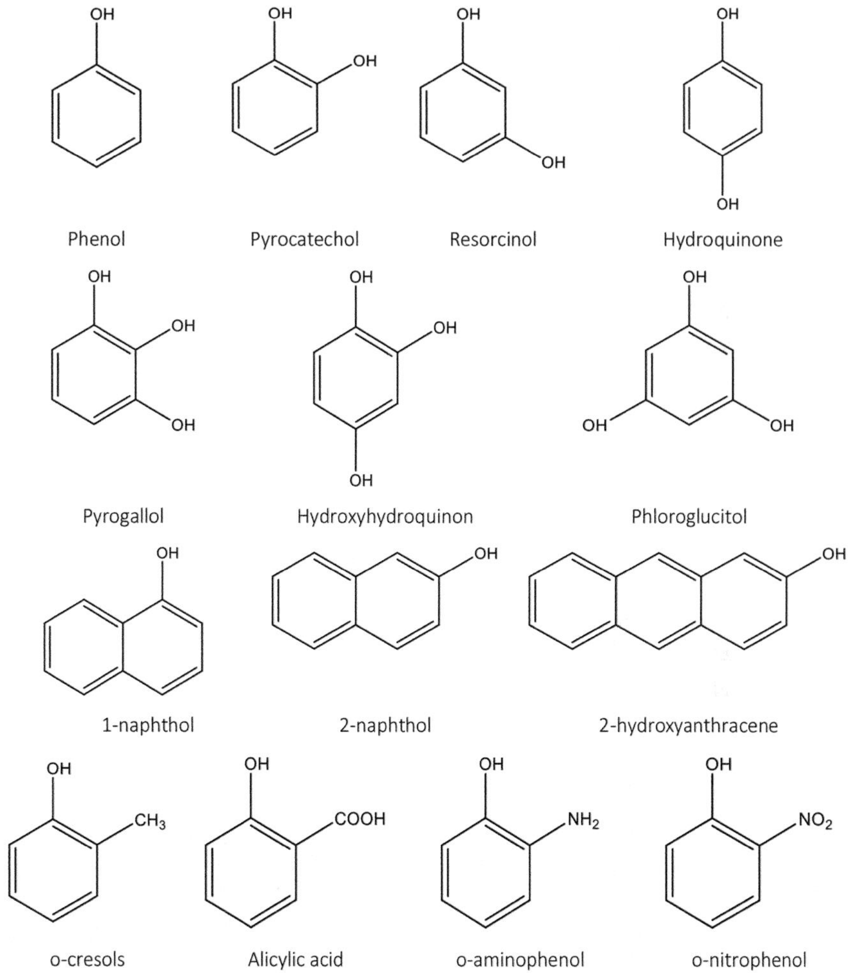

FIGURE 8.2 Examples of simple phenolic molecules.

The chemical structure of flavonoids is a phenyl benzopyran skeleton, which contains two benzene rings (ring A and ring B) linked to a heterocyclic pyran ring (ring C). Flavonoids can be divided into six families, according to differences in the pyran ring, including flavones, flavonols, flavanones, flavanonol, isoflavones, and flavan-3-ols. They are different from the level of oxidation and pattern of substitution of the C ring (Figures 8.3 and 8.4) (Kumar et al., 2013).

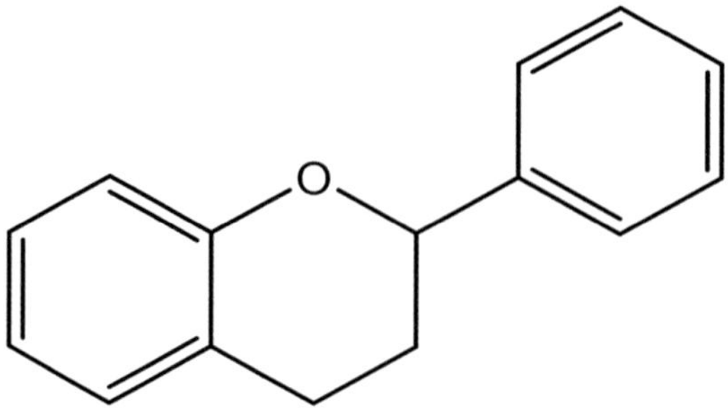

FIGURE 8.3 Basic structure of flavonoids.

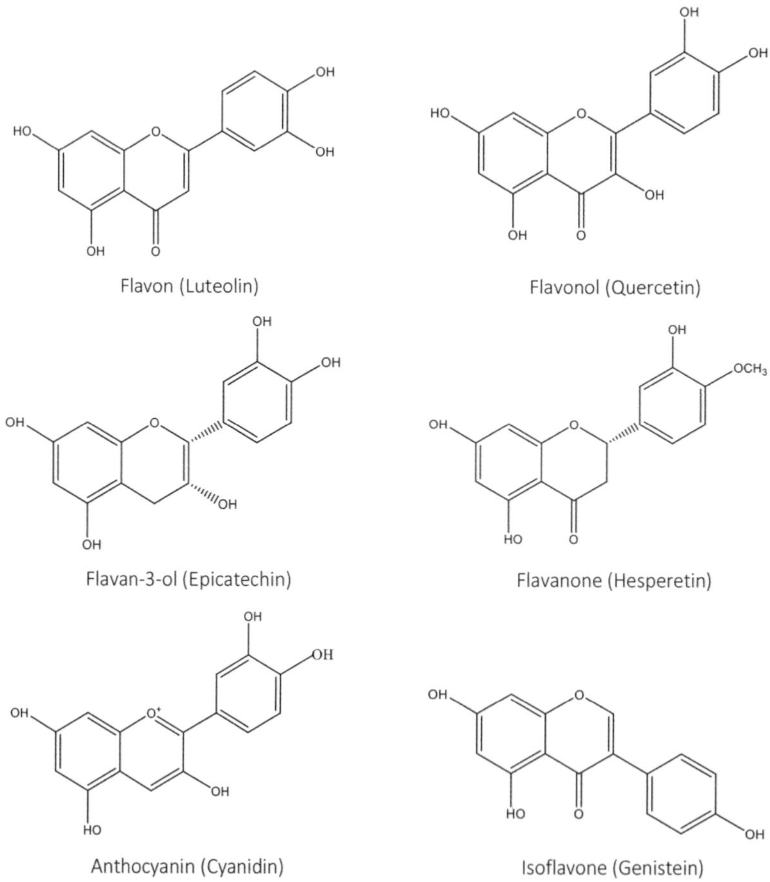

FIGURE 8.4 Structure backbone of six families of flavonoids group and examples.

8.3.2.2 NONFLAVONOIDS

The chemical structures of the nonflavonoids group are abundant. Most of them are generally smaller and simpler than flavonoids structures, a few ones are complex and have high molecular weights (Laura et al., 2019). Nonflavonoid compounds include four main families such as phenolic acids, volatile phenols, stilbenes, and other polyphenols (Rentzsch et al., 2009). Among them, phenolic acid is the most important nonflavonoids in vegetables and fruits. Phenolic acids can be divided into two families including hydroxybenzoic acids (HBAs) and hydroxycinnamic acids (HCAs) (Figure 8.5) (Laura et al., 2019).

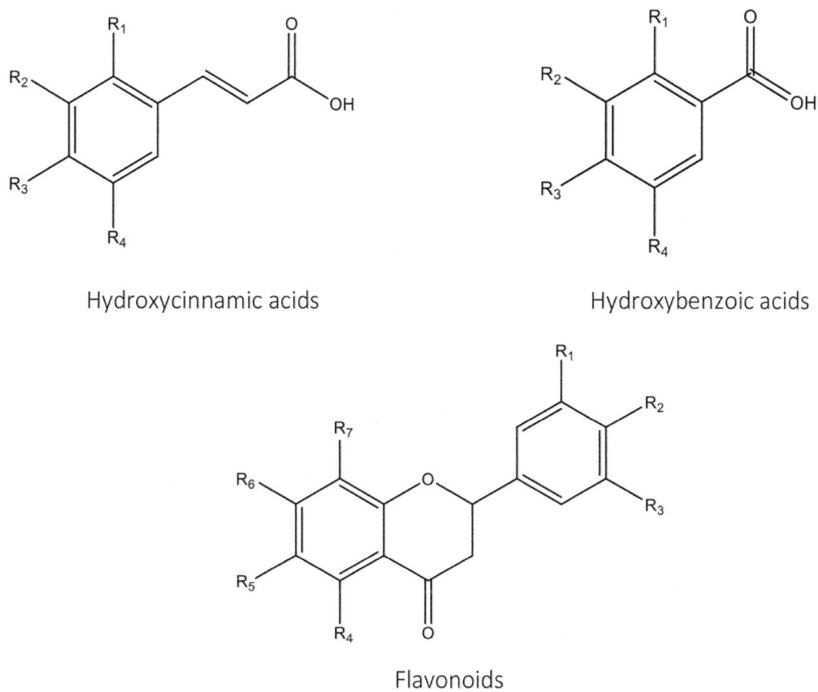

FIGURE 8.5 General structure of HBAs, HCAs, and flavonoids.

HBAs are the most popular nonflavonoids, which are found in many vegetables and fruits such as nuts, berries, chicory, tea, and some spices (Laura et al., 2019).

8.4 THE EFFECTIVE AGAINST TYPE 2 DIABETES (T2D) OF PHENOLIC COMPOUNDS IN SOME PLANTS

It is easy to find phenolic compounds in many fruits and vegetables such as grapes, berries, and tomatoes. It is said that PCs in plants are antioxidants and respond for defense reactions. From the standpoint of human physiology, PCs also are essential in the defense responses in humans. They promote the health benefits of taking part in anti-oxidation, anti-aging, anti-inflammatory, and anti-proliferative activities. Therefore, using plant foods, which have a high concentration of phenolic compounds, is very beneficial (Lin et al., 2016).

Polyphenols, especially flavonoids, phenolic acids, and tannins, can inhibit α-amylase and α-glucosidase, which are responsible for the metabolism of carbohydrates to glucose. They adjust carbohydrates and lipid metabolism, control sugar concentration in blood, attenuate dyslipidemia, and insulin resistance, improve β-cell function, increase insulin secretion, ...Thus, the PCs can control the early stage of T2D. In addition, PCs also have the ability to moderate long-term diabetes complications, including retinopathy, nephropathy, neuropathy, and cardiovascular disease (Lin et al., 2016). However, different plants give different PCs structures, so their biological effects are also distinctive. This review focuses on some plants that have phenolic compounds effective against T2D.

8.4.1 PUNICA GRANATUM (POMEGRANATE)

Punica grantum, is also known as Pomegranate, Carthaginian Apple, Granada. It is planted all over the world that includes 500 cultivars. For a long time ago, Pomegranate juice is well known for its healthy compounds, because of its high concentration of antioxidant activity and total PCs content. The level of flavonoids in *P. grantum* is very high; it includes flavanols, flavonols, anthocyanins, hydrolysable, and condensing tannins, organic acids, and phenolic acids. In the Middle East and India, *P. grantum* is used as a hyperglycemia and diabetes method (Saxena et al., 2004). In the medieval age, *P. grantum* also was used to prepare holy and make medicinal drinks in Persian (Jandari et al., 2020).

A lot of studies indicated that total phenolic compounds of peel extract are approximately 5.9-fold higher than that of juice extract. A seed extract has 2.1-fold more total PCs than that of juice extract (Kalaycıoğlu et al., 2017).

Being claimed to supply several health benefits, pomegranate has many effects such as anti-inflammatory and chemo-preventive effects. Moreover, it has been shown that potential role in obesity, diabetes, and cardiovascular disease treatment (Jurenka, 2008).

In the treatment of T2D, *P. grantum* decreases the lipid peroxidation and oxidative stress in order to affect T2D. The efficiency of this fruit is shown through many different mechanisms, such as enhancing some enzymes to react antioxidant, inhibiting reactive oxygen species (ROS), generating metal chelating activity, and inhibiting or activating certain transcriptional factors, including peroxisome proliferator-activated receptor γ (PPAR-γ*) and nuclear factor κappa B* (κB) (Banihani et al., 2013).

Ahmad Esmaillzadeh and his co-worker conducted a study that assessed the effect of the level of pomegranate juice on the T2D patient, who are hyperlipidemia (cholesterol >/= 5.2 mmol/L or triacylglycerol >/= 2.3 mmol/L), in 2004. They realized if the patients drink 40-gram pomegranate juice/day for 8 weeks, the levels of total cholesterol and LDL in their serum will be significantly reduced (Esmaillzadeh et al., 2004). If type 2 diabetic patients with hyperlipidemia consume pomegranate juice for three months, their ability to oxidize N-linoleoyl in their blood also will be decreased remarkably, and the oxysterol/total cholesterol ratio in their blood will be decreased by approximately 93% too (Szuchman et al., 2008).

Because of *Punica grantum*'s benefit, it will be useful to drink 20–50 ml juice or 200–500 mg pills of that fruit. Besides, pomegranate can be put together with raspberry (*Rubus idaeus*), goji berries (*Lycium barbarum*), apple (*Malus domestica*), … to using easily (Dragan et al., 2015).

8.4.2 CYNARA SCOLYMUS L. (ARTICHOKE)

Cynara scolymus or artichoke is a perennial herb of the Compositae family, which is from the Mediterranean area and till now is wide-ranging all over the world (Zhu et al., 2004). *C. scolymus* has been used medically since the 4th BCE (Mahboubi, 2018). The first time *C. scolymus* was transplanted in China during the 1990s from Italy and Spain. Then, it was started cultivating widely for commercial purposes by the Kunming Yandi Agricultural Development Co., Ltd., in southwest China since 2001 (Zhu et al., 2004).

In traditional medicine, *C. scolymus* has been used as a cholagogues and fat metabolizer. There are a lot of chemical structures in the parts of this plant, such as potassium, vitamin C, cynarin, orthophenole derivates, magnesium, folate, and dietary fiber in artichoke globe (Lin et al., 2016); flavonoids (scolymoside, luteolin, and cynaroside), hydroxycinnamic acids (caffeic acid, ferulic acid, chlorogenic acid, dicaffeolquinic acids), cynarin (1,5-dicaffeolquinic acid) in the leaves of artichoke (Azzini et al., 2007). Recent studies indicated that flavonoids of *C. scolymus* participate in upregulating nitric oxide synthesis in endothelial cells. Endothelial nitric oxide synthase (eNOS) can generate both nitric oxide (NO) and superoxide, regulate the dilating and shrinking of blood vessels. It is believed that eNOS has antithrombotic, anti-atherosclerotic, and antihypertensive properties, it is vital in vascular function (Lin et al., 2016).

Facts, a study has shown that polyphenolic compounds, which were extracted from artichoke leaves, can extract lower sugar concentration in blood (Fantini et al., 2011). They also have been determined that it can increase eNOS activity, so, influence NO synthesis and prevent patients from long-term diabetes complications (Lin et al., 2016).

8.4.3 THE COMMON BEAN (PHASEOLUS VULGARIS L.)

The common bean (*Phaseolus vulgaris* L.) is a plant of the Leguminosae (Fabaceae) family, genus Phaseolus (Graham et al., 1997). *P. vulgaris* is the main legume for human consumption and is the world's third most important food legume after soybeans and peanuts (Lin et al., 2008). The common bean is a rich source of protein, carbohydrates, vitamins, minerals, and dietary fiber for the people in developing and developed nations, especially for the poorer people of Africa and Latin America (Broughton et al., 2003).

Many studies in the dietetics and prevention of diabetes fields showed that bean seed coats extract has a high concentration of natural polyphenolic compounds, which can control disorders in these diseases. Results of researches are confirmed by *in vivo* studies on the experimental models of diabetes in lab animals (Lee et al., 2014; Inoue et al., 2011; Yuji et al., 2013; Jang et al., 2013; Yao et al., 2008; Wang et al., 2014) and clinical studies (Kianbakht et al., 2013; Hoggard et al., 2013; Sidorova et al., 2017).

A study indicated that the common bean has a high concentration of phenolic compounds. This bean contained hydrocinnamic acid derivatives and lots of kinds of flavonoids such as flavonols, flavanones, flavones, isoflavones, bioflavonoids, isoflavones, anthocyanins, coumestans, and pterocarpans (Lin et al., 2008). All of them are useful in reducing glycemia, lipid, and carbohydrate absorption. *P. vulgaris can regulate* serum levels of plasma insulin, HDL-cholesterol, glutathione, vitamin C, and the antiatherogenic index in diabetic rats normal. The other study also determined that cooked common beans can prevent pancreatic beta-cell from damaging (Pari et al., 2003). So, it is one of the most well-known hypoglycemic herbal drugs.

8.4.4 *TERMINALIA CHEBULA RETZ*

Terminalia chebula Retz is a native plant of India, it is grown widely all-over southern Asia (Kolla et al., 2017). The dried fruit of *T. chebula* is also known as Xi-Qin-Ge or Zhang-Qin-Ge in China, and black myrobalan in English (Ali Asgar, 2013).

In ethnobotanical, the *T. chebula* fruit is useful in dental disease, ophthalmology … Paste of this fruit is as anti-inflammatory, analgesic, and healing the pain (Bag et al., 2013). Its powder has been used in bleeding and ulceration in gums. It is also used in nervous weakness, nervous irritability, and renal conditions (Bag et al., 2013).

Many studies are showing that it is possible to extract bioactive compounds from different parts of *T. chebula*, such as hydrolysable tannins, phenol compounds, fructose, amino acids, … Therefore, *T. chebula* has antidiabetic and protective activity. Dosing effect of *T. chebula* extracts dependent blood glucose concentration of streptozotocin-induced diabetic rats (Kannan et al., 2012). The dried fruit of *T. chebula* has high quantities of phenolic compounds, which were investigated by Gao et al. in 2007. They found three active ellagitannins in the extracted dried fruit, chebulanin, chebulagic acid, and chebulinic acid. Then, scientists showed the strong α-glucosidase inhibitory activity of three compounds in rat intestinal, with the IC_{50} values of 690 μM, 97 μM, and 36 μM, respectively. Because the delayed absorption process of glucose via inhibition of α-glucosidase activity is one of the treatment plans for preventing T2D, *T. chebula* could be a choice to control blood glucose level in type 2 diabetes (Ali Asgar, 2013).

8.4.5 *FINGER MILLET (ELEUSINE CORACANA L.)*

Finger millet (*Eleusine coracana L.*) belongs to the Poaceae family of the monocotyledon group. It is one of the minor cereals, which originated in Ethiopia, but cultivated in India and many of the African countries (Chandra et al., 2016). *E. coracana* is like a power source of nutrients, supplies calories and protein to the population in these countries (O'Kennedy et al., 2006).

However, not many Indians fully understand the health-benefiting nutrients of this plant (Chandra et al., 2016). Especially, the millet peel contains a high concentration of phenolic compounds (10% w/w). The characterized of PCs extracted from the millet peel are made out by Shobana et al. through HPLC and ESI-MS. This study indicated that many different types of PCs in *E. coracana*, including phenolic acids and polyphenols, such as gallic acid, gentisic acid, caffeic acid, protocatechuic acid, syringic acid, ferulic acid, kaempferol, naringenin, luteolin glycoside, phloroglucinol, apigenin, … (Shobana et al., 2009). These PCs showed strong inhibition towards α-glucosidase and pancreatic amylase and the IC_{50} values for them respectively were 16.9 and 23.5 μg (Table 8.1).

TABLE 8.1 Phenolic Compounds Identified in Finger Millet

Class	Basic Skeleton	Compounds	References
Hydroxybenzoic acid derivatives	C6–C1	Gallic acid, protocatechuic acid, p-hydroxybenzoic acid, vanillic acid, syringic acid	McDonough et al. (1986); Subba Rao and Muralikrishna (2002); Chethan and Malleshi (2007).
Hydroxycinnamic acid derivatives	C6–C3	Ferulic acid, trans-cinnamic acid, p-coumaric acid, caffeic acid, sinapic acid	Chethan et al. (2008); Shobana et al. (2009)
Flavonoids	C6–C3–C6	Quercetin Proanthocyanidins	Chethan et al. (2008); Dykes and Rooney (2006); Chandrasekara and Shahidi (2010)

8.4.6 PAKHANBHED (BERGENIA CILIATA)

Bergenia ciliate belongs to the family *Saxifragaceae*, and it is a Nepalese herb. In traditional medicine in Nepal, it is used against various diseases (Ali Asgar, 2013). A total of 185 research papers, which is like a reliable scientific data source, indicated that *B. ciliate* is used to treat 104 different diseases, especially, *B. ciliate* is well known for the therapy of kidney disorders (Ahmad et al., 2018). Besides, rhizomes of the plant are determined that it can be used to treat lung disease, etc.

A lot of compounds, which are useful in medical treatment, are found from the root of *B. ciliata*, such as saponins, flavonoids, glycosides, and terpenoids. However, there is very little research on *B. ciliate.* In 2008,

Bhandari et al. announced the anti-diabetes activity of *B. ciliate.* They isolated the active compounds and exam this action by an *in vitro* model. For the first time, two active compounds, (–)-3-*O*-galloylepicatechin and (–)-3-*O*-galloylcatechin were extracted and isolated from this plant. This research demonstrated that these active substances have the strong inhibition of porcine pancreatic α-amylase and rat intestinal α-glucosidase activity in a dose-dependent. It is also the first time, the anti-diabetes potential of *B. ciliata* was reported and becoming beneficial evidence for the following studies (Table 8.2) (Bhandari et al., 2008).

TABLE 8.2 Key Enzyme Inhibitors from Plant Origins Involved in Starch Digestion

Plant	Enzyme Inhibitor	Enzyme Targets	References
Finger millet (*E. coracana*)	Naringenin, kaempferol, luteolin glycoside, apigenin, (+)-catechin/(–)-epicatechin, daidzein, caffeic acid, ferulic acid, syringic acid	Pancreatic α-amylase, α-glucosidase	Shobana et al. (2009)
Pakhanbhed (*B. ciliata*)	(–)-3-*O*-galloylepicatechin, (–)-3-*O*-galloylcatechin	Pancreatic α-amylase, maltase	Bhandari et al. (2008)
Raspberries (*Rubus idaeus* L.)	Ellagic acid, catechin, pelargonidin-3-rutinoside, cyanidin-diglucoside	α-Glucosidase	Zhang et al. (2010)
Persimmon (*Diospyros kaki*)	Proanthocyanidin [(–)-epigallocatechin, (–)-epigallocatechin-3-*O*-gallate, (–)-epicatechin, (+)-catechin, (–)-epicatechin-3-*O*-gallate]	Pancreatic α-amylase	Kawakami et al. (2010)

8.4.7 RASPBERRIES (RUBUS IDAEUS L.)

Rubus idaeus L. is in the *Idaeobatus* subgenus, the Rubus genus. It is a very hardy perennial species that is distributed throughout temperate North America and Eurasia, but it is only found on mountains in Southern Europe (Tutin et al., 1968). Raspberries are well known as a great source of antioxidants because they contain a high level of phenolic compounds. Many studies were shown the characterization of this fruit. Guyliana D. Noratto et al. reported that eating red raspberry for 6 weeks can decrease oxidative stress in obese diabetic mice (Noratto et al., 2017). Wolfe et al. indicated that a raspberry in the USA extract has the highest antioxidant activity (Wolfe et al., 2008). Zhang et al. assessed the inhibitory effect of PCs extracted from raspberry on starch digestive enzymes (Zhang et al., 2010). The *R. idaeus* extracts are ellagic acid, cyanidin-diglucoside, pelargonidin-3-rutinoside, and catechin, which were verified as the strong α-glucosidase inhibitors compounds, they can against yeast α-glucosidase, porcine pancreatic α-amylase, and porcine pancreatic lipase. However, no substance in the *R. idaeus* extracts occurred against α-amylase and lipase (Zhang et al., 2010). Besides, the anthocyanin substance was found from extracts of raspberries, too (McDougall et al., 2005). McDougall et al. reported that anthocyanin substance can inhibitive strongly α-glucosidase activity, both salivary and pancreatic amylase, but this ability was related to their tannin content. Therefore, the inhibitory α-glucosidase affection from PCs in the raspberry extract is very beneficial in the treatment of T2D.

8.4.8 PERSIMMON (DIOSPYROS KAKI)

Diospyros kaki, a common name is persimmon, originated from China and was introduced to Japan in the seventh century and became a popular plant species in Japan. It is a tropical fruit and is used daily in the diet (Woolf et al., 2011). Persimmon is a great source of nutrients, including carbohydrates, vitamins, dietary fibers, minerals, carotenoids, especially, phenolic compounds, and other bioactive phytochemicals (Matheus et al., 2020; Keservani et al., 2010a, b).

Kawakami et al. extracted a lot of phenolic compounds from persimmon leaf. The major components were proanthocyanidins, covering (+)-catechin, (–)-epigallocatechin, (–)-epigallocatechin-3-O-gallate, (–)-epicatechins, and (–)-epicatechin-3-O-gallate (Kawakami et al., 2010). Although the PCs concentration of *D. kaki* leaf tea remarkably inhibitsα-amylase activity, it has no effect on maltase. Research results showed that the percentage of PCs powder to inhibit porcine pancreas α-amylase at the concentration 24 μg/ml was 24.6 ± 3.2, and at the concentration 48 μg/ml was 45.2 ± 2.8 (Kawakami et al., 2010). In a similar study, Shin et al. (2020) showed that another effect of persimmon related to diabetes complications. This research was conducted in the high-fat diet-induced mouse. It determined that the ethanolic extract of *D. kaki* reduced fasting glucose level in blood and improved diabetic cognitive impairment (Shin et al., 2020). So, proanthocyanidin compound from Persimmon (*Diospyros kaki*) extracts contributes significantly to control hyperglycemia by the inhibition of α-amylase.

8.5 CONCLUSION

T2D and its complications are on the rise and have become a great burden on society, so managing hyperglycemia after meals are very important. This review has shown potential treatment of this disease of phenolic compounds, which are found in many plant species such as *Punica granatum*, *Cynara scolymus* L., *Punica granatum*, *Bergenia ciliate*, etc. Phenolic compounds against hyperglycemia-induced chronic diseases through both antioxidant and inhibition of starch digestion. Human researches have indicated that a correlation between PCs concentrations and their effects on sugar-metabolizing enzymes. Besides, PCs can interact with other dietary components, such as protein may inhibit polyphenol activity in the gastrointestinal tract. It is necessary to conduct long-term human studies, which would help to determine the efficacy, optimal dosage, and safety of inhibitors and the relationship of PCs and other dietary components.

KEYWORDS

- **flavonoids**
- **hydroxybenzoic acids**
- **hydroxycinnamic acids**
- **phenolic compounds**
- **plant foods**
- **polyphenol compounds**
- **type 2 diabetes**

REFERENCES

Ahmad, M., Butt, M. A., Zhang, G., Sultana, S., Tariq, A., & Zafar, M., (2018). Bergenia ciliata: A comprehensive review of its traditional uses, phytochemistry, pharmacology and safety. *Biomedicine & Pharmacotherapy, 97*, 708–721.

Ali, A. M., (2013). Anti-diabetic potential of phenolic compounds: A review. *International Journal of Food Properties, 16*(1), 91–103.

Association, A. D., (2019). Classification and diagnosis of diabetes: Standards of medical care in diabetes—2019. *Diabetes Care, 42*(Supplement 1), S13–S28.

Azzini, E., Bugianesi, R., Romano, F., D. Di Venere, M. S., Durazzo, A., Foddai, M., Catasta, G., Linsalata, V., & Maiani, G., (2007). Absorption and metabolism of bioactive molecules after oral consumption of cooked edible heads of *Cynara scolymus* L. (cultivar Violetto di Provenza) in human subjects: A pilot study. *British Journal of Nutrition, 97*(5), 963–969.

Bag, A., Bhattacharyya, S. K., & Chattopadhyay, R. R., (2013). The development of *Terminalia chebula* Retz. (Combretaceae) in clinical research. *Asian Pacific Journal of Tropical Biomedicine, 3*(3), 244–252.

Banihani, S., Swedan, S., & Alguraan, Z., (2013). Pomegranate and type 2 diabetes. *Nutrition Research, 33*(5), 341–348.

Bhandari, M. R., Jong-Anurakkun, N., Hong, G., & Kawabata, J., (2008). α-Glucosidase and α-amylase inhibitory activities of Nepalese medicinal herb Pakhanbhed (Bergenia ciliata, Haw.). *Food Chemistry, 106*(1), 247–252.

Broughton, W. J., Hernandez, G., Blair, M., Beebe, S., Gepts, P., & Vanderleyden, J., (2003). Beans (Phaseolus spp.)–model food legumes. *Plant and Soil, 252*(1), 55–128.

Care, D., (2020). Standards of medical care in diabetes 2020. *Diabetes Care, 43*, S1–S224.

Chandra, D., Chandra, S., & Sharma, A., (2016). Review of finger millet (*Eleusine coracana* (L.) Gaertn): A powerhouse of health benefiting nutrients. *Food Science and Human Wellness 5*(3), 149–155.

Chandrasekara, A., & Shahidi, F., (2010). Content of insoluble bound phenolics in millets and their contribution to antioxidant capacity. *Journal of Agricultural and Food Chemistry, 58*(11), 6706–6714.

Chatterjee, S., Khunti, K., & Davies, M. J., (2017). Type 2 diabetes. *The Lancet, 389*(10085), 2239–2251.

Chethan, S., & Malleshi, N., (2007). Finger millet polyphenols: Optimization of extraction and the effect of pH on their stability. *Food Chemistry, 105*(2), 862–870.

Chethan, S., Dharmesh, S. M., & Malleshi, N. G., (2008). Inhibition of aldose reductase from cataracted eye lenses by finger millet (*Eleusine coracana*) polyphenols. *Bioorganic & Medicinal Chemistry, 16*(23), 10085–10090.

Chethan, S., Sreerama, Y., & Malleshi, N., (2008). Mode of inhibition of finger millet malt amylases by the millet phenolics. *Food Chemistry, 111*(1), 187–191.

Dragan, S., Andrica, F., Serban, M. C., & Timar, R., (2015). Polyphenols-rich natural products for treatment of diabetes. *Current Medicinal Chemistry, 22*(1), 14–22.

Dykes, L., & Rooney, L. W., (2006). Sorghum and millet phenols and antioxidants. *Journal of Cereal Science, 44*(3), 236–251.

Esmaillzadeh, A., Tahbaz, F., Gaieni, I., Alavi-Majd, H., & Azadbakht, L., (2004). Concentrated pomegranate juice improves lipid profiles in diabetic patients with hyperlipidemia. *Journal of Medicinal Food, 7*(3), 305–308.

Fantini, N., Colombo, G., Giori, A., Riva, A., Morazzoni, P., Bombardelli, E., & Carai, M. A., (2011). Evidence of glycemia-lowering effect by a *Cynara scolymus* L. extract in normal and obese rats. *Phytotherapy Research, 25*(3), 463–466.

Gao, H., Huang, Y. N., Xu, P. Y., & Kawabata, J., (2007). Inhibitory effect on α-glucosidase by the fruits of *Terminalia chebula* Retz. *Food Chemistry, 105*(2), 628–634.

Graham, P., & Ranalli, P., (1997). Common bean (*Phaseolus vulgaris* L.). *Field Crops Research, 53*(1–3), 131–146.

Jandari, S., Hatami, E., Ziaei, R., Ghavami, A., & Yamchi, A. M., (2020). The effect of pomegranate (*Punica granatum*) supplementation on metabolic status in patients with type 2 diabetes: A systematic review and meta-analysis. *Complementary Therapies in Medicine*, 102478.

Jurenka, J., (2008). Therapeutic applications of pomegranate (*Punica granatum* L.): A review. *Alternative Medicine Review, 13*(2).

Kalaycıoğlu, Z., & Erim, F. B., (2017). Total phenolic contents, antioxidant activities, and bioactive ingredients of juices from pomegranate cultivars worldwide. *Food Chemistry, 221*, 496–507.

Kannan, V. R., Rajasekar, G. S., Rajesh, P., Balasubramanian, V., Ramesh, N., Solomon, E. K., Nivas, D., & Chandru, S., (2012). Anti-diabetic activity on ethanolic extracts of fruits of *Terminalia chebula* Retz. alloxan induced diabetic rats. *American Journal of Drug Discovery and Development, 2*(3), 135–142.

Kawakami, K., Aketa, S., Nakanami, M., Iizuka, S., & Hirayama, M., (2010). Major water-soluble polyphenols, proanthocyanidins, in leaves of persimmon (*Diospyros kaki*) and their α-amylase inhibitory activity. *Bioscience, Biotechnology, and Biochemistry, 74*(7), 1380–1385.

Keservani, R. K., & Sharma A. K., (2014). Flavonoids: Emerging trends and potential health benefits. *J. Chin. Pharm. Sci, 23*(12), 815–822.

Keservani, R. K., Kesharwani, R. K., Sharma, A. K., Vyas, N., & Chadoker, A., (2010b). Nutritional supplements: An overview. *Int. J. of Curr. Pharm. Rev. and Res., 1*(1), 59–75.

Keservani, R. K., Kesharwani, R. K., Vyas, N., Jain, S., Raghuvanshi, R., & Sharma, A. K., (2010a.). Nutraceuticals and functional food as future food. *Der Pharm. Lett., 2*(1), 106–116.

Kolla, J. N., Kulkarni, N. M., Kura, R. R., & Theepireddy, S., K. R., (2017). *Terminalia chebula* Retz.–an important medicinal plant. *Herba Polonica, 63*(4), 45–56.

Kumar, S., & Pandey, A. K., (2013). Chemistry and biological activities of flavonoids: An overview. *The Scientific World Journal, 2013*.

Laura, A., Moreno-Escamilla, J. O., Rodrigo-García, J., & Alvarez-Parrilla, E., (2019). Phenolic compounds. *Postharvest Physiology and Biochemistry of Fruits and Vegetables* (pp. 253–271). Elsevier.

Lin, D., Xiao, M., Zhao, J., Li, Z., Xing, B., Li, X., Kong, M., et al., (2016). An overview of plant phenolic compounds and their importance in human nutrition and management of type 2 diabetes. *Molecules, 21*(10), 1374.

Lin, L. Z., Harnly, J. M., Pastor-Corrales, M. S., & Luthria, D. L., (2008). The polyphenolic profiles of common bean (*Phaseolus vulgaris* L.). *Food Chemistry, 107*(1), 399–410.

Mahboubi, M., (2018). Cynara scolymus (artichoke) and its efficacy in management of obesity. *Bulletin of Faculty of Pharmacy, 56*(2), 115–120. Cairo University.

Matheus, J. R. V., Andrade, C. J. D., Miyahira, R. F., & Fai, A., E. C., (2020). Persimmon (*Diospyros kaki* L.): Chemical properties, bioactive compounds and potential use in the development of new products: A review. *Food Reviews International*, 1–18.

McDonough, C., Rooney, L., & Earp, C., (1986). Structural characteristics of *Eleusine corocana* (finger millet) using scanning electron and fluorescence microscopy. *Food Structure 5*(2), 9.

McDougall, G. J., Shpiro, F., Dobson, P., Smith, P., Blake, A., & Stewart, D., (2005). Different polyphenolic components of soft fruits inhibit α-amylase and α-glucosidase. *Journal of Agricultural and Food Chemistry, 53*(7), 2760–2766.

Muhammad, S., Khan, B. A., Akhtar, N., Mahmood, T., Rasul, A., Hussain, I., Khan, H., & Badshah, A., (2012). The morphology, extractions, chemical constituents and uses of *Terminalia chebula*: A review. *Journal of Medicinal Plants Research, 6*(33), 4772–4775.

Noratto, G. D., Chew, B. P., & Atienza, L. M., (2017). Red raspberry (*Rubus idaeus* L.) intake decreases oxidative stress in obese diabetic (db/db) mice. *Food Chemistry, 227*, 305–314.

O'Kennedy, M. M., Grootboom, A., & Shewry, P., (2006). Harnessing sorghum and millet biotechnology for food and health. *Journal of Cereal Science, 44*(3), 224–235.

Pari, L., & Venkateswaran, S., (2003). Effect of an aqueous extract of *Phaseolus vulgaris* on plasma insulin and hepatic key enzymes of glucose metabolism in experimental diabetes. *Die Pharmazie-an International Journal of Pharmaceutical Sciences, 58*(12), 916–919.

Patil, R., Patil, R., Ahirwar, B., & Ahirwar, D., (2011). Current status of Indian medicinal plants with antidiabetic potential: A review. *Asian Pacific Journal of Tropical Biomedicine, 1*(2), S291–S298.

Rentzsch, M., Wilkens, A., & Winterhalter, P., (2009). Non-flavonoid phenolic compounds. *Wine Chemistry and Biochemistry* (pp. 509–527). Springe.

Risérus, U., Willett, W. C., & Hu, F. B., (2009). Dietary fats and prevention of type 2 diabetes. *Progress in Lipid Research, 48*(1), 44–51.

Roberts, A. M., & Roberts, A., (2010). *The Complete Human Body*. Dorling Kindersley Ltd.

Sabu, M., & Kuttan, R., (2002). Anti-diabetic activity of medicinal plants and its relationship with their antioxidant property. *Journal of Ethnopharmacology, 81*(2), 155–160.

Saxena, A., & Vikram, N. K., (2004). Role of selected Indian plants in management of type 2 diabetes: A review. *The Journal of Alternative & Complementary Medicine, 10*(2), 369–378.

Shin, E. J., Kim, J. M., Kang, J. Y., Park, S. K., Han, H. J., Kim, H. J., Kim, C. W., Lee, U., & Heo, H. J., (2020). Ameliorative effect of persimmon (Diospyros kaki) in cognitively impaired diabetic mice. *Journal of Food Biochemistry*, e13581.

Shobana, S., Sreerama, Y., & Malleshi, N., (2009). Composition and enzyme inhibitory properties of finger millet (*Eleusine coracana* L.) seed coat phenolics: Mode of inhibition of α-glucosidase and pancreatic amylase. *Food Chemistry, 115*(4), 1268–1273.

Sidorova, Y., Shipelin, V., Mazo, V., Zorin, S., Petrov, N., & Kochetkova, A., (2017). Hypoglycemic and hypolipidemic effect of *Vaccinium myrtillus* L. leaf and *Phaseolus vulgaris L.* seed coat extracts in diabetic rats. *Nutrition, 41*, 107–112.

Subba, R. M., & Muralikrishna, G., (2002). Evaluation of the antioxidant properties of free and bound phenolic acids from native and malted finger millet (Ragi, *Eleusine coracana* Indaf-15). *Journal of Agricultural and Food Chemistry, 50*(4), 889–892.

Szuchman, A., Aviram, M., Musa, R., Khatib, S., & Vaya, J., (2008). Characterization of oxidative stress in blood from diabetic vs. hypercholesterolaemic patients, using a novel synthesized marker. *Biomarkers, 13*(1), 119–131.

Tutin, T., Heywood, V., Burges, N., Moore, D., Valentine, D., Walters, S., & Webb, D., (1968). *Flora Europaea* (Vol. 2–5). Cambridge University Press, Cambridge.

Wolfe, K. L., Kang, X., He, X., Dong, M., Zhang, Q., & Liu, R. H., (2008). Cellular antioxidant activity of common fruits. *Journal of Agricultural and Food Chemistry, 56*(18), 8418–8426.

Woolf, A., & Ben-Arie, R., (2011). Persimmon (*Diospyros kaki* L.). *Postharvest Biology and Technology of Tropical and Subtropical Fruits* (pp. 166–194e). Elsevier.

World Health Organization, (2016). *Global report on diabetes*. World Health Organization, 1–83.

Yahia, E. M., & Carrillo-Lopez, A., (2018). *Postharvest Physiology and Biochemistry of Fruits and Vegetables*. Woodhead Publishing.

Yang, Q., Zhou, L., Liu, C., Liu, D., Zhang, Y., Li, C., Shang, Y., Wei, X., Li, C., & Wang, J., (2018). Brain iron deposition in type 2 diabetes mellitus with and without mild cognitive impairment—An *in vivo* susceptibility mapping study. *Brain Imaging and Behavior 12*(5), 1479–1487.

Zhang, L., Li, J., Hogan, S., Chung, H., Welbaum, G. E., & Zhou, K., (2010). Inhibitory effect of raspberries on starch digestive enzyme and their antioxidant properties and phenolic composition. *Food Chemistry, 119*(2), 592–599.

Zhu, X., Zhang, H., & Lo, R., (2004). Phenolic compounds from the leaf extract of artichoke (*Cynara scolymus* L.) and their antimicrobial activities. *Journal of Agricultural and Food Chemistry, 52*(24), 7272–7278.

CHAPTER 9

Role of Vitamins in the Metabolic Syndrome

DURGESH SINGH

Department of Zoology, Sadanlal Sanwaldas Khanna Mahila Mahavidyalay (A Constituent College of University of Allahabad), Attarsuiya, Prayagraj, Uttar Pradesh – 211003, India, E-mail: ddurgesh12@gmail.com

ABSTRACT

Metabolic syndrome has been defined as a cluster or constellation of metabolic abnormalities which leads to various health ailments such as high glucose level, cardiovascular disease, abdominal obesity, diabetes, etc. Metabolic syndromes have been frequently reported worldwide, and the main reason behind their prevalence has been suggested to be the modified lifestyle which involves decreased physical activity and an unhealthy diet. Vitamins are organic compounds that are essential for the proper development of the body. These micronutrients are essential, and their absence in the diet leads to various diseases. Apart from their active role in the development of the body, vitamins possessing antioxidant and anti-inflammatory activity have been widely used in treating or managing metabolic syndrome. Supplementation of diet with vitamins has been reported to help in removal of reactive oxygen species and hence provide protection against cardiovascular disease, cancer, neurological, and immunological disorders. A healthy diet, including antioxidant and anti-inflammatory vitamins along with proper physical activity, is the key toward avoiding and managing metabolic syndrome.

The Metabolic Syndrome: Dietary Supplements and Food Ingredients. Raj K. Keservani, MPharm, Durgavati Yadav, PhD, Rajesh K. Kesharwani, PhD, Sippy Singh, DPhil & Kumar Sandeep, PhD (Eds.)

9.1 INTRODUCTION

Vitamins are organic compounds required in a very minute amount in diet for proper growth and development of the body. These organic compounds are required on a daily basis in small quantity, which varies from organism to organism and depends on the rate of metabolism of a person. Around 13 reported vitamins have been categorized into two groups, namely water-soluble (B complex and C) and fat-soluble vitamins (A, D, E, and K). Deficiencies of these vitamins in the body can lead to many diseases (Albahrani and Greaves, 2016). The common diseases caused by deficiency of these vitamins are: Vitamin B complex deficiency – beriberi, chelosis, pellagra, pernicious anemia, etc.; vitamin C deficiency – scurvy; vitamin A deficiency – night blindness; vitamin D deficiency – osteomalacia and rickets; vitamin E – accredited oxidative stress; vitamin K – hemorrhage. Two of the fat-soluble vitamins: A and D have been found associated with medical conditions like cancer, Type II diabetes mellitus and disorders related to the immune system (Stephensen, 2001; Holick, 2004; Keservani et al., 2015).

"Good health is an investment in economic growth," and one of the best ways to manage good health and prevent health issues is to focus on proper nutrition (Brenseke et al., 2013). There has been burgeoning interest in studying and managing metabolic syndrome (MetS) by using anti-oxidative and anti-inflammatory agents (Wong et al., 2017, 2018). MetS is a combination of metabolic abnormalities which includes major risk factors such as cardiovascular diseases (CDVs), abdominal obesity, type 2 diabetes (T2D), high glucose level, etc. (Alberti et al., 2009). MetS is one of the major public health concerns these days, and the main reason is associated with lifestyle changes, including less healthy eating habits and reduced physical activities (Saklayen, 2018; Engin, 2017). The history of MetS starts back from 1920 when a Swedish physician Kylin observed correlation between of hypertension, gout, and hyperglycemia. Vague in 1947 suggested a close relationship of CDVs and T2D mellitus with visceral obesity. Further, in 1965, in an abstract, a syndrome was highlighted which included symptoms like obesity, hypertension, and hyperglycemia. Reaven (1988) gave a single term SYNDROME X to all the symptoms which generated due to diabetes and CDVs which was later renamed in coming years as The Deadly Quartet in 1989; the insulin resistance syndrome in 1992 and finally to metabolic syndrome (MetS) (Kaur, 2014).

Researches so far indicate a crucial role of micronutrients, especially vitamins having anti-oxidative and anti-inflammatory properties in the regulation of metabolic disorder or syndromes. As the oxidative stress and inflammation are correlated with health issues like obesity (Jarouliya and Keservani, 2019), insulin resistance and metabolic disorders, hence it has been proposed that agents with anti-oxidative properties can overcome the oxidative stress and are potent enough to prevent health complications related to oxidative damage (Avignon et al., 2012). Some of the vitamins exhibiting such properties have been discussed in subsections.

9.1.1 VITAMIN A

Usually referred to as Retinol or Retinoic acid is best known for its role in maintaining vision. Apart from it, it also promotes immune function and is also reported to reverse the inflammation by decreasing adipocytokines (Mozos and Marginean, 2015). Green leafy vegetables, pigmented fruits are good sources of vitamin A. Intestinal juices and bile salts help in the absorption of vitamin A in the gut region (D' Ambroso et al., 2011; Keservani et al., 2010a). The retinyl esters (preformed vitamin A form animal source) are hydrolyzed into retinol in intestine while small amount of retinoid is converted to retinoic acid which is absorbed into blood (Ross, 2005). Nearly 70% of retinoid present in diet is absorbed by the liver and converted into retinyl palmitate (D' Ambroso et al., 2011) which is further rehydrolyzed into retinol by enzymes like retinyl ester hydrolase.

Vitamin A has an important role in regulating immune function, and its deficiency can cause drying up of skin as well as its keratinization, disrupt neutrophil development, decreased number of natural killer cells, and hence leads to increased risk of infections (de Pee and Dary, 2002). On the contrary, hypervitaminosis of this vitamin can cause toxicity within the body leading to blurred vision, nausea, headache, etc. (Ross, 2005). Mozos and Marginean (2015) in their study reported that retinol-binding protein and insulin sensitivity are inversely related and in case of obese and diabetic patients there is an increase in retinol-binding protein. Whayne and Maulik (2012) propounded that vitamin A along with other micronutrients having antioxidant property lead to reduced cardiovascular related deaths. Vitamin A has also been reported to inhibit the growth of tumorous cells *in vitro* and can control other functions like apoptosis, cell division, etc. (Niles, 2000).

9.1.2 VITAMIN D

This hormone plays an important role in calcium and phosphorus metabolism. Vitamin D is synthesized in the body from 7-dehydrocholesterol by the action of sunlight. The endogenous synthesis of vitamin D depends upon the thickness and color of skin, period of exposure to sunlight (Albahrani and Greaves, 2016). During exposure to UV radiations of sunlight, C9–C10 of 7-dehydrocholesterol breaks down into pre-vitamin D3 within skin. Excess of pre-vitamin D3 is converted into lumisterol and tachysterol (Kochupillai, 2008) which can be converted back into pre-vitamin D3 whenever there is a fall in its level in skin. Vitamin D produced in the body circulates through blood by binding to vitamin D binding protein and reaches the liver (Bikle, 2009). In the liver, vitamin D is converted into calcidiol and then reaches to kidney via blood circulation where it is converted into biologically active form calcitriol. The latter binds to vitamin D receptor present on different cells, thereby stimulating calcium absorption, cell differentiation, insulin secretion, inducing maturation of preosteoclasts into osteoclasts which helps in bone reabsorption (Holick, 2007). Vitamin D deficiency may lead to rickets in children and osteomalacia in adults, while low level result in bone fractures (Holick, 2006; Cranney et al., 2007) and breast cancer. This hormone shows exerts several important effects such as anti-inflammatory, anti-apoptotic, and anti-fibrotic. Two major forms of vitamin D are cholecalciferol (derived from animal) and ergocalciferol (derived from plants).

Gradillas-García et al. (2015) have reported that low level of vitamin D in serum can lead to a higher risk of MetS. Apart from this, deficiency of vitamin D has been reported to pose threat to autoimmune diseases (Illescas-Montes et al., 2019), and even variation in vitamin D receptor expression has been correlated with occurrence of MetS (Karonova et al., 2018). Some of the interesting studies have correlated higher risk of MetS in people living away from the equator (Mutt et al., 2019); vitamin D deficiency in postmenopausal women and higher risk of MetS (Schmitt, 2018); inverse relationship between level of vitamin D and presence of MetS (Huang et al., 2015); and high ratio of MetS in cardiology unit in patients suffering vitamin D deficiency (Barbalho et al., 2018). On the contrary, many researchers objected to the direct correlation of vitamin d level and the occurrence of metabolic disorders like MetS (Zhengetal, 2020; Wang et al., 2020).

Amongst different views of scientists, the potential of vitamin D remains one of the promising micronutrients in controlling metabolic disorders because of its pleiotropic role in various actions which includes mitochondrial oxidation, improved arterial stiffness, improved phospholipid metabolism, inflammatory cytokines, etc. (Faraji and Alizadeh, 2020). Supplementation of vitamin D in patients has been reported to lead to reduced blood pressure, reduction in triglyceride, BMI, and waist circumference (Melguizo-Rodríuez et al., 2021).

9.1.3 VITAMIN E

This fat-soluble vitamin also known as tocopherol, is well known for its antioxidant properties, in addition to these some non-antioxidant activities have also been reported (Traber, 2011). Vitamin E is majorly absorbed in the small intestine and the process is enhanced in the presence of fat and food matrix (Albahrani and Greaves, 2016). This vitamin is metabolized in the liver and alpha-tocopherol is resecreted in blood by the action of a protein named α-tocopherol, defect in this protein has been reported to lead to deficiencies related to vitamin E (Eggermont, 2006). Vitamin E helps in removing reactive oxygen species (ROS) and has been reported to provide protection against cardiovascular disease, cancer, neurological, and immunological disorders (Clarke et al., 2008; Dror and Allen, 2011). Deficiency of vitamin E has been correlated with malabsorption disorders.

9.1.4 VITAMIN K

This fat-soluble vitamin is present in diet in two variable forms namely: menaquinones and phylloquinone, out of which the former is found in animal products, whereas the latter is rich in green leafy vegetables. High vitamin K intake has been associated with better lipoprotein profile, low body mass index (BMI) (Braam et al., 2004) and better sensitivity towards insulin (Ibarrola-Jurado et al., 2012). Vitamin K insufficiency has been reported to lead to high level of uncarboxylated Gla protein in the body. High phylloquinone intake has been suggested to cause low risk of high HDL cholesterol and hyperglycemia (Pan and Jackson, 2009), whereas high intake of menaquinone has been suggested to cause reduction in MetS occurrence leading to decrease in triacylglycerol and waist circumference (Dam et al., 2015). Knapen et al. (2012) reported an increase in body weight (BW), BMI, and fat mass (FM) in people with poor intake of vitamin K.

9.1.5 VITAMIN B

Vitamin B9 (Folate) and Vitamin B7 (Biotin) are water-soluble vitamins which have been correlated with several metabolic disorders (Keservani et al., 2010a, b). Folate deficiencies have been suggested to pose a high risk for occurrence of CDVs. Hummel et al. (2007) reported that in patients suffering from obesity and high weight issues, the serum level of folate has been recoded to be low. Biotin and folate together have been studied to suppress the proinflammatory cytokines in cell cultures (Mozos and Marginean, 2015). Folic acid supplementation helps in improving insulin resistance and decrease in homocysteine plasma level (Gargari et al., 2011). Folate deficiency has the ability to elevate obesity by various modes. Vitamin B1 (Thiamine) is necessary for the proper functioning of many tissues and organs. In obese females, reduced level of plasmatic thiamine as well as thiamine in erythrocyte has been observed (Patrini et al., 2004). Vitamin B6 (Pyridoxine) acts as a cofactor that helps in the breakdown of glycogen, transamination, and decarboxylation which has been suggested to affect weight (Nachtigal et al., 2005). Serum level of pyridoxine and BMI are negatively correlated (Aasheim et al., 2008). Vitamin B12 is a cofactor responsible for synthesis of methionine to homocysteine and its deficiency leads to hyperhomocysteinemia which in turn is correlated with insulin resistance and risk for cardiovascular disease (Pinhas-Hamiel et al., 2006).

9.1.6 VITAMIN C

This water-soluble vitamin present in plants and animals is well known for its antioxidant and anti-inflammatory properties. This vitamin has been widely accepted and proposed to be beneficial in treating MetS as high plasma vitamin C has been reported to result in decreased BMI, waist circumference, blood glucose, and triglycerides (Wong et al., 2020). Oxidative stress caused by the formation of reactive oxygen species is also an indicative signal for the occurrence of MetS. Vitamin C is potent enough to remove free radicals from the body thereby protecting the body from the damage caused by free radicals and hence helps in reducing oxidative damage or stress (Padayatty et al., 2003; Singh et al., 2017; Keservani et al., 2016). Bilbies et al. (2012) studied the effects of vitamin C supplementation in treating MetS rat models and found that there was a remarkable decrease in glucose, insulin resistance, weight gain, and triglyceride level. Similarly, ascorbate supplemented Werner syndrome mice showed decreased glucose levels, and low visceral fat weight (Shamanna et al., 2017). There have been many studies which tested the effect of vitamin C in managing MetS both alone as well as in combination with other antioxidant compounds. All the studies showed the positive response of this vitamin in treating MetS. Studies conducted in association with National Health and Nutrition Examination Survey have validated that concentration of vitamin C in serum and MetS occurrence show inverse relationship (Wong et al., 2020). If the animals are fed with a combination of vitamin C, polyphenols, and other antioxidant compounds, then it helps in combating oxidative stress, which in turn can help in managing metabolic disorders. The current recommended daily allowance for vitamin C has referred to be in the range of 40 mg/day in order to maintain a balance between oxidative stress and protection by antioxidant micronutrients (Carr and Lykkesfeldt, 2020). Increased level of Vitamin C leads to oxidation of fat in the body and hence lowers the case of adiposity, apart from this it also activates glycolysis (Wong et al., 2020). Vitamin C deficiency may lead to endotoxemia and increased inflammation.

9.1.6.1 PATHOGENIC MECHANISM OF METABOLIC SYNDROME

The pathogenesis of MetS is very complex. Environmental factors, life style and lack of physical activity are considered as major causes for MetS. Among these factors, high calorie intake leading to visceral adiposity has been considered as the primary reason for the onset of pathways causing MetS (Matsuzawa et al., 2011). Some of the important mechanisms that can lead to MetS are: insulin resistance, neurohormonal activation and inflammation.

Insulin resistance causes increase in amount of free fatty acid which is one the important way in pathogenesis of MetS. Insulin resistance within adipose tissues inhibits insulin mediated lipolysis resulting in the amount of free fatty acid that in turn inhibits insulin's antilipolytic activity (Boden and Shulman, 2002). Free fatty acid inhibits activation of protein kinase in muscle whereas increases kinase activation in liver resulting in reduced glucose uptake and increased gluconeogenesis and lipogenesis respectively leading to hyperinsulinemic condition. Hence, overall insulin secretion decreases (Tooke and Hannemann, 2000). Insulin resistance causes an increase in the release of proinflammatory cytokines, serum viscosity from adipose tissue which increases the risk for CDVs (Juhan-Vague et al., 2003).

Adipokines released by visceral tissues have been reported to be related with CDVs and MetS (Wallace et al., 2001). Obese condition has been reported to increase leptin, an adipokine which also increases the risk of CDVs. Renin-angiotensin system activation acts as important neurohormonal pathway resulting in the development of MetS. All the pathways in MetS narrows down to a common pathway of inflammation. Inflammation is one the important pathogenesis in MetS. An increase in adipose tissue mass has been correlated with an increase in the release of tumor necrosis factor by macrophages. An increase in the level of tumor necrosis factor has also been correlated with major components of MetS: obesity and insulin resistance (Tsigos et al., 1999). High level of C-reactive proteins (CRPs) has been correlated with MetS (Bastard et al., 2000).

9.1.6.2 MANAGEMENT OF METS

Apart from the use of vitamins; healthy lifestyle, pharmacotherapy, and the use of nutraceuticals can help us in the management of MetS. Dietary manipulations including low intake of calories, cholesterol, simple sugar, trans fat, etc., along with moderate exercise can be beneficial in managing MetS. Pharmacological aspects, including the use of insulin sensitizer, use of antiplatelet can be useful in decreasing the risk of diabetes. Various natural products derived from herbs, plant extract, spices have proved to be beneficial in combating MetS. Some of the nutraceuticals like garlic, curcumin, cinnamon, neem, resveratrol, etc., have shown promising effects in managing patients with MetS. Some of the nutraceuticals with their MetS combating activities have been summed up in Table 9.1.

TABLE 9.1 List of the Important Nutraceuticals Playing Role in Managing MetS

Nutraceutical	Source	MetS Combating Activities
Allicin	Garlic	• Possess antioxidant and antithrombotic activity • Raw garlic improves insulin sensitivity • Lowers cholesterol and triglyceride
Curcumin	Turmeric	• Active ingredient diferuloylmethane has anti-inflammatory and antioxidative property • Downregulates tumor necrosis factor • Decreases expression of pro-inflammatory cytokines • Inhibits Wnt/β-catenin pathway which is related to obesity • Increases adiponectin expression

TABLE 9.1 *(Continued)*

Nutraceutical	Source	MetS Combating Activities
Polyphenol	Cinnamon	• Possess antithrombotic activity • Results in insulin sensitization • Possess lipid lowering power • Anti-inflammatory and antioxidant property • Improves blood pressure • Improves glucose transport and insulin signaling
Neem extract	Neem	• Increases glucose tolerance • Regenerates pancreatic beta cells thereby improving insulin secretion
Resveratrol (polyphenol)	Grapes, nuts	• Improves insulin sensitivity • Increases glucose tolerance • Improves weight and BMI
Quercetin	Onion, berries, tea	• Anti-inflammatory and antioxidant property • Lowers blood pressure • Lowers insulin resistance • Improves waist circumference • Decreases adipogenesis • Lowers blood glucose and cholesterol level
Omega fatty acid (polyunsaturated fatty acid)	Fish oil	• Increases fatty acid oxidation • Inhibits lipogenesis

Source: Rochlani et al. (2017).

9.1.6.3 NANOSTRUCTURED PHYTOMEDICINES AND METS

Medicinal plants have been widely used in managing MetS but it faces problem of low bioavailability and diminished bioactivity (Tajmohammadi et al., 2018). However, these problems can be overcome, and the bioavailability, stability, and solubility of these compounds from medicinal plants can be enhanced by formulating the compounds in nanostructured form. The beneficial effects of nanostructured phytochemicals have been studied in relation to diabetes and many other diseases, and the efficacy and safety of drugs have been observed to improve (Ganesan et al., 2017; Kesharwani et al., 2018). Nano-drug delivery systems like nanomicelles, liposomes, etc., help in targeting drugs to specific sites. The nanoformulations of conventional drugs used for treating MetS have been found to have better efficacy and reduced side effects. Different types of nanoparticles have been used for drug delivery system like: Polymer nanoparticles (100–1,000 nm) which easily decompose into biocompatible compounds easily encapsulates drug into the target tissue; Dronerimes (3D polymer system, 10–20 nm) which has improved physical and chemical nature; Ceramic nanoparticles (less than 50 nm) consisting of silica, aluminum, calcium phosphate with increased biocompatibility and stability; Micelles (less than 100 nm) consisting of amphiphilic macromolecules; Liposomes which are combination of lipid molecule in aqueous solution and are non-toxic, biodegradable, and lack immune response; Nanopumps which are capable of pumping insulin and balance blood sugar; and Chitosan combined golden nanoparticles which good carrier of insulin.

The nanostructured herbal extracts have anti-diabetic properties. Marrubin obtained from *Leonotis leonurus* induces secretion of insulin, upregulates glucose transporter type-2 gene and increase insulin sensitivity (Mnonopi et al., 2012). Peepal tress possess antioxidant, anti-inflammatory, antidiabetic, and neuroprotective activity (Singh et al., 2011). α-eleostearic acid found in bitter gourd has anti-diabetic, anti-inflammatory property but has low efficiency and bioavailability whereas nanoemulsion containing bitter gourd oil has been observed to increase gastrointestinal absorption which diminishes blood sugar and increases antioxidant enzymes (Kumar et al., 2012; Paul et al., 2014). Nanoemulsion of garlic oil has been reported to show therapeutic effect against HFD-dyslipidemia (Cheraghpour et al., 2019). There has been growing interest in using biocompatible nanoparticles of herbal extracts because it is cost-effective, eco-friendly, and highly efficient and stable (Rane et al., 2020; Rane and Tadavi, 2020).

9.2 CONCLUSION

Looking into the complications of MetS and various mechanisms involved in pathophysiology, efficient strategies are essential for the prevention and treatment of the disease. Obese patients have low levels of vitamins in comparison to non-obese individuals. Increase in adipose tissue leads to inflammation and various health issues and when this condition is accompanied with vitamin deficiency then situation can become even worse. Vitamins have high potential in treating or managing metabolic disorders. These vitamins exhibiting antioxidative and anti-inflammatory properties protect the body against oxidative damage to the body which if left unattended, can lead to metabolic syndrome. MetS is a global epidemic that poses a high risk to health. Unhealthy eating habits and low physical activities result in metabolic syndrome (Wong et al., 2020). Proper diet containing micronutrients rich in antioxidant properties (Vitamins) have been associated with reduced MetS status. Proper supplementation of these vitamins can potentially reverse metabolic syndrome.

Greater awareness is required regarding the importance and benefits of vitamins. It is essential to promote a better understanding of the consequences of poor dietary habits especially lack or deficiency of vitamins which may result in an increased risk of metabolic disease. Modification in lifestyle is the initial step towards managing MetS, hence it is advised to incorporate a proper healthy diet and suggestive physical activities in daily routine in order to maintain good health.

KEYWORDS

- **cardiovascular disease**
- **insulin resistance**
- **metabolic syndrome**
- **nutraceuticals**
- **reactive oxygen species**
- **vitamins**

REFERENCES

Aasheim, E. T., et al., (2008). Vitamin status in morbidly obese patients: A cross-sectional study. *The Ame. J of Cli. Nut., 87*(2), 362–369.

Albahrani, A. A., & Greaves, R. F., (2016). Fat-soluble vitamins: Clinical indications and current challenges for chromatographic measurement. *Clin. Biochem Rev., 37*(1), 27–47.

Alberti, K. G., Eckel, R. H., Grundy, S. M., et al., (2009). Harmonizing the metabolic syndrome: A joint interim statement of the international diabetes federation task force on epidemiology and prevention; national heart, lung, and blood institute; American heart association; world heart federation; international atherosclerosis society; and international association for the study of obesity. *Circ., 120*(16), 1640–1645.

Avignon, A., Hokayem, M., Bisbal, C., & Lambert, K., (2012). Dietary antioxidants: Do they have a role to play in the ongoing fight against abnormal glucose metabolism? *Nut., 28*(7, 8), 715–721.

Barbalho, S. M., Tofano, R. J., De Campos, A. L., et al., (2018). Association between vitamin D status and metabolic syndrome risk factors. *Diab. Metab. Synd., 12*, 501–507.

Bastard, J. P., Jardel, C., Bruckert, E., et al., (2000). Elevated levels of interleukin 6 are reduced in serum and subcutaneous adipose tissue of obese women after weight loss. *J. Clin. Endocrinol. Metab., 85*(9), 3338–3342.

Bikle, D., (2009). Nonclassic actions of vitamin D. *J. Clin. Endocrinol. Metab*., 94, 26–34.

Bilbis, L. S., Muhammad, S. A., Saidu, Y., et al., (2012). Effect of vitamins, A, C, and E supplementation in the treatment of metabolic syndrome in albino rats. *Biochem. Res. In., 2012,* 678582–678582.

Boden, G., & Shulman, G. I., (2002). Free fatty acids in obesity and type 2 diabetes: Defining their role in the development of insulin resistance and beta-cell dysfunction. Eur. J. Clin. Invest., 32(Suppl. 3), 14–23.

Braam, L., McKeown, N., Jacques, P., et al., (2004). Dietary phylloquinone intake as a potential marker for a heart-healthy dietary pattern in the Framingham offspring cohort. *J Am Diet Asso.,* 104(9), 1410–1414.

Brenseke, B., Prater, M. R., Bahamonde, J., & Gutierrez, J. C., (2013). Current thoughts on maternal nutrition and fetal programming of the metabolic syndrome. *J. Pre., 2013*, 368461.

Carr, A. C., & Lykkesfeldt, J., (2020). Discrepancies in global vitamin C recommendations: A review of RDA criteria and underlying health perspectives. *Crit. Rev. Food Sci. Nut.,* 1–14.

Cheraghpour, M., Imani, H., Ommi, S., Alavian, S. M., Karimi-Shahrbabak, E., Hedayati, M., et al., (2019). Hesperidin improves hepatic steatosis, hepatic enzymes, and metabolic and inflammatory parameters in patients with nonalcoholic fatty liver disease: A randomized, placebo-controlled, double-blind clinical trial. *Phytother. Res., 33*, 2118–2125.

Clarke, M. W., Burnett, J. R., & Croft, K. D., (2008). Vitamin E in human health and disease. *Crit. Rev. Clin. Lab. Sc., 45*, 41750.

Cranney, A., Horsley, T., O'Donnell, S., Weiler, H., Puil, L., Ooi, D., et al., (2007). *Effectiveness and Safety of Vitamin D in Relation to Bone Health* (p. 235). US Department of Health and Human Services, Public Health Service, Agency for Healthcare Research and Quality.

D'Ambrosio, D. N., Clugston, R. D., & Blaner, W. S., (2011). Vitamin A metabolism: An update. *Nutri., 3*, 63–103.

Dam, V., Dalmeijer, G. W., Vermeer, C., et al., (2015). Association between vitamin K and the metabolic syndrome: A 10-year follow-up study in adults. *J Clin. Endocrinol. Metab., 100*(6), 2472–2479.

De Pee, S., & Dary, O., (2002). Biochemical indicators of vitamin A deficiency: Serum retinol and serum retinol binding protein. *J. Nutr., 132*(Suppl), 2895S–2901S.

Dror, D. K., & Allen, L. H., (2011). Vitamin E deficiency in developing countries. *Food Nutr. Bull., 32*, 124–143.

Eggermont, E., (2006). Recent advances in vitamin E metabolism and deficiency. *Eur J. Pediatr., 165*, 429–434.

Engin, A., (2017). The definition and prevalence of obesity and metabolic syndrome. In: Engin, A. B., & Engin, A., (eds.), *Obesity and Lipotoxicity: Advances in Experimental Medicine and Biology* (pp. 1–17). Springer International Publishing: Cham, Switzerland. ISBN 978-3-319-48382-5.

Faraji, S., & Alizadeh, M., (2020). Mechanistic effects of vitamin D supplementation on metabolic syndrome components in patients with or without vitamin D Deficiency. *J. Obes. Metab. Synd., 29*, 270–280.

Ganesan, P., Arulselvan, P., & Choi, D. K., (2017). Phytobioactive compound based nano delivery systems for the treatment of type 2 diabetes mellitus– current status. *Int. J. Nanomedicine, 12*, 1097–1111.

Gargari, B. P., Aghamohammadi, V., & Aliasgharzadeh, A., (2011). Effect of folic acid supplementation on biochemical indices in overweight and obese men with type 2 diabetes. *Dia. Res. and Clin. Prac., 94*(1), 33–38.

Gradillas-García, A., Álvarez, J., Rubio, J. A., & De Abajo, F. J., (2015). Relationship between vitamin D deficiency and metabolic syndrome in adult population of the community of Madrid. *Endocrinol. Nut., 62*, 180–187.

Holick, M. F., (2004). Sunlight and vitamin D for bone health and prevention of autoimmune diseases, cancers, and cardiovascular disease. *Am. J. Clin. Nutr., 80*(Suppl), 1678S–1688S.

Holick, M. F., (2006). Resurrection of vitamin D deficiency and rickets. *J. Clin. Invest., 116*, 2062–2072.

Holick, M. F., (2007). Vitamin D deficiency. *N. Engl. J. Med., 357*, 266–281.

Huang, C. Y., Chang, H. H., Lu, C. W., Tseng, F. Y., Lee, L. T., & Huang, K. C., (2015). Vitamin D status and risk of metabolic syndrome among non-diabetic young adults. *Clin. Nut., 34*, 484–489.

Hummel, M., Standl, E., & Schnell, O., (2007). Chromium in metabolic and cardiovascular disease. *Horm. Metab Res.*, *39*(10), 743–751.

Ibarrola-Jurado, N., Salas-Salvadó, J., Martinez-González, M. A., & Bulló, M., (2012). Dietary phylloquinone intake and risk of type 2 diabetes in elderly subjects at high risk of cardiovascular disease. *Am. J. Clin. Nut., 96*(5), 1113–1118.

Illescas-Montes, R., Melguizo-Rodríguez, L., Ruiz, C., & Costela-Ruiz, V. J., (2019). Vitamin D and autoimmune diseases. *Life Sc., 233*, 116744.

Jarouliya, U., & Keservani, R. K., (2019). Pathways leading to child obesity. In: Bagchi, D., (ed.), *Global Perspectives on Childhood Obesity: Current Status, Consequences and Prevention* (2nd edn., pp. 137–146). Academic Press, Elsevier; Chapter 12. ISBN: 9780128128404.

Juhan-Vague, I., Alessi, M. C., Mavri, A., et al., (2003). Plasminogen activator inhibitor-1, inflammation, obesity, insulin resistance and vascular risk. *J. Thromb. Haemost., 1*(7), 1575–1579.

Karonova, T., Grineva, E., Belyaeva, O., et al., (2018). Relationship between vitamin D status and vitamin D receptor gene polymorphisms with markers of metabolic syndrome among adults. *Front. Endocrinol., 9*, 448.

Kaur, J., (2014). A comprehensive review on metabolic syndrome. *Cardiol. Res. and Prac.*, 21. Article ID 943162. http://dx.doi.org/10.1155/2014/943162.

Keservani, R. K., Kesharwani, R. K., Sharma, A. K., & Ahmad, M. F., (2015). Stress, Parkinson's, Alzheimer's disease: Role of dietary supplements. In: Bagchi, D., Preuss, H. G., & Swaroop, A., (eds.), *Nutraceutical and Functional Foods in Human Life and Disease Prevention* (pp. 241–253). Boca Raton, Fl. Taylor and Francis, Ch 14. ISBN: 9781482237214.

Keservani, R. K., Kesharwani, R. K., Sharma, A. K., Vyas, N., & Chadoker, A., (2010b). Nutritional supplements: An overview. *Int. J. of Curr. Pharm. Rev. and Res., 1*(1), 59–75.

Keservani, R. K., Kesharwani, R. K., Vyas, N., Jain, S., Raghuvanshi, R., & Sharma, A. K., (2010a). Nutraceuticals and functional food as future food. *Der Pharm. Lett., 2*(1), 106–116.

Keservani, R. K., Sharma, A. K., & Kesharwani, R. K., (2016). An overview and therapeutic applications of nutraceutical & functional foods. In: Keservani, R. K., Sharma, A. K., & Kesharwani, R. K., (eds.), *Recent Advances in Drug Delivery* Technology (pp. 160–201). IGI Global Publisher, USA, Ch 6. ISBN:9781522507543.

Kesharwani, P., Gorain, B., Low, S. Y., Tan, S. A., Ling, E. C. S., Lim, Y. K., et al., (2018). Nanotechnology based approaches for anti-diabetic drugs delivery. *Diabetes Res. Clin. Pract., 136*, 52–77. doi: 10.1016/j.diabres.2017.11.018.

Knapen, M. H., Schurgers, L. J., Shearer, M. J., Newman, P., Theuwissen, E., & Vermeer, C., (2012). Association of vitamin K status with adiponectin and body composition in healthy subjects: Uncarboxylated osteocalcin is not associated with fat mass and body weight. *Br. J. Nutr., 108*(6), 1017–1024.

Kochupillai, N., (2008). The physiology of vitamin D: Current concepts. *Indian J. Med. Res., 127*, 256–262.

KumarDey, T., Ghosh, S., Ghosh, M., Koley, H., & Dhar, P., (2012). Comparative study of gastrointestinal absorption of EPA & DHA rich fish oil from nano and conventional emulsion formulation in rats. *Food Res. Int., 49*, 72–79.

Matsuzawa, Y., Funahashi, T., & Nakamura, T., (2011). The concept of metabolic syndrome: Contribution of visceral fat accumulation and its molecular mechanism. *J. Atheroscler Thromb., 18*(8), 629–639.

Melguizo-Rodríguez, L., Costela-Ruiz, V. J., García-Recio, E., et al., (2021). Role of vitamin D in the metabolic syndrome. *Nutrients, 13*, 830.

Mnonopi, N., Levendal, R. A., Mzilikazi, N., & Frost, C., (2012). Marrubiin, a constituent of leonotisleonurus, alleviates diabetic symptoms. *Phytomedicine, 19*, 488–493. doi: 10.1016/j.phymed.2011.12.008.

Mozos, I., & Marginean, O., (2015). Links between vitamin D deficiency and cardiovascular diseases. *Biomed Res. Int*., 109275.

Mutt, S. J., Jokelainen, J., Sebert, S., Auvinen, J., Järvelin, M. R., Keinänen-Kiukaanniemi, S., & Herzig, K. H., (2019). Vitamin D status and components of metabolic syndrome in older subjects from Northern Finland (Latitude 65° North). *Nutrients*, 11.

Nachtigal, M., et al., (2005). Dietary supplements and weight control in a middle-age population. *J of Alte. & Compl. Med*., *11*(5), 909–915.

Niles, R. M., (2000). Vitamin A and cancer. *Nutri., 16*, 573–576.

Padayatty, S. J., Katz, A., Wang, Y., et al., (2003). Vitamin C as an antioxidant: Evaluation of its role in disease prevention. *J. Am. Coll. Nut., 22*, 18–35.

Pan, Y., & Jackson, R. T., (2009). Dietary phylloquinone intakes and metabolic syndrome in US young adults. *J. Am. Coll. Nut., 28*(4), 369– 379.

Patrini, C., Griziotti, A., & Ricciardi, L., (2004). Obese individuals as thiamin storers. *Inter J of Obes*., *28*(7), 920–924.

Paul, D., Dey, T. K., Mukherjee, S., Ghosh, M., & Dhar, P., (2014). Comparative prophylactic effects of α-eleostearic acid rich nano and conventional emulsions in induced diabetic rats. *J. Food Sci. Technol., 51*, 1724–1736. doi: 10.1007/ s13197-014-1257-2.

Pinhas-Hamiel, O., et al., (2006). Obese children and adolescents: A risk group for low vitamin B12 concentration. *Arch of Ped. & Adoles. Med., 160*(9), 933–936.

Rane, B. R., Patil, A. S., Keservani, R. K., & Jain, A. S., (2020). Novel approaches in herbal formulation. In: Kesharwani, R. K., Keservani, R. K., & Sharma, A. K., (eds.), *Enhancing the Therapeutic Efficacy of Herbal Formulations Through Novel Drug Delivery Systems* (pp. 43–68). IGI Global International Publisher, Pennsylvania, USA, Chapter 2. ISBN13,9781799844532.

Rane, B. R., Tadavi, S. A., & Keservani, R. K., (2020). Naturopathy. In: Sharma, A. K., Keservani, R. K., & Gautam, S. P., (eds.), *Herbal Product Development* (pp. 321–347). Apple Academic Press, Taylor & Francis Group, Chapter 12. ISBN: 9781771888776.

Reaven, G. M., (1988). Role of insulin resistance in human disease. *Diabetes, 37*(12), 1595–1607.

Rochlani, Y., Pothineni, N. V., Kovelamudi, S., & Mehta, J. L., (2017). Metabolic syndrome: Pathophysiology, management, and modulation by natural compounds. *Ther. Adv. Cardiovasc. Dis*., *11*(8), 215–225.

Ross, C., (2005). Vitamin A. In: Coates, P. M., Blackman, M. R., Cragg, G. M., Levine, M., Moss, J., & White, J. D., (eds.), *Encyclopedia of Dietary Supplements* (p. 713). New York: Marcel Dekker.

Saklayen, M. G., (2018). The global epidemic of the metabolic syndrome. *Curr. Hypertens. Re.*, 20.

Schmitt, E. B., Nahas-Neto, J., Bueloni-Dias, F., Poloni, P. F., Orsatti, C. L., & Petri-Nahas, E. A., (2018). Vitamin D deficiency is associated with metabolic syndrome in postmenopausal women. *Maturitas*., *107*, 97–102.

Shamanna, R. A., Croteau, D. L., Lee, J. H., et al., (2017). Recent advances in understanding Werner syndrome. *F1000Re., 6*, 1779. doi: 10.12688/f1000research.12110.1. PMID: 29043077; PMCID: PMC5621106.

Singh, D., Singh, B., & Goel, R. K., (2011). Traditional uses, phytochemistry and pharmacology of *Ficus religiosa*: A review. *J. Ethnopharmacol., 134*, 565–583. doi: 10.1016/j.jep.2011.01.046.

Singh, P., Kesharwani, R. K., & Keservani, R. K., (2017b). Antioxidants and vitamins: Roles in cellular functions and metabolism. In: Bagchi, D., (ed.), *Sustained Energy for Enhanced Human Function and Activity.* Elsevier. Ch 24. ISBN: 9780128054130.

Stephensen, C. B., (2001). Vitamin, A., infection, and immune function. *Annu Rev Nutr., 21*, 167–192.

Tajmohammadi, A., Razavi, B. M., & Hosseinzadeh, H., (2018). *Silybum marianum* (milk thistle) and its main constituent, silymarin, as a potential therapeutic plant in metabolic syndrome: A review. *Phytother. Res., 32*, 1933–1949.

Tooke, J. E., & Hannemann, M. M., (2000). Adverse endothelial function and the insulin resistance syndrome. *J. Intern. Med., 247*(4), 425–431.

Traber, M. G., (2011). Vitamin E- alpha-tocopherol. In: Herrmann, W., & Obeid, R., (eds.), *Vitamins in the Prevention of Human Disease* (p. 45782). Berlin: Walter de Gruyter.

Tsigos, C., Kyrou, I., Chala, E., et al., (1999). Circulating tumor necrosis factor-alpha concentrations are higher in abdominal versus peripheral obesity. *Metabolism, 48*(10), 1332–1335.

Wallace, A. M., McMahon, A. D., Packard, C. J., et al., (2001). On behalf of the WOSCOPS executive committee. Plasma leptin and the risk of cardiovascular disease in the west of Scotland coronary prevention study (WOSCOPS). *Circulation, 104*(25), 3052–3056.

Wang, N., Wang, C., Chen, X., Wan, H., Chen, Y., Chen, C., Han, B., & Lu, Y., (2020). Vitamin, D., Prediabetes and type 2 diabetes: Bidirectional mendelian randomization analysis. *Eur. J. Nut.,* 59, 1379–1388.

Whayne, T. F. Jr, & Maulik, N., (2012). Nutrition and the healthy heart with an exercise boost. *Can. J. Physiol. Pharmaco., 90*(8), 967–976.

Wong, S. K., Chin, K. Y., & Ima-Nirwana, S., (2020). Vitamin C: A review on its role in the management of metabolic syndrome. *Int. J. Med. Sc.,* 17(11), 1625–1638. doi: 10.7150/ijms.47103.

Wong, S. K., Chin, K. Y., Suhaimi, F. H., et al., (2017). Vitamin E as a potential interventional treatment for metabolic syndrome: Evidence from animal and human studies. *Front Pharmacol., 8*, 444.

Wong, S. K., Chin, K. Y., Suhaimi, F. H., et al., (2018). The effects of palm tocotrienol on metabolic syndrome and bone loss in male rats induced by high-carbohydrate high-fat diet. *J. Funct. Food, 44*, 246–254.

Zheng, J. S., Luan, J., Sofianopoulou, E., et al., (2020). The association between circulating 25-hydroxyvitamin D metabolites and type 2 diabetes in European populations: A meta-analysis and mendelian randomization analysis. *PLoS Med.,* 17.

CHAPTER 10

Pathophysiology and Application of Herbs in the Metabolic Syndrome

BHUSHAN R. RANE,[1] RAJESH A. AHIRRAO,[2] NAYAN A. GUJARATHI,[3] ASHISH S. JAIN,[1] and RAJ K. KESERVANI[4]

[1]Shri D. D. Vipsute College of Pharmacy and Research Center, Panvel, Maharashtra – 410206, India

[2]P. G. College of Pharmaceutical Science and Research, Chaupale, Nandurbar, Maharashtra – 425412, India

[3]Shri Vile Parle Kelvani Mandal's Institute of Pharmacy, Dhule – 424001, Maharashtra, India

[4]Faculty of B. Pharmacy, CSM Group of Institutions, Prayagraj, Uttar Pradesh – 212111, India

ABSTRACT

Metabolic syndrome is defined as a collection of interrelated processes such as physiological, biochemical, clinical, and metabolic factors that cause the body's temperature to rise, the threat of issues associated with the cardiovascular system, diabetes mellitus (type 2), and all-cause mortality. Insulin resistance, visceral adiposity, atherogenic dyslipidemia, endothelial dysfunction (ED), hereditary predisposition, high blood pressure, hypercoagulable state, and chronic stress are some of the elements that make up the metabolic syndrome. For this demographic, lifestyle modification is still the first line of defense. Modern lifestyle modification therapy combines dietary and exercise guidelines with behavioral techniques. If lifestyle changes alone are not enough to reduce risk factors, pharmacological treatment should

The Metabolic Syndrome: Dietary Supplements and Food Ingredients. Raj K. Keservani, MPharm, Durgavati Yadav, PhD, Rajesh K. Kesharwani, PhD, Sippy Singh, DPhil & Kumar Sandeep, PhD (Eds.)

be considered. Metabolic syndrome can be treated with a variety of herbal medicines. This study summarizes the evidence on the diagnosis of metabolic syndrome and herbal treatment methods for each of the risk variables that make up the syndrome.

10.1 INTRODUCTION TO METABOLIC SYNDROME

The metabolic condition is a group of the most risky cardiovascular failure hazard factors: diabetes, as well as increased fasting plasma glucose and stomach weight, elevated cholesterol and high blood pressure (Alberti et al., 2005, 2006; www.idf.org) 1–4 It is assessed that more than 25% of the world's grown-up population have the metabolic disorder and they are twice as prone to kick the bucket from and multiple times as liable to suffer from a heart attack or a stroke contrasted and individuals without the syndrome.

Metabolic syndrome is a combination of symptoms that includes of individual weight, hyperglycemia, hypertension, and hyperlipidemia (Eckel et al., 2005). There have been distinctive determination standards for metabolic condition after it was first being characterized according to the World Health Organization (WHO) in 1998 (Alberti et al., 1998). The most usually acknowledged definition utilizes the rules proposed by the WHO, EGIR which is an acronym for Electronic Government Information Resources (European group of insulin resistance) in 1999, APP-III (Adult Treatment Panel III) in 2001, AACE is an initial for the American Association for the (American Association of Clinical Endocrinology) IN 2003 AHA/NHLBI stands for American Heart Association (AHA)/National Heart, Lung, and Blood Institute. As per the different agencies, metabolic condition is actually analyzed as per the conditions in Table 10.1.

TABLE 10.1 Clinical Measures for Diagnosis of Metabolic Syndrome as per Different Agencies

Clinical Measures	AACE (2003)	ATPIII (2001)	EGIR (1999)	IDF (2005)	WHO (1998)
Body weight	BMI ≥25 Kg/m^2	WC 102 cm for men and 88 cm for women	WC 94 cm for men and 80 cm for ladies	Increased WC (as per population and any 2 of the other measures)	Men waist to hip Ratio > 0.90; women with a waist-to-hip ratio greater than 0.85 and/or a BMI greater than 30 kg/m^2

TABLE 10.1 *(Continued)*

Clinical Measures	AACE (2003)	ATPIII (2001)	EGIR (1999)	IDF (2005)	WHO (1998)
Blood pressure	≥130/85 mmHg	≥130/85 mmHg	≥140/90 mmHg or hypertension	≥130 and 90 mmHg systolic and diastolic respectively or hypertension	≥140/90 mmHg
Glucose	Is it IGT or IFG? (but not diabetes)	A concentration of 110 mg/dL (includes diabetes)	Is it IGT or IFG? (but not diabetes)	≥100 mg/dL (includes diabetes)	IGT, IFG, or T2DM
Insulin resistance	IGT or IFG plus any of the other measures based on clinical judgment	None of the above, but any three of the following other measures	Any two of the abovementioned conditions plus plasma insulin > 75th percentile measures	None	IGT, any two of the following: IFG, T2DM, or decreased insulin sensitivity other measures
Lipids	Triglycerides ≥ 150 mg/dL	Triglycerides ≥ 150 mg/dL	Triglycerides ≥ 150 mg/dL	Triglycerides ≥ 150 mg/Dl or on TGs	Triglycerides ≥ 150 mg/dL
	HDL-C levels of 40 mg/dL in men and 50 mg/dL in women	In men, HDL-C is less than 40 mg/dL, or <50 mg/dL in women	Men and women with HDL-C levels of less than 39 mg/dL	In men, HDL-C is less than 40 mg/dL, or <50 mg/dL in women	Men's HDL-C is 35 mg/dL, whereas women's HDL-C is 39 mg/dL
Other factors	Other features of insulin resistance	–	–	–	Microalbuminuria is defined as a urinary excretion rate of more than 20 mg/min or a creatinine level of albumin ration of >30 mg/g

Source: Alberti et al. (1998); Balkau et al. (1999); Cleeman et al. (2001); Einhorn et al. (2003); www.idf.org.

However, the measure for abdomen circumference is marginally unique for every nation and race. Each metabolic threat factor is related with each other, and together the threat factors advance atherosclerotic cardiovascular illness (NCEP, 2002).

The primary basic danger factors for metabolic disorder are obesity and insulin resistance (Carr et al., 2004; Ferrannini et al., 1991). Therefore, forestalling atherosclerotic cardiovascular illness by controlling the abdomen area and insulin resistance is the way to overseeing metabolic condition. Clinically, every treatment to treat patients suffering from hyperglycemia, hypertension, and hyperlipidemia is recommended on the basis of the patient's state. Nonetheless, ACE inhibitor that is a medication utilized for hypertension, including enalapril and captopril, may cause unfavorable occasions like cough, expanded serum creatinine, migraine, and skin rash (Luo et al., 2015; Bangalore et al., 2010). Metformin, a drug used to treat T2D, has also been demonstrated to be effective. It can prompt gastrointestinal side effects and lactic acidosis (Ferrannini et al., 1991). Therefore, herbal drugs showing proof of security and adequacy can be elective medicines for metabolic sicknesses.

Despite the fact that a few remains audits of herbal drugs for obesity hypertension, and T2D Mellitus, efficient survey of metabolic condition has not been directed at this point (Sui et al., 2012; Hasani et al., 2009; Chen et al., 2015; Yeh et al., 2003). This investigation, notwithstanding, audits a solitary sickness, yet in addition metabolic disorder all in all. The point of this examination is to assess the adequacy and wellbeing of herbal drugs to assist with overseeing metabolic syndrome.

Furthermore, individuals with metabolic condition have a five times more serious danger of creating T2D. They would add to the 230 million individuals overall who as of now have diabetes 6a, quite possibly the most well-known persistent illnesses worldwide and the fourth or fifth driving reason for death in the developed countries. The bunching of cardiovascular infection (CVD) hazard factors that symbolize the metabolic disorder is presently viewed as the main impetus for another CVD scourge. The metabolic syndrome is a significant and raising general wellbeing and clinical test in the aftermath of urbanization, excess energy use, growing stoutness, and immobile life proclivities over the world. Metabolic syndrome started as a hypothesis rather than a conclusion. The metabolic illness got its start in 1920, when a Swedish doctor named Kylin discovered a link between hypertension, high blood glucose (hyperglycemia), and gout. Later that year, in 1947, Vague portrayed that the instinctive obesity was regularly connected with the metabolic issues found in CVD (Shaw and Chisholm, 2003; Kylin, 1923).

Following this, in 1965, a theoretical abstract was presented at the annual meeting of the European Association for the Study of Diabetes by Avogaro and Crepaldi which again portrayed a disorder which contained Obesity, hypertension, and hyperglycemia. Following Reaven's Banting Lecture in 1988, the field essentially moved forward. He depicted "a bunch of hazard factors for diabetes and cardiovascular sickness" and gave it the name "Syndrome X." His primary commitment was a presentation of the idea of the insulin resistance (Avogaro and Crepaldi, 1965; Reaven, 1988).

Whatever the case may be, he omitted obesity or intrinsic beefiness from the definition which was subsequently added as a vital abnormality. In 1989, Kaplan renamed the disorder "The Deadly Quartet" for a combination of obesity, diabetes, hypertriglyceridemia, and hypertension Regardless, it was renamed "The Insulin Resistance Syndrome" again in 1992. A few gatherings have endeavored to foster demonstrative models for the conclusion of the metabolic syndrome (Kaplan, 1989; Haffner et al., 1992; Eckel, 2005). A diabetic expert from the WHO spearheaded the effort bunch in 1998 to give a meaning of the metabolic syndrome (Alberti and Zimmet, 1998). As a result, in 1999, the European Group for Insulin Resistance Research (EGIR) reacted with a revision in the WHO definition (Malhotra, 2001). The Adult Treatment Panel (NCEP/ATP) of the National Cholesterol Education Program (NCEP) published their criteria in 2001 (Cleeman, 2001). In this vein, the American Association of Clinical Endocrinologists (AACE) presented its viewpoints on the meaning of the condition in 2003 (Einhorn et al., 2003). The multiplication of definitions recommended that a solitary binding together definition was attractive (Ford, 2004).

10.2 EPIDEMIOLOGY

Overall pervasiveness of metabolic syndrome goes from 10% to 84%, contingent upon the area, metropolitan or provincial climate, synthesis (sex, age, race, and identity) of the populace contemplated, and the meaning of the disorder utilized (Desroches and Lamarche, 2007; Kolovou et al., 2007).

Higher financial status, stationary way of life, and high weight were essentially connected with metabolic syndrome. Cameron et al. have inferred that the distinctions in hereditary foundation, diet, levels of actual work, smoking, family background of diabetes, and schooling all impact the predominance of the metabolic syndrome and its parts (Cameron et al., 2004).

The metabolic syndrome was found to be prevalent in 5% of subjects of normal weight, 22% of overweight people, and 60% of obese people in the National Health and Nutrition Examination Survey (NHANES). It increases with age (10% in those aged 20–29, 20% in people aged 40–49), and 45% in people matured 60–69). The pervasiveness of MetS (in light of NCEP-ATP III models, 2001) across the globe, men's rates increased from 8% to 43% and women's rates increased from 7% to 56% (Cameron et al., 2004; Park et al., 2003; Ford et al., 2002).

Park et al. saw that there is an expansion in the predominance of metabolic syndrome from 20 years of age through the 6th and 7th decade of life for men and women, individually. Ponholzer et al. revealed that there is high commonness of metabolic syndrome among postmenopausal ladies, which varies between 32.6% and 41.5% (Park et al., 2003; Ponholzer et al., 2008). A Framingham Heart Study report demonstrated that a weight increment of ≥2.25 kg over a time of 16 years was associated with a 45% increased risk of fostering the metabolic syndrome (Wilson et al., 1999), and it has been displayed by Palaniappan et al. that every 11 cm expansion in midsection periphery (WC) is related with a changed 80% expanded danger of fostering the syndrome within five years (Palaniappan et al., 2004). The metabolic modifications happen all the while more habitually than would be normal by some coincidence and the simultaneousness of a few components increments cardiovascular danger far beyond the danger related with the individual factors alone (Reilly and Rader, 2003).

10.3 CAUSES OF METABOLIC SYNDROME

The underlying cause of metabolic disease continues to confound experts, but insulin resistance and obesity are both regarded as significant causes (Qiao et al., 2004; Carr et al., 2004). Genetics, physical inactivity, maturing, a proinflammatory state and hormonal changes may likewise have a causal influence, despite the fact that their work may vary depending on ethnic group (Saad et al., 1991; Anderson et al., 2001). Insulin resistance happens when cells in the body organs like liver, skeletal muscle, and fat/adipose tissue) become less delicate, and in the end, impervious insulin is a protein produced by beta cells in the pancreas that helps the body deal with glucose. Glucose can presently do not be consumed by the tissues yet stays in the blood, setting off the requirement for increasingly more insulin (hyperinsulinemia) to be created trying to deal with the glucose. The creation of steadily expanding measures of insulin debilitates and it is possible that the beta cells will be destroyed in the end. When the pancreas is presently not capable of creating sufficient insulin then an individual becomes hyperglycemic (a blood glucose level that is too high) and will be determined to have T2D. Indeed, even before this occurs, harm is happening to the body, including a development of fatty oils which further weakens insulin affectability.

Insulin resistance and metabolic disease are linked to obesity. Hypertension, high blood cholesterol, and low HDL-c are all linked to obesity and hyperglycemia, and is autonomously connected with higher CVD risk (Qiao et al., 2004; Zimmet et al., 2001; Carey et al., 1997).

The danger of genuine wellbeing outcomes as T2D, coronary illness, and a scope of different conditions, including a few types of malignancy, has been displayed to ascend with an expansion in weight, yet it is an overabundance of fat ratio in the mid-region, estimated basically by abdomen size, that is more characteristic of the metabolic disorder profile (Ohlson et al., 1985; Pouliot et al., 1994). According to the International Obesity Task Force (IOTF), 1.7 billion of the total planet populaces is as of now at an elevated danger of weight-related, non-transmittable sicknesses, for example, T2D (Table 10.2) (Rexrode et al., 1998).

TABLE 10.2 To Understand Obesity – Waist Circumference Region Wise

Country Ethnic Group	Waist Circumference	
	Male	Female
Chinese	≥ 90 cm	≥ 80 cm
Europids	≥ 94 cm	≥ 80 cm
Japanese	≥ 90 cm	≥ 80 cm
South Asians (Indian, Malaysia, and Chinese)	≥ 90 cm	≥ 80 cm
Ethnic south and central Americans	≥ 90 cm	≥ 80 cm
Sub Saharan Africans	≥ 94 cm	≥ 80 cm
Populations from the eastern Mediterranean and the middle east (Arabs)	≥ 94 cm	≥ 80 cm
USA (for clinical purpose)	102 cm	88 cm

Extra metabolic measurements also need to be performed to investigate the metabolic syndrome. To confirm the general body fat distribution, a few parameters, need to be investigated like central fat distribution, adipose tissue biomarkers (leptin), adiponectin, and liver fat content.

For many people have elevated blood pressure, i.e., vascular dysregulation which is authenticated by measurement of ED and microalbuminuria.

Atherogenic dyslipidemia verification the increased triglyceride and low HDL levels need to be checked by estimating the proportion of small LDL particles. Various patients show resistance to insulin which is confirmed by some test like fasting insulin/proinsulin levels, elevated free fatty acids determination and M value from clamp. OGTT is checked to verify Dysglycemia. Proinflammatory condition and prothrombic state is proved after testing C-reactive protein (CRP), TNF, alpha, IL-6 and fibrinolytic factors fibrinogen level respectively. Many times, hormonal levels are also need to checked if any abnormality is observed (IDF, 2004).

10.4 PATHOPHYSIOLOGY

Metabolic syndrome is a condition of persistent little mark inflammation as a result of multifaceted interaction between hereditary and the surrounding environment. Insulin resistance, visceral adiposity, hypercoagulable condition, ED, atherogenic dyslipidemia, hereditary predisposition, high blood pressure, and chronic stress are just a few of the symptoms that make up the syndrome.

10.4.1 ABDOMINAL OBESITY

Obesity is mostly caused by a high-calorie diet and no exercise or physical activity. Adipocytes, stromal preadipocytes, immune cells, and endothelium make up adipose tissue, and it is possible to react quickly and energetically to changes in nutrient overload during adipocytes hypertrophy and hyperplasia (Halberg et al., 2008). With fatness and progressive adipocytes swelling, the blood supply to adipocytes may be reduced with resultant hypoxia (Cinti et al., 2005). It results into leads to an extra making of active chemicals such as adipocytokines, tumor necrosis factor-alpha (TNFα) and interleukin-6 (IL-6), plasminogen activator inhibitor-1 and CRP (Lau et al., 2005). After all the inflammation observed in adipose tissue which increases and leads to whole systemic inflammation finally leads to progress of obesity and related issues (Trayhurn and Wood, 2004). Adipocytokines use endocrine, autocrine, and paracrine signals to trigger a variety of events, such as insulin resistance (Charmandari, 2005), arousal (Ferrannini, 1991), energy biotransformation, blood coagulation, and inflammatory reactions (Ferrannini, 1991) are just a few examples which are responsible to hasten atherosclerosis, plaque crack, and atherothrombosis (Saleem et al., 2009; Tsimikas et al., 2009; Jacobs et al., 2009). It is concluded that the storage-related adipose tissue and movement of lipids as well as releases the various cytokines.

10.4.2 FREE FATTY ACID (FFA)

Internal belly fat levels are associated to splanchnic FFA levels, white adipose tissues in the upper body make up the majority of moving FFA. It could be to blame for the deposition of fat in the liver result in abdominal obesity. Additionally, insulin resistance develops when skeletal muscle is exposed to high amounts of FFA because insulin-mediated glucose absorption is hampered. Whereas, a constant introduction of the pancreas to the increased FFA damages a function of β-cell pancreas (Miles and Jensen, 2005; Boden et al., 2001).

10.4.3 TNFα

It is a mediator in adipocytes which act locally to decrease the sensitivity of insulin adipocytes. Plasma TNFα is completely related with the weight and triglycerides, while, constructive relationship existing between the plasma TNFα and HDL-C (Lau et al., 2005; Xydakis et al., 2004).

10.4.4 C-REACTIVE PRITEIN (CRP)

An increase in CRP is linked with an insulin resistance, BMI, and hyperglycemia. CRP increases with the number of the metabolic syndrome chemicals. Most of the times it is observed in human being with obesity insulin-resistant issues levels of CRP is increased while in obese insulin-sensitive subjects no change observed. An elevated level of CRP leads to heart issues in obese patients (SotoGonz′alez et al., 2006; Deepa et al., 2006; Guldiken et al., 2007; McLaughlin et al., 2002; Ridker et al., 2003).

10.4.5 INTERLEUKIN-6 (IL-6)

Skeletal muscles and adipose tissues release the IL-6 in individuals. It has both an inflammatory and an anti-inflammatory action. Receptor of IL-6 is present in the hypothalamus in the brain which manages hunger and energy ingestion. It is a systemically released chemical adipokine which is responsible for insulin sensitivity as well as main factor responsible for production of CRP in liver. It is also able to suppress the activity of lipoprotein lipase (Pedersen et al., 2003; Stenl et al., 2003; Diamant et al., 2005).

10.4.6 PLASMINOGEN ACTIVATOR INHIBITOR-1 (PAI-1)

It is an inhibitor of protease which released by adipocytes from the abdomen, platelets, and endothelium. Its effects observed because of inhibition of tissue plasminogen activator (tPA) which reveals that it is one of the main factors for atherothrombosis and fibrinolysis (Lau et al., 2005; Alessi and Juhan, 2006).

10.4.7 ADIPONECTIN

It is responsible for the metabolism of glucose and lipid; it increases insulin sensitivity. Foodstuff ingestion and body mass regulated by adiponectin and it protects against a persistent swelling (Liu and Liu, 2010). The rate of glucose production in the liver (endogenously) is reduced by adiponectin because it is responsible for inhibition of liver gluconeogenic enzymes. Glucose transport in muscles is increased, and it enhances the oxidation of fatty acid (Eckel et al., 2005). It is inversely connected with cardio vascular disease factors like blood pressure, LDL-C, and triglycerides (Kazumi et al., 2002).

10.4.8 LEPTIN

It is a chemical (adipokine) released in the body which regulates of satiety and caloric intake (Lau et al., 2010). Patients with obesity and weight increase are said to have leptin resistance, leptin level is increased and it decreases in weight loss conditions. Leptin receptors are present in the brain parts, i.e., hypothalamus as well as the brain stems, it is in charge of neuroendocrine functioning, it manages satiety and regulate requirement of energy in body. The hypothalamus' ventromedial and dorsomedial regions also trigger the sympathetic nervous system and regulate blood pressure (Hutley and Prins, 2005; Carlyle et al., 2002).

10.4.9 INSULIN RESISTANCE

Qualities of the insulin sensitive aggregate incorporate an ordinary body weight (BW) without stomach or instinctive corpulence, being reasonably dynamic, and devouring an eating regimen low in saturated fats (Bravata et al., 2004; Carey et al., 1996; Mayer et al., 1998; Vessby et al., 2001). On the other hand, insulin-resistant people exhibit an impeded glucose digestion or an unusual reaction to a glucose challenge as a source of resistance, a raised fasting glucose levels or potentially unmistakable hyperglycemia, or a decrease in insulin activity after intravenous organization of insulin (euglycemic clip strategy) with diminished insulin-mediated glucose freedom and additionally decreases in the concealment of endogenous glucose creation. It is characterized as a pathologic state in which a typical insulin fixation doesn't sufficiently create an ordinary insulin reaction in the fringe target tissues, for example, fat, muscle, and liver. The pancreatic beta cell secretes more insulin in this situation (i.e., hyperinsulinemia) to conquer the hyperglycemia among insulin-safe people. Despite the fact that hyperinsulinemia may make up for insulin protection from some natural activities of insulin, that is, support of normoglycemia, notwithstanding, it might cause an overexpression of insulin action in some ordinarily delicate tissues. This complement of some insulin activities combined with a protection from different activities of insulin brings about the clinical indications of metabolic syndrome (Mugo et al., 2005).

The beta cells of the pancreas are impotent in the long run to create adequate insulin to address the demolishing tissue insulin obstruction prompts hyperglycemia and obvious type 2 debates Mellitus (Petersen and Shulman, 2006). An insulin obstruction prompts the vascular irregularities that incline to atherosclerosis. Regardless of the fact that insulin-safe people need not be clinically stout, they in any case usually have an unusual fat appropriation that is portrayed by a dominating chest area fat. Notwithstanding the general commitments of instinctive fat and stomach subcutaneous fat to insulin obstruction, an example of stomach (or chest area) stoutness connects all the more unequivocally with the insulin opposition and the metabolic syndrome than brings down body corpulence (Jensen et al., 1989).

10.4.10 DYSLIPIDEMIA

This dyslipidemia is a serious condition portrayed by a range of subjective lipid anomalies reflecting annoyances in the construction, digestion, and natural exercises of both pro- and anti-atherosclerotic lipoproteins which incorporates a height of apolipoprotein B (apo B)-containing lipoproteins are a type of lipoprotein that contains apo B, raised TGs, expanded degrees of little particles of LDL, and low degrees of HDLC. Insulin resistance causes atherogenic dyslipidemia in a variety of ways.

To begin with, insulin ordinarily smothers lipolysis in adipocytes, so a debilitated insulin flagging increments lipolysis, coming about in expanded FFA levels. In the liver, FFAs fill in as a substrate for the blend of triglycerides.

FFAs additionally balance out the creation of apoB, the major lipoprotein of extremely low-density lipoprotein (VLDL) particles, bringing about a more VLDL creation. Second, insulin normally corrupts apoB via PI3K-dependent pathways, therefore insulin resistance directly increases VLDL synthesis. Third, insulin directs the movement of lipoprotein lipase, the rate-restricting and significant arbiter of VLDL freedom. Along these lines, insulin resistance and hypertriglyceridemia are the aftereffect of both an expansion in VLDL creation and a reduction in VLDL freedom.

VLDL is used to remainder lipoproteins and small thick LDL, both of these factors can hasten the establishment of an atheroma. The triglycerides in VLDL are moved cholesterol ester transport protein to HDL (CETP) in return for cholesteryl esters, bringing about the triglycerides-enhanced HDL and cholesteryl ester enriched VLDL particles. Further, the triglycerides-advanced HDL is a superior substrate for hepatic lipase, as a result, it has been cleared quickly from the flow, passing on a low HDL particle to take an interest in a converse cholesterol transport from the vasculature. Accordingly, in the liver of insulin-resistant patients, the FFA transition is high, the capacity and combination of triglycerides are extended, and abundance triglycerides are discharged as VLDL (Lewis and Steiner, 1996). Generally, it is accepted that the dyslipidemia related with insulin opposition is an immediate result of expanded VLDL discharge by the liver (Ginsberg et al., 2005). These inconsistencies are intently associated with an expanded oxidative pressure and an endothelial brokenness, along these lines building up the proinflammatory idea of macrovascular atherosclerotic infection.

10.4.11 HYPERTENSION

Fundamental hypertension is oftentimes connected with the few metabolic irregularities, of which heftiness, glucose narrow mindedness, and the most well-known type of dyslipidemia is dyslipidemia (Ferrannini and Natali, 1991). According to research, both hyperglycemia and hyperinsulinemia activate the Renin-Angiotensin System (RAS) via increasing the expression of angiotensinogen. Angiotensin II (AT-II) and the AT1 receptor, which may help people with insulin resistance improve their hypertension (Malhotra et al., 2001). There is additionally proof Insulin resistance and hyperinsulinemia are two terms for the same thing leads to SNS enactment, and, accordingly, the kidneys increment sodium reabsorption, the heart increments cardiovascular yield, and supply routes respond with vasoconstriction bringing about hypertension (Morse et al., 2005). It has been as of late found that adipocytes likewise produce aldosterone because of ATII (Briones et al., 2012). In such manner, the adipocyte might be viewed as a small-scale renin-angiotensin aldosterone framework.

10.4.12 HEREDITARY FACTORS

The incredible varieties in the helplessness and period of beginning in people with a fundamentally the same as hazard profile propose a major collaboration among hereditary and environmental elements (Ordovas, 2007). It is perceived that a few groups who are not fat by conventional measures all things considered are insulin resistant and have unusual degrees of metabolic danger factors.

Models are found in people with 2 diabetic guardians or 1 parent and a second- or third-degree relative; the equivalent is valid for some people of South Asian identity (Perseghin et al., 1997; Abate et al., 2004). Extensive people and ethnic varieties additionally exist in the clinical example of metabolic danger factors in large/insulin-resistant subjects (Martin et al., 2003). All things considered, the statement of each metabolic danger factor falls somewhat under its own hereditary control, which impacts the reaction to various ecological openings. For instance, an assortment of polymorphisms in qualities influencing lipoprotein digestion is related to the deterioration of dyslipidemia among fat individuals (Laakso, 2004).

Essentially, a hereditary inclination to the flawed insulin discharge when joined with insulin obstruction can cause abnormally high amounts of glucose in the blood (Poulsen et al., 2005). As indicated by the frugal genotype speculation proposed by Neel in 1962, people living in an unforgiving climate with an unsound food supply would amplify their likelihood of endurance on the off chance that they could expand a capacity of surplus energy. In such circumstances, hereditary selection would prefer genotypes that conserve energy. In any case, the chose hereditary varieties that were supported during ailing health would become horrible when sustenance improved. This theory accepts that the normal hereditary variations of frugal qualities incline to metabolic syndrome. Another frugal aggregate theory was presented in 1992 by Hales and Barker (Hales and Barker, 1992). As indicated by this theory, infants who encountered an intrauterine unhealthiness may have adjusted to a helpless sustenance by diminishing energy consumption and becoming "frugal." These metabolic variations are advantageous when people are ineffectively sustained during youth and grown-up life; in any case, with an expanded food admission, these transformations are as of now not valuable and would prompt an expanded danger of metabolic syndrome in a later life. Backing for this theory comes from the noticed relationship of Low birth weight combined with a delayed start advancement of insulin obstruction and T2D mellitus in a few populaces (Hales et al., 1997).

10.4.13 ENDOTHELIAL FUNCTION

It is described by a debilitated endothelium-subordinate vasodilatation, a decreased blood vessel consistency, and a sped-up cycle of atherosclerosis (Kraemer et al., 2008). Different elements like oxidative pressure, hyperglycemia, progressed glycation items, FFAs, fiery cytokines, or adipokinase cause a powerlessness of endothelium to serve its typical physiological and defensive instruments. Hansson has shown that safe cells assume a significant part in every one of the phases of the atherosclerotic cycle (Hansson, 2005); furthermore, a decrease in NO, a critical controller of endothelial homeostasis, and an expansion in receptive oxygen species bring about an endothelial brokenness and a proatherogenic vascular bed (Harrison, 1997).

10.4.14 HYPERCOAGULABLE STATE

Raised circulating cytokines and strong stage reactants are signs of a pro-inflammatory condition (e.g., CRP). A prothrombotic condition also implies abnormalities in the procoagulant factors. Specifically, an increase in fibrinogen, factor VII, and factor VIII just as the antifibrinolytic factor (PAI-1), platelet scraped spots, and EDs. Grundy (2004) has shown that a fibrinogen, an intense stage reactant protein like CRP, ascends because of a cytokine-rich environment. This suggests that prothrombotic and proinflammatory states may be linked physiologically.

10.4.15 DIET

According to Aljada et al. research, a high dietary fat intake is linked to a higher risk of heart disease is related with an oxidative stress and an enactment of the proinflammatory transcription factor, that is, atomic factor kappa-beta (NFκB) (Aljada et al., 2004). Interestingly, in comparison to an eating routine rich in foods grown from the ground, an eating routine rich in foods grown from the ground has no aggravation provoking limit and a high-fat eating regimen regardless of whether it has similar calories content (Dandona et al., 2005).

10.4.16 CONSTANT STRESS AND GLUCOCORTICOID (GC) ACTION

Constant hypersecretion of stress mediators, like cortisol, in people with a hereditary inclination presented to a lenient climate, may prompt the instinctive fat collection because of persistent hypercortisolism, low development chemical discharge, and hypogonadism. GCs increment the exercises of compounds engaged with unsaturated fat combination and advance the emission of lipoproteins; actuate the hepatic gluconeogenic pathway (Charmandari et al., 2005; Wang et al., 1995; Argaud et al., 1996); enhance preadipocyte-to-adipocyte differentiation, which could result in a higher muscle-to-fat ratio mass; hinder an insulin-invigorated amino take-up by adipocytes; and increment lipolysis or lipid oxidation which prompts the fringe insulin obstruction. A decent connection was seen between plasma cortisol levels, all out urinary GC metabolites, and the quantity of highlights of the metabolic syndrome among these patients. Both the discharge rate and the fringe leeway of cortisol in these patients were decidedly related with the systolic pulse, and fasting glucose and insulin (Hauner et al., 1989; Grunfeld and Jones, 1986; Guillaume et al., 1993; Andrew et al., 2002). These hormonal modifications may prompt a receptive insulin hypersecretion, an expanding instinctive heftiness, and sarcopenia, coming about in T2DM, hypertension, and dyslipidemia (Figure 10.1) (Chrousos, 2009).

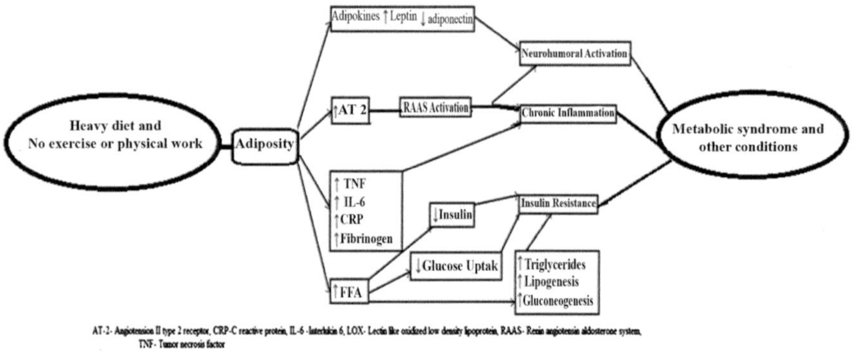

FIGURE 10.1 Different pathophysiological mechanisms which result into metabolic syndrome.

10.5 TREATMENT FOR METABOLIC SYNDROME

Metabolic syndrome is a condition of ongoing less inflammation with the significant systemic impacts. Clinical distinguishing proof and the management of patients with the metabolic issues are essential to start endeavors to sufficiently carry out the medicines to decrease their danger of ensuing sicknesses (Wong, 2005). Successful preventive methodologies incorporate way of life changes, fundamentally weight reduction, diet, and work out, and the treatment contains the suitable utilization of pharmacological specialists to diminish the particular danger factors. Pharmacological treatment ought to be considered for those whose hazard factors are not sufficiently diminished with the preventive measures and way of life changes (Deen, 2004). The clinical management of metabolic syndrome is complicated by the fact that there is no perceived strategy to forestall or work in general disorder, the foundation of which is basically insulin opposition (Reaven, 1988).

Accordingly, most doctors treat every segment of metabolic syndrome independently, laying a specific accentuation on those parts that are effectively amiable to the medication treatment. Truth be told, it is simpler to recommend a medication to bring down pulse, blood glucose, or fatty substances instead of starting a drawn-out system to change individuals' way of life (practice more and eat better) with the expectation that they will eventually get in shape and will in general have a lower circulatory strain, blood glucose, and fatty oils. For the treatment of hazard factors of metabolic syndrome, the doctor ought to follow the National Cholesterol Education Program's current treatment guidelines (Table 10.3).

TABLE 10.3 Various Active Components from Herbal Drugs and Their Action

Active Component	Name of Herbal Drug	Biological Source	Role of Category or its Action
Allicin	Garlic	*Allium sativum*	Anti-oxidant, anti-inflammatory, increase insulin sensitivity, lower overall cholesterol levels and triglycerides and increase adiponectin
Berberine	Huang Lian	*Rhizoma coptidis*	Increase insulin sensitivity, decrease triglycerides, decrease body weight, decrease systolic blood pressure
Bergamol essential oil	Bergamol orange	*Citrus bergamia*	Antioxidant, anti-inflammatory, decrease lectin-like low density lipoprotein, decrease reactive oxygen species formation
Cumin aldehyde	Cumin	*Cuminum cyminum*	Decrease lipid, decrease BSL
Curcumin	Turmeric	*Curcuma longa*	Antioxidant, anti-inflammatory, increase insulin sensitivity, decrease obesity, decrease leptin, and increase adiponectin
Galactomannan Saponins	Fenugreek	*Trigonella foenum*	Increase insulin sensitivity, decrease total cholesterol and triglycerides, decrease body weight, decrease postprandial blood glucose
Gingerols Parasols Sojails	Ginger	*Zingiber officinale*	Anti-inflammatory, decrease systolic blood pressure, decrease cyclooxygenase-2, decrease 5-lipooxygenease
Neem oil	Neem	*Azadirecta indica*	Increase insulin secretion, decrease postprandial hyperglycemia
Polyphenols	Cinnamon	*Cinnamomum verum*	Anti-inflammatory, increase insulin sensitivity, antithrombotic, fasting blood glucose, blood pressure, and body composition are all improved
PUFA	Omega fatty acids	Fish oils	Decrease lipogenesis, increase fatty acid oxidation, regulate peroxisome proliferator-activated receptor gamma

TABLE 10.3 *(Continued)*

Active Component	Name of Herbal Drug	Biological Source	Role of Category or its Action
Quercetin	Onions	*Allium cepa*	Anti-oxidant, anti-inflammatory, increase lipolysis, decrease adipogenesis, decrease cholesterol, decrease BSL
Resveratrol 3,5,4-trihydroxy trans stilbene	Grapes	*Vitus vinifera*	Increase lipolysis, decrease adipogenesis, decrease insulin resistance, decrease BMI
Sulforaphane	Broccoli	*Brassica oleracea*	Anti-oxidant, anti-inflammatory, activates nuclear factor erythroid 2 – related factor 2

10.5.1 CURCUMIN

Curcumin is a turmeric derivative (Curcuma longa), a flavor normally utilized in south-east Asia. The dynamic fixing, diferuloylmethane, is known to have anti-inflammatory in addition to antioxidative properties. Curcumin is a spice that has been used for centuries and displayed to smother NF-kB initiation, which blunts the incendiary course by decrease in the declaration of favorable to fiery cytokines, downregulates TNF-α articulation, and stifles articulation type-1 plasminogen activator inhibitor, which is answerable for a prothrombotic state (Aggarwal, 2010; Singh and Aggarwal, 1995; Pendurthi and Rao, 2000). Curcumin additionally hinders the Wnt/β-catenin pathway connected to corpulence and enacts peroxisome proliferator-actuated receptor gamma in hepatic stellate cells. Curcumin is likewise known to hinder leptin flagging accordingly expanding adiponectin articulation. Obesity has negative repercussions, but insulin sensitivity has positive outcomes just as blunting of the fiery pathways fill in as advantageous impacts in Metabolic syndrome.

10.5.2 GARLIC

Garlic (Allium sativum), an ordinarily utilized as condiment and spices, is likewise known to have therapeutic worth as a result because of its anti-inflammatory and anti-thrombotic effects Padiya and colleagues discovered that in fructose-fed animals, crude garlic improves insulin sensitivity and may have comparative impacts in humans. Reinhart and associates, in an examined randomized placebo-controlled controlled trials when comparing the effects of garlic on lipid profiles, it was discovered that garlic consumption lowers total cholesterol and fatty substance levels. In another investigation, Gomez-Arbelaez, and the effect of matured garlic extract on adiponectin levels in people with metabolic syndrome was demonstrated by companions. Utilization of matured garlic extract for 12 weeks was displayed to improve adiponectin levels. Garlic's organosulfur compounds in its subsidiaries have a mitigating effect. These mixtures have a cell reinforcement activity due to thiol bunches that assist with battling ROS-intervened aggravation making garlic a promising normal treatment for metabolic syndrome (Padiya et al., 2011; Pendurthi and Rao, 2000; Reinhart et al., 2009; Gomez et al., 2013).

10.5.3 CINNAMON

Cinnamon obtained from a bark of plant *Cinnamomum verum,* it is used in spices also as a flavoring agent, is commonly used in Ayurvedic, Chinese, and Unani traditional medicines. Extracts of cinnamon possess anti-inflammatory, antithrombotic, insulin-sensitizing, lowering of lipid and antioxidant properties which are useful in the management of metabolic syndrome. Polyphenols present in cinnamon have the ability to lower glucose like insulin and many investigations reported its use in glycemia and in lowering of lipid. Ziegenfuss and colleagues showed in a randomized placebo-controlled experiment that make use of an aqueous cinnamon extract was related with progress in fasting blood glucose, blood pressure, and body composition in people as a result of metabolic syndrome (Ziegenfuss et al., 2006).

10.5.4 BERBERINE

Berberine is an alkaloid derived from the *Rhizoma coptidis* plant that is used in China for its antibacterial effects and has been linked to antidiabetic properties. Studies in insulin-resistant models shows that the utilization of berberine is helpful in BW issue, it also helps to lower fatty oil levels, and in insulin sensitivity. Berberine works by upregulating genes involved in energy metabolism and downregulating genes involved in lipogenesis. It has an insulin-sensitizing activity, like metformin and thiazolidinediones, interceded by adenosine monophosphate-related protein kinase actuation in adipocytes. In persons with metabolic syndrome, studies have found a reduction in waist circumference, triglyceride levels, and systolic blood pressure, especially in women (Yang et al., 2012; Lee et al., 2006; Perez et al., 2013).

10.5.5 NEEM

Azadirachta indica, otherwise called the neem tree, also known as the neem tree, is well-known for its medicinal properties, restorative worth. Neem extricate is related with an expanded glucose tolerance by diminishing intestinal and pancreatic glucosidase movement, which works on postprandial hyperglycemia. An increment in glucose-6-phosphate dehydrogenase movement which expands glucose usage has likewise been observed beforehand. Furthermore, Neem extract additionally assist with the recovery of pancreatic beta cells, further developing insulin emission (Bhat et al., 2009).

10.5.6 BERGAMOT OIL

Bergamot fundamental oil is a citrus fundamental oil obtained from *Citrus bergamia,* and possesses anticancer, antimicrobial, anti-inflammatory, and antianxiety properties. With regards to metabolic syndrome, the antioxidant impacts of bergamot oil are generally significant. Mollace and coworkers considered the impact of the cancer prevention agent segment of bergamot oil on LOX-1 articulation and effects of free radical generation on rat carotid injury. Other normal mixtures, for example, flavonoids, 6-shogaol procyanidins, and procyanidins have additionally been related with cell reinforcement properties that might be advantageous for the administration of Metabolic syndrome (Mollace et al., 2008; Keservani and Sharma, 2014).

10.5.7 RESVERATROL

Resveratrol (3,5,4′-trihydroxystilbene) is a chemical polyphenol which is present in grapes and almonds, as well as servants like wine. It works as a sirtuin pathway activator which controls a few cell capacities identified with digestion, oxidation, and aging. It effectively affects cell energy homeostasis, for example, diminishing adipogenesis and expanding lipolysis, through different instruments, and furthermore restrains cyclooxygenase with a subsequent cancer prevention agent action. The utilization of resveratrol in metabolic syndrome has been concentrated in animal models just as in people. Clinical examinations in patients with insulin resistance and nonalcoholic fatty liver illness have shown promising results. Truth be told, utilization of resveratrol has displayed to further develop insulin sensitivity, glucose tolerance, and in general weight and BMI in people who have metabolic syndrome (Bremer, 2014; Chen et al., 2015; Mendez et al., 2014).

10.5.8 QUERCETIN

Quercetin, a plant-determined flavonoid found in vegetables and natural products like onions, fruits, and tea, has been accounted for to have anti-inflammatory and antioxidant metabolic impacts. Quercetin acts by means of mitochondrial pathways engaged with adipokinesis and lipolysis that influence the improvement of obesity (Leiherer et al., 2016). Rivera and associates considered the impact of quercetin in obese Zucker rodents. In this examination, rodents that got quercetin had lower circulatory strain and compared to rodents that didn't, those who did had lower cholesterol levels and less insulin blockage. Higher portions of quercetin additionally delivered a calming result in the instinctive fat tissue. Pfeuffer and partners contemplated the metabolic ramifications of quercetin use in people, explicitly men with the apolipoprotein E genotype have a higher risk of heart disease and found that quercetin further develops metabolic boundaries like abdomen periphery, postprandial blood glucose, and lipids, yet additionally an increase in flammable markers, such as TNF (Rivera et al., 2008; Pfeuffer et al., 2013).

10.5.9 OMEGA FATTY ACIDS

Omega-3 long-chain polyunsaturated unsaturated fats (PUFAs) have been broadly concentrated in numerous epidemiological investigations, explicitly according to their defensive job among disorders linked to the metabolic syndrome (Wang et al., 2006). Eicosapentaenoic acid (EPA) and docosahexaenoic acid (DHA), two specific PUFAs found in abundance in fish oils, have received a lot of attention, prompting major cultural tips for prevention Lipogenesis has been shown to be inhibited by 48 PUFAs and initiate unsaturated fat oxidation in liver and fat tissue through guideline of key transcription factors, for example, the sterol regulatory component and peroxisome proliferator-activated receptors restricting protein (Flachs et al., 2005; Guo et al., 2005; Gillies et al., 2012; Wu and Juurlink, 2001). Despite a great effect on different segments of metabolic syndrome, long haul information on counteraction of difficult cardiovascular outcomes with utilization of PUFAs have not been affirmed. Notwithstanding, solid patho-physiological associates of fundamental systems are empowering and need further investigation.

10.5.10 SULFORAPHANE

Sulforaphane, a phytochemical isolated from Brassica family crops such as broccoli, is another phytochemical, has likewise as of late, been displayed to have expected valuable impacts in metabolic syndrome, because of its cancer prevention agent and calming properties. Sulforaphane actuates atomic factor erythroid 2-related factor 2, antioxidant transcription factor. Sulforaphane has been proven in animal studies to have a protective role against many disorders such as hypertension, hyperlipidemia, and diabetes, all of which are important components of the metabolic syndrome (Table 10.4) (Wu and Juurlink, 2001; Desouza et al., 2012; Song et al., 2009).

TABLE 10.4 Herbs and Their Effects on the Components of Metabolic Syndrome and Related Disorders

Components	Plants/Substances	Mechanism
Abnormal obesity	Caffeine	Adrenergic stimulation
	Tea (catechins)	Adrenergic stimulation
Impaired glucose tolerance	Berries (anthocyanins, flavonoids, stilbenes)	Slow carbohydrate digestion and absorption, increasing insulin production via the AMPK pathway, and reducing oxidative stress in cells
	Pomegranate is a type of pomegranate that (ellagitannins, anthocyanins)	Inhibition of -glycosidase, which increases the activity of the glucose transporter type 4 and slows carbohydrate digestion and absorption; stimulation of the incretin response. Insulin signaling is improved, and the gut microbiota is modulated.
	Cocoa	The lowering of TNF improves IR.
	Curcuma (curcumin)	Carbohydrate digestion and absorption are slowed; Anti-inflammatory action helps to improve IR.
	Olive oil (polyphenols)	

TABLE 10.4 *(Continued)*

Components	Plants/Substances	Mechanism
Hypertriglyceridemia	Olive oil (MUFAs, polyphenols)	Catabolism of VLDL is being increased.
Low levels of HDL-C	Curcuma (curcumin)	Improvement of HDL metabolism
	Olive oil (polyphenols)	
High blood pressure	Cocoa (catechins, proanthocyanidins, flavonols)	Endothelial NO synthase is activated, whereas the angiotensin-converting enzyme is inhibited.
	Pomegranate (ellagitannins)	The activity of the angiotensin-converting enzyme (ACE) in the blood is reduced.
	Olive oil (polyphenols)	NO levels are being raised.
		Increasing NO levels
Associated Conditions		
Fatty liver	Milk thistle curcuma (curcumin)	Anti-oxidant and anti-inflammatory effects; inhibition of fibrogenesis
Chronic inflammation	Catechins in tea, berries (anthocyanins, flavonoids, stilbenes)	Antioxidant activity, signaling modulation of inflammatory response pathways
		Suppression of NF-kB
	Pomegranate is a type of pomegranate that (ellagitannins, anthocyanins)	Peroxisome proliferator-activated receptor pathway modulation
		Modulation of toll-like receptor
	Catechins, proanthocyanidins, and flavonoids) are found in cocoa, flavonols)	Reduction of TNFα; COX inhibition and LOS
		COX inhibition; inflammatory gene expression reduction
	Curcuma (curcumin)	
	Olive oil (polyphenols)	
LDL-C/small LdL	Phytosterols, tea catechins, berries (anthocyanins, flavonoids, stilbenes)	Reduced lipid absorption in the intestine, increased expression of LDL receptors, and preservation of lipids against oxidative damage
		By activating AMPK, cholesterol production is inhibited.
		Intestinal cholesterol absorption is inhibited.

10.6 CONCLUSION

Metabolic syndrome is a worldwide epidemic and a set up hazard factor for atherosclerotic and nonatherosclerotic cardio vascular diseases. Herbal medicines showed helpful impacts on controlling the abdomen periphery, blood glucose, blood lipids, and pulse in this efficient survey. This implies natural medications can possibly be correlative and elective drugs for metabolic disorders. Be that as it may, all the more excellent preliminaries are expected to demonstrate the viability and security of home-grown meds.

Huge varieties exist in the demonstrative standards and meaning of metabolic syndrome, which address a transient change in the comprehension of this infection. Different upgrades coming full circle in a condition of ongoing inflammation appear to be the fundamental pathophysiological drivers for metabolic syndrome. First and foremost, the presence of just a small bunch of metabolic syndromes that have been displayed to convincingly affect long-term results settles on the decision of treatment evaluations. Also, the persistent idea of the segments of metabolic syndrome justifies delayed and frequently endless utilization of different medications like statins, prompting an expanded weight of drug-related antagonistic impacts and patient resistance. In this unique situation, the improvement of nutraceuticals that are promptly accessible and with negligible incidental effects may address a space of guarantee in the advancement of novel treatments (Keservani et al., 2010a, b).

KEYWORDS

- **adult treatment panel**
- **cardiovascular failure**
- **cardiovascular infection**
- **diabetes mellitus (type 2)**
- **herbs**
- **metabolic syndrome**

REFERENCES

Abate, N., Chandalia, M., Snell, P. G., & Grundy, S. M., (2004). Adipose tissue metabolites and insulin resistance in nondiabetic Asian Indian men. *Journal of Clinical Endocrinology and Metabolism, 89*(6), 2750–2755.

Aggarwal, B. B., (2010). Targeting inflammation-induced obesity and metabolic diseases by curcumin and other nutraceuticals. *Annu. Rev Nutr., 30*, 173–199.

Alberti, K. G. M. M., & Zimmet, P. Z., (1998). Definition, diagnosis and classification of diabetes mellitus and its complications. Part 1: Diagnosis and classification of diabetes mellitus. Provisional report of a WHO consultation. *Diabetic Medicine, 15*(7), 539–553.

Alberti, K. G., & Zimmet, P. Z., (1998). Definition, diagnosis and classification of diabetes mellitus and its complications. Part 1: Diagnosis and classification of diabetes mellitus provisional report of a WHO consultation. *Diabetic Medicine, 15*(7), 539–553.

Alberti, K. G., Zimmet, P., & Shaw, J., (2005). IDF epidemiology task force consensus group. The metabolic syndrome new worldwide definition. *Lancet, 366*, 1059–1062.

Alberti, K. G., Zimmet, P., & Shaw, J., (2006). Metabolic syndrome-a new worldwide definition. A consensus statement from the international diabetes federation. *Diabet. Med., 23*(5), 469–480.

Alessi, M. C., & Juhan-Vague, I., (2006). PAI-1 and the metabolic syndrome: Links, causes, and consequences. *Arteriosclerosis, Thrombosis, and Vascular Biology, 26*(10), 2200–2207.

Aljada, A., Mohanty, P., Ghanim, H., et al., (2004). Increase in intranuclear nuclear factor κB and decrease in inhibitor κB in mononuclear cells after a mixed meal: Evidence for a proinflammatory effect. *The American Journal of Clinical Nutrition, 79*(4), 682–690.

Anderson, P. J., Critchley, J. A. J. H., Chan, J. C. N., et al., (2001). Factor analysis of the metabolic syndrome: Obesity vs insulin resistance as the central abnormality. *International Journal of Obesity, 25*, 1782.

Andrew, R., Gale, C. R., Walker, B. R., Seckl, J. R., & Martyn, C. N., (2002). Glucocorticoid metabolism and the metabolic syndrome: Associations in an elderly cohort. *Experimental and Clinical Endocrinology and Diabetes, 110*(6), 284–290.

Argaud, D., Zhang, Q., Pan, W., Maitra, S., Pilkis, S. J., & Lange, A. J., (1996). Regulation of rat liver glucose-6-phosphatase gene expression in different nutritional and hormonal states: Gene structure and 5'-flanking sequence. *Diabetes, 45*(11), 1563–1571.

Avogaro, P., & Crepaldi, G., (1965). Essential hyperlipidemia, obesity and diabetes. *Diabetologia, 1*, 137.

Balkau, B., & Charles, M. A., (1999). Comment on the provisional report from the WHO consultation: European group for the study of insulin resistance (EGIR). *Diabetic Medicine, 16*(5), 442, 443.

Bangalore, S., Kumar, S., & Messerli, F. H., (2010). Angiotensin-converting enzyme inhibitor-associated cough: Deceptive information from the physicians' desk reference. *The American Journal of Medicine, 123*(11), 1016–1030.

Bhat, M., Kothiwale, S. K., Tirmale, A. R., et al., (2009). Antidiabetic properties of *Azadirachta indica* and *Bougainvillea spectabilis: In vivo* studies in murine diabetes model. *Evid. Based Complement Alternat. Med., 2011*, 561625.

Boden, G., Lebed, B., Schatz, M., Homko, C., & Lemieux, S., (2001). Effects of acute changes of plasma free fatty acids on intramyocellular fat content and insulin resistance in healthy subjects. *Diabetes, 50*(7), 1612–1617.

Bravata, D. M., Wells, C. K., Concato, J., Kernan, W. N., Brass, L. M., & Gulanski, B. I., (2004). Two measures of insulin sensitivity provided similar information in a U.S. population. *Journal of Clinical Epidemiology, 57*(11), 1214–1217.

Bremer, A. A., (2014). Resveratrol use in metabolic syndrome. *Metab. Syndr. Relat. Disord., 12*(10), 493–495.

Briones, A. M., Cat, A. N. D., Callera, G. E., et al., (2012). Adipocytes produce aldosterone through calcineurin-dependent signaling pathways: Implications in diabetes mellitus-associated obesity and vascular dysfunction. *Hypertension, 59*(5), 1069–1078.

Cameron, A. J., Shaw, J. E., & Zimmet, P. Z., (2004). The metabolic syndrome: Prevalence in worldwide populations. *Endocrinology and Metabolism Clinics of North America, 33*(2), 351–375.

Carey, D. G., Jenkins, A. B., Campbell, L. V., Freund, J., & Chisholm, D. J., (1996). Abdominal fat and insulin resistance in normal and overweight women: Direct measurements reveal a strong relationship in subjects at both low and high risk of NIDDM. *Diabetes, 45*(5), 633–638.

Carey, V. J., Walters, E. E., Colditz, G. A., et al., (1997). Body fat distribution and risk of noninsulin-dependent diabetes in women: The nurses' health study. *Am. J. Epidemiol., 145*, 614–619.

Carlyle, M., Jones, O. B., Kuo, J. J., & Hall, J. E., (2002). Chronic cardiovascular and renal actions of leptin: Role of adrenergic activity. *Hypertension, 39*(2), 496–501.

Carr, D. B., Utzschneider, K. M., Hull, R. L., et al., (2004). Intra-abdominal fat is a major determinant of the national cholesterol education program adult treatment panel III criteria for the metabolic syndrome. *Diabetes, 53*(8), 2087–2094.

Carr, D. B., Utzschneider, K. M., Hull, R. L., et al., (2004). Intraabdominal fat is a major determinant of the national cholesterol education program adult treatment panel III criteria for the metabolic syndrome. *Diabetes, 53*(8), 2087–2094.

Charmandari, E., Tsigos, C., & Chrousos, G., (2005). Endocrinology of the stress response. *Annual Review of Physiology, 67*, 259–284.

Chen, S., Zhao, X., Ran, L., et al., (2015). Resveratrol improves insulin resistance, glucose and lipid metabolism in patients with non-alcoholic fatty liver disease: A randomized controlled trial. *Dig. Liver Dis., 47*(3), 226–232.

Chen, Z., Wang, L., Yang, G., Xu, H., & Liu, J., (2015). Chinese herbal medicine combined with conventional therapy for blood pressure variability in hypertension patients: A systematic review of randomized controlled trials. *Evidence-Based Complementary and Alternative Medicine, 2015*, 16. Article ID 582751.

Chrousos, G. P., (2009). Stress and disorders of the stress system. *Nature Reviews Endocrinology, 5*(7), 374–381.

Cinti, S., Mitchell, G., Barbatelli, G., et al., (2005). Adipocyte death defines macrophage localization and function in adipose tissue of obese mice and humans. *Journal of Lipid Research, 46*(11), 2347–2355.

Cleeman, J. I., (2001). Executive summary of the third report of the national cholesterol education program (NCEP) expert panel on detection, evaluation, and treatment of high blood cholesterol in adults (adult treatment panel III). *Journal of the American Medical Association*, 285(19), 2486–2497.

Dandona, P., Aljada, A., Chaudhuri, A., Mohanty, P., & Garg, R., (2005). Metabolic syndrome: A comprehensive perspective based on interactions between obesity, diabetes, and inflammation. *Circulation, 111*(11), 1448–1454.

De Souza, C. G., Sattler, J. A., De Assis, A. M., et al., (2012). Metabolic effects of sulforaphane oral treatment in streptozotocin-diabetic rats. *J. Med. Food, 15*(9), 795–801.

Deen, D., (2004). Metabolic syndrome: Time for action. *The American Family Physician, 69*(12), 2875–2887.

Deepa, R., Velmurugan, K., Arvind, K., et al., (2006). Serum levels of interleukin 6, C-reactive protein, vascular cell adhesion molecule 1, and monocyte chemotactic protein 1 in relation to insulin resistance and glucose intolerance-the Chennai urban rural epidemiology study (CURES). *Metabolism, 55*(9), 1232–1238.

Desroches, S., & Lamarche, B., (2007). The evolving definitions and increasing prevalence of the metabolic syndrome. *Applied Physiology, Nutrition and Metabolism, 32*(1), 23–32.

Diamant, M., Lamb, H. J., Van De, R. M. A., et al., (2005). The association between abdominal visceral fat and carotid stiffness is mediated by circulating inflammatory markers in uncomplicated type 2 diabetes. *Journal of Clinical Endocrinology and Metabolism, 90*(3), 1495–1501.

Eckel, R. H., Grundy, S. M., & Zimmet, P. Z., (2005). The metabolic syndrome. *The Lancet, 365*(9468), 1415–1428.

Einhorn, D., Reaven, G. M., Cobin, R. H., et al., (2003). American college of endocrinology position statement on the insulin resistance syndrome. *Endocrine Practice, 9*(3), 237–252.

Ferrannini, E., & Natali, A., (1991). Essential hypertension, metabolic disorders, and insulin resistance. *The American Heart Journal, 121*(4), 1274–1282.

Ferrannini, E., Haffner, S. M., Mitchell, B. D., & Stern, M. P., (1991). Hyperinsulinaemia: The key feature of a cardiovascular and metabolic syndrome. *Diabetologia, 34*(6), 416–422.

Flachs, P., Horakova, O., Brauner, P., et al., (2005). Polyunsaturated fatty acids of marine origin upregulate mitochondrial biogenesis and induce beta-oxidation in white fat. *Diabetologia*, *48*, 2365–2375.

Ford, E. S., (2004). Prevalence of the metabolic syndrome in US populations. *Endocrinology and Metabolism Clinics of North America, 33*(2), 333–350.

Ford, E. S., Giles, W. H., & Dietz, W. H., (2002). Prevalence of the metabolic syndrome among US adults: Findings from the third national health and nutrition examination survey. *Journal of the American Medical Association, 287*(3), 356–359.

Gill, H., Mugo, M., Whaley-Connell, A., Stump, C., & Sowers, J. R., (2005). The key role of insulin resistance in the cardiometabolic syndrome. *The American Journal of the Medical Sciences, 330*(6), 290–294.

Gillies, P. J., Bhatia, S. K., Belcher, L. A., et al., (2012). Regulation of inflammatory and lipid metabolism genes by eicosapentaenoic acid-rich oil. *J. Lipid Res., 53*, 1679–1689.

Ginsberg, H. N., Zhang, Y. L., & Hernandez-Ono, A., (2005). Regulation of plasma triglycerides in insulin resistance and diabetes. *Archives of Medical Research, 36*(3), 232–240.

Gomez-Arbelaez, D., Lahera, V., Oubina, P., et al., (2013). Aged garlic extract improves adiponectin levels in subjects with metabolic syndrome: A double-blind, placebo-controlled, randomized, crossover study. *Mediators Inflamm., 2013*, 285795.

Grundy, S. M., (2004). Obesity, metabolic syndrome, and cardiovascular disease. *Journal of Clinical Endocrinology and Metabolism, 89*(6), 2595–2600.

Grunfeld, C., & Jones, D. S., (1986). Glucocorticoid-induced insulin resistance *in vitro:* Inhibition of insulin-stimulated methylaminoisobutyric acid uptake. *Hormone and Metabolic Research, 18*(3), 149–152.

Guillaume-Gentil, C., Assimacopoulos-Jeannet, F., & Jeanrenaud, B., (1993). Involvement of non-esterified fatty acid oxidation in glucocorticoid-induced peripheral insulin resistance *in vivo* in rats. *Diabetologia, 36*(10), 899–906.

Guldiken, S., Demir, M., Arikan, E., et al., (2007). The levels of circulating markers of atherosclerosis and inflammation in subjects with different degrees of body mass index:

Soluble CD40 ligand and high-sensitivity C-reactive protein. *Thrombosis Research, 119*(1), 79–84.

Guo, W., Xie, W., Lei, T., et al., (2005). Eicosapentaenoic acid, but not oleic acid, stimulates beta-oxidation in adipocytes. *Lipids, 40*, 815–821.

Haffner, S. M., Valdez, R. A., Hazuda, H. P., Mitchell, B. D., Morales, P. A., & Stern, M. P., (1992). Prospective analysis of the insulin resistance syndrome (syndrome X). *Diabetes, 41*(6), 715–722.

Halberg, N., Wernstedt-Asterholm, I., & Scherer, P. E., (2008). The adipocyte as an endocrine cell. *Endocrinology and Metabolism Clinics of North America, 37*(3), 753–768.

Hales, C. N., & Barker, D. J. P., (1992). Type 2 (non-insulin-dependent) diabetes mellitus: The thrifty phenotype hypothesis. *Diabetologia, 35*(7), 595–601.

Hales, C. N., Desai, M., & Ozanne, S. E., (1997). The thrifty phenotype hypothesis: How does it look after 5 years? *Diabetic Medicine, 14*(3), 189–195.

Hansson, G. K., (2005). Mechanisms of disease: Inflammation, atherosclerosis, and coronary artery disease. *The New England Journal of Medicine, 352*(16), 1685–1695.

Harrison, D. G., (1997). Cellular and molecular mechanisms of endothelial cell dysfunction. *Journal of Clinical Investigation, 100*(9), 2153–2157.

Hasani-Ranjbar, S., Nayebi, N., Larijani, B., & Abdollahi, M., (2009). A systematic review of the efficacy and safety of herbal medicines used in the treatment of obesity. *World Journal of Gastroenterology, 15*(25), 3073–3085.

Hauner, H., Entenmann, G., Wabitsch, M., et al., (1989). Promoting effect of glucocorticoids on the differentiation of human adipocyte precursor cells cultured in a chemically defined medium. *Journal of Clinical Investigation, 84*(5), 1663–1670.

Hu, G., Qiao, Q., Tuomilehto, J., et al., (2004). Plasma insulin and cardiovascular mortality in non-diabetic European men and women: A meta-analysis of data from eleven prospective studies. The DECODE insulin study group. *Diabetologia, 47*, 1245–1256.

Hutley, L., & Prins, J. B., (2005). Fat as an endocrine organ: Relationship to the metabolic syndrome. *The American Journal of the Medical Sciences, 330*(6), 280–289.

Jacobs, M., M. Van. G. M. J. M., Van, D. K. C. J. H., et al., (2009). Low-grade inflammation can partly explain the association between the metabolic syndrome and either coronary artery disease or severity of peripheral arterial disease: The CODAM study. *European Journal of Clinical Investigation, 39*(6), 437–444.

Jensen, M. D., Haymond, M. W., Rizza, R. A., Cryer, P. E., & Miles, J. M., (1989). Influence of body fat distribution on free fatty acid metabolism in obesity. *Journal of Clinical Investigation, 83*(4), 1168–1173.

Kaplan, N. M., (1989). The deadly quartet. Upper-body obesity, glucose intolerance, hypertriglyceridemia, and hypertension. *Archives of Internal Medicine, 149*(7), 1514–1520.

Kazumi, T., Kawaguchi, A., Sakai, K., Hirano, T., & Yoshino, G., (2002). Young men with high-normal blood pressure have lower serum adiponectin, smaller LDL size, and higher elevated heart rate than those with optimal blood pressure. *Diabetes Care, 25*(6), 971–976.

Keservani, R. K., & Sharma, A. K., (2014). Flavonoids: Emerging trends and potential health benefits. *J. Chin. Pharm. Sci., 23*(12), 815–822.

Keservani, R. K., Kesharwani, R. K., Sharma, A. K., Vyas, N., & Chadoker, A., (2010b). Nutritional supplements: An overview. *Int. J. of Curr. Pharm. Rev. and Res., 1*(1), 59–75.

Keservani, R. K., Kesharwani, R. K., Vyas, N., Jain, S., Raghuvanshi, R., & Sharma, A. K., (2010a). Nutraceuticals and functional food as future food. *Der. Pharm. Lett., 2*(1), 106–116.

Kolovou, G. D., Anagnostopoulou, K. K., Salpea, K. D., & Mikhailidis, D. P., (2007). The prevalence of metabolic syndrome in various populations. *The American Journal of the Medical Sciences, 333*(6), 362–371.

Kraemer-Aguiar, L. G., Laflor, C. M., & Bouskela, E., (2008). Skin microcirculatory dysfunction is already present in normoglycemic subjects with metabolic syndrome. *Metabolism, 12*, 1740–1746.

Kylin, E., (1923). Studies on the hypertension-hyperglycemia-hyperuricemia syndrome. *Central Journal for Internal Medicine, 44*, 105–127.

Laakso, M., (2004). Gene variants, insulin resistance, and dyslipidaemia. *Current Opinion in Lipidology, 15*(2), 115–120.

Lau, D. C. W., Dhillon, B., Yan, H., Szmitko, P. E., & Verma, S., (2005). Adipokines: Molecular links between obesity and atherosclerosis. *The American Journal of Physiology-Heart and Circulatory Physiology, 288*(5), H2031–H2041.

Lee, Y. S., Kim, W. S., Kim, K. H., et al., (2006). Berberine, a natural plant product, activates AMP-activated protein kinase with beneficial metabolic effects in diabetic and insulin-resistant states. *Diabetes, 55*(8), 2256–2264.

Leiherer, A., Stoemmer, K., Muendlein, A., et al., (2016). Quercetin impacts expression of metabolism- and obesity-associated genes in SGBS adipocytes. *Nutrients, 8*(5), 10.3390/nu8050282.

Lewis, G. F., & Steiner, G., (1996). Acute effects of insulin in the control of VLDL production in humans: Implications for the insulin resistant state. *Diabetes Care, 19*(4), 390–393.

Liu, M., & Liu, F., (2010). Transcriptional and post-translational regulation of adiponectin. *Biochemical Journal, 425*(1), 41–52.

Luo, J. Q., He, F. Z., Wang, Z. M., et al., (2015). SLCO1B1 variants and angiotensin-converting enzyme inhibitor (enalapril)—Induced cough: A pharmacogenetic study. *Scientific Reports, 5*, Article ID 17253.

Malhotra, A., Kang, B. P. S., Cheung, S., Opawumi, D., & Meggs, L. G., (2001). Angiotensin II promotes glucose-induced activation of cardiac protein kinase C isozymes and phosphorylation of troponin I. *Diabetes, 50*(8), 1918–1926.

Marsh, A. J., Fontes, M. A. P., Killinger, S., Pawlak, D. B., Polson, J. W., & Dampney, R. A. L., (2003). Cardiovascular responses evoked by leptin acting on neurons in the ventromedial and dorsomedial hypothalamus. *Hypertension, 42*(4), 488–493.

Martin, L. J., North, K. E., Dyer, T., Blangero, J., Comuzzie, A. G., & Williams, J., (2003). Phenotypic, genetic, and genome-wide structure in the metabolic syndrome. *BMC Genetics, 4*(1), Article S95.

Mayer-Davis, E. J., D'Agostino, Jr. R., Karter, A. J., et al., (1998). Intensity and amount of physical activity in relation to insulin sensitivity the insulin resistance atherosclerosis study. *Journal of the American Medical Association, 279*(9), 669–674.

McLaughlin, T., Abbasi, F., Lamendola, C., et al., (2002). Differentiation between obesity and insulin resistance in the association with C-reactive protein. *Circulation, 106*(23), 2908–2912.

Mendez-Del, V. M., Gonzalez-Ortiz, M., Martinez-Abundis, E., et al., (2014). Effect of resveratrol administration on metabolic syndrome, insulin sensitivity, and insulin secretion. *Metab. Syndr. Relat. Disord., 12*(10), 497–501.

Miles, J. M., & Jensen, M. D., (2005). Counterpoint: Visceral adiposity is not causally related to insulin resistance. *Diabetes Care, 28*(9), 2326–2328.

Mollace, V., Ragusa, S., Sacco, I., et al., (2008). The protective effect of bergamot oil extract on lecitine-like oxy LDL receptor-1 expression in balloon injury-related neointima formation. *J. Cardiovasc. Pharmacol. Ther., 13*(2), 120–129.

Morse, S. A., Zhang, R., Thakur, V., & Reisin, E., (2005). Hypertension and the metabolic syndrome. *The American Journal of the Medical Sciences, 330*(6), 303–310.

NCEPE (2002). Third report of the National Cholesterol Education Program (NCEP) expert panel on detection, evaluation, and treatment of high blood cholesterol in adults (Adult Treatment Panel III) final report. *Circulation, 106*(25), 3143–3421.

Neel, J. V., (1962). Diabetes mellitus: A "thrifty" genotype rendered detrimental by progress? *The American Journal of Human Genetics, 14*, 353–362.

Ohlson, L. O., Larsson, B., Svardsudd, K., et al., (1985). The influence of body fat distribution on the incidence of diabetes mellitus: 13.5 years of follow-up of the participants in the study of men born in 1913. *Diabetes, 34*, 1055–1058.

Ordovas, J. M., (2007). Genetic links between diabetes mellitus and coronary atherosclerosis. *Current Atherosclerosis Reports, 9*(3), 204–210.

Padiya, R., Khatua, T. N., Bagul, P. K., et al., (2011). Garlic improves insulin sensitivity and associated metabolic syndromes in fructose fed rats. *Nutr. Metab. (Lond.), 8*(53), 7075-8-53.

Palaniappan, L., Carnethon, M. R., Wang, Y., et al., (2004). Predictors of the incident metabolic syndrome in adults: The insulin resistance atherosclerosis study. *Diabetes Care, 27*(3), 788–793.

Park, Y. W., Zhu, S., Palaniappan, L., Heshka, S., Carnethon, M. R., & Heymsfield, S. B., (2003). The metabolic syndrome: Prevalence and associated risk factor findings in the US population from the third national health and nutrition examination survey, 1988–1994. *Archives of Internal Medicine, 163*(4), 427–436.

Pedersen, B. K., Steensberg, A., Fischer, C., et al., (2003). Searching for the exercise factor: Is IL-6 a candidate? *Journal of Muscle Research and Cell Motility, 24*(2, 3), 113–119.

Pendurthi, U. R., & Rao, L. V., (2000). Suppression of transcription factor Egr-1 by curcumin. *Thromb. Res., 97*(4), 179–189.

Perez-Rubio, K. G., Gonzalez-Ortiz, M., Martinez-Abundis, E., et al., (2013). Effect of berberine administration on metabolic syndrome, insulin sensitivity, and insulin secretion. *Metab. Syndr. Relat. Disord., 11*(5), 366–369.

Perseghin, G., Ghosh, S., Gerow, K., & Shulman, G. I., (1997). Metabolic defects in lean nondiabetic offspring of NIDDM parents: A cross-sectional study. *Diabetes, 46*(6), 1001–1009.

Petersen, K. F., & Shulman, G. I., (2006). Etiology of insulin resistance. *The American Journal of Medicine, 119*(5), S10–S16.

Pfeuffer, M., Auinger, A., Bley, U., et al., (2013). Effect of quercetin on traits of the metabolic syndrome, endothelial function and inflammation in men with different APOE isoforms. *Nutr. Metab. Cardiovasc Dis., 23*(5), 403–409.

Ponholzer, A., Temml, C., Rauchenwald, M., Marszalek, M., & Madersbacher, S., (2008). Is the metabolic syndrome a risk factor for female sexual dysfunction in sexually active women? *International Journal of Impotence Research, 20*(1), 100–104.

Pouliot, M. C., Després, J. P., Lemieux, S., et al., (1994). Waist circumference and abdominal sagittal diameter: Best simple anthropometric indexes of abdominal visceral adipose tissue accumulation and related cardiovascular risk in men and women. *Am. J. Cardiol., 73*, 460–468.

Poulsen, P., Levin, K., Petersen, I., Christensen, K., Beck- Nielsen, H., & Vaag, A., (2005). Heritability of insulin secretion, peripheral and hepatic insulin action, and intracellular glucose partitioning in young and old Danish twins. *Diabetes, 54*(1), 275–283.

Reaven, G. M., (1988). Role of insulin resistance in human disease. *Diabetes, 37*(12), 1595–1607.

Reilly, M. P., & Rader, D. J., (2003). The metabolic syndrome: More than the sum of its parts? *Circulation, 108*(13), 1546–1551.

Reinhart, K. M., Talati, R., White, C. M., et al., (2009). The impact of garlic on lipid parameters: A systematic review and meta-analysis. *Nutr. Res. Rev., 22*(1), 39–48.

Rexrode, K. M., Carey, V. J., Hennekens, C. H., et al., (1998). Abdominal adiposity and coronary heart disease in women. *JAMA, 280*, 1843–1848.

Ridker, P. M., Buring, J. E., Cook, N. R., & Rifai, N., (2003). C-reactive protein, the metabolic syndrome, and risk of incident cardiovascular events: An 8-year follow-up of 14 719 initially healthy American women. *Circulation, 107*(3), 391–397.

Rivera, L., Moron, R., Sanchez, M., et al., (2008). Quercetin ameliorates metabolic syndrome and improves the inflammatory status in obese Zucker rats. *Obesity (Silver Spring), 16*(9), 2081–2087.

Saad, M. F., Lillioja, S., Nyomba, B. L., et al., (1991). Racial differences in the relation between blood pressure and insulin resistance. *New England Journal of Medicine, 324*, 733–739.

Saleem, U., Khaleghi, M., Morgenthaler, N. G., et al., (2009). Plasma carboxy-terminal vasopressin (copeptin): A novel marker of insulin resistance and metabolic syndrome. *Journal of Clinical Endocrinology and Metabolism, 94*(7), 2558–2564.

Shaw, J. E., & Chisholm, D. J., (2003). Epidemiology and prevention of type 2 diabetes and the metabolic syndrome. *Medical Journal of Australia, 179*(7), 379–383.

Singh, S., & Aggarwal, B. B., (1995). Activation of transcription factor NF-kappa B is suppressed by curcumin (diferuloylmethane) [corrected]. *J. Biol. Chem., 270*(42), 24995–25000.

Song, M. Y., Kim, E. K., Moon, W. S., et al., (2009). Sulforaphane protects against cytokine- and streptozotocin-induced β-cell damage by suppressing the NF-κB pathway. *Toxicol. Appl. Pharmacol., 235*(1), 57–67.

SotoGonz´alez, A., BellidoGuerrero, D., Bu˜noSoto, M., P´ertegaD´ıaz, S., Martinez-Olmos, M., & Vidal, O., (2006). Metabolic syndrome, insulin resistance and the inflammation markers C-reactive protein and ferritin. *European Journal of Clinical Nutrition, 60*(6), 802–809.

Stenl¨of, K., Wernstedt, I., Fj¨allman, T., Wallenius, V., Wallenius, K., & Jansson, J. O., (2003). Interleukin-6 levels in the central nervous system are negatively correlated with fat mass in overweight/obese subjects. *Journal of Clinical Endocrinology and Metabolism, 88*(9), 4379–4383.

Sui, Y., Zhao, H. L., Wong, V. C. W., et al., (2012). A systematic review on the use of Chinese medicine and acupuncture for treatment of obesity. *Obesity Reviews, 13*(5), 409–430.

Trayhurn, P., & Wood, I. S., (2004). Adipokines: Inflammation and the pleiotropic role of white adipose tissue. *British Journal of Nutrition, 92*(3), 347–355.

Tsimikas, S., Willeit, J., Knoflach, M., et al., (2009). Lipoprotein-associated phospholipase A2 activity, ferritin levels, metabolic syndrome, and 10-year cardiovascular and non-cardiovascular mortality: Results from the Bruneck study. *European Heart Journal, 30*(1), 107–115.

Vague, J., (1947). Sexual differentiation. A factor affecting the forms of obesity. *Presse Medicale, 30*, S39, S40.
Vessby, B., Uusitupa, M., Hermansen, K., et al., (2001). Substituting dietary saturated for monounsaturated fat impairs insulin sensitivity in healthy men and women: The KANWU study. *Diabetologia, 44*(3), 312–319.
Wang, C. N., McLeod, R. S., Yao, Z., & Brindley, D. N., (1995). Effects of dexamethasone on the synthesis, degradation, and secretion of apolipoprotein B in cultured rat hepatocytes. *Arteriosclerosis, Thrombosis, and Vascular Biology, 15*(9), 1481–1491.
Wang, C., Harris, W. S., Chung, M., et al., (2006). *n*–3 fatty acids from fish or fish-oil supplements, but not alpha-linolenic acid, benefit cardiovascular disease outcomes in primary- and secondary prevention studies: A systematic review. *Am J. Clin. Nutr., 84*, 5–17.
Wilson, P. W. F., Kannel, W. B., Silbershatz, H., & D'Agostino, R. B., (1999). Clustering of metabolic factors and coronary heart disease. *Archives of Internal Medicine*, vol. 159, no. 10, pp. 1104– 1109.
Wong, N. D., (2005). Intensified screening and treatment of the metabolic syndrome for cardiovascular risk reduction. *Preventive Cardiology, 8*(1), 47–54.
Wu, L., & Juurlink, B. H. J., (2001). The impaired glutathione system and its up-regulation by sulforaphane in vascular smooth muscle cells from spontaneously hypertensive rats. *J. Hypertens., 19*(10), 1819–1825.
www.idf.org/metabolic_syndrome, website of the International Diabetes Federation .
Xydakis, A. M., Case, C. C., Jones, P. H., et al., (2004). Adiponectin, inflammation, and the expression of the metabolic syndrome in obese individuals: The impact of rapid weight loss through caloric restriction. *Journal of Clinical Endocrinology and Metabolism, 89*(6), 2697–2703.
Yang, J., Yin, J., Gao, H., et al., (2012). Berberine improves insulin sensitivity by inhibiting fat store and adjusting adipokines profile in human preadipocytes and metabolic syndrome patients. *Evid. Based Complement Alternat. Med., 2012*, 363845.
Yeh, G. Y., Eisenberg, D. M., Kaptchuk, T. J., & Phillips, R. S., (2003). Systematic review of herbs and dietary supplements for glycemic control in diabetes. *Diabetes Care, 26*(4), 1277–1294.
Ziegenfuss, T. N., Hofheins, J. E., Mendel, R. W., et al., (2006). Effects of a water-soluble cinnamon extract on body composition and features of the metabolic syndrome in pre-diabetic men and women. *J. Int. Soc. Sports Nutr., 3*, 45–53.
Zimmet, P., Alberti, K. G. M. M., & Shaw, J., (2001). Global and societal implications of the diabetes epidemic. *Nature, 414*, 782–787.

CHAPTER 11

The Metabolic Syndrome: A Concerning Area for Future Research

EKNATH D. AHIRE,[1] KHEMCHAND R. SURANA,[2] VIJAYRAJ N. SONAWANE,[2] SWATI G. TALELE,[3] GOKUL S. TALELE,[4] SANJAY J. KSHIRSAGAR,[1] SHUBHAM J. KHAIRNAR,[5] and NILIMA A. THOMBRE[1]

[1]*Department of Pharmaceutics, MET's Institute of Pharmacy, Bhujbal Knowledge City (BKC), Affiliated to Savitribai Phule Pune University, Adgaon, Nashik, Maharashtra – 422003, India, Tel.: +91-9588431965, Fax: (02552)-303-515, E-mail: eknathahire05@gmail.com (E. D. Ahire)*

[2]*Department of Pharmaceutical Chemistry, Shreeshakti Shaikshanik Sanstha, Divine College of Pharmacy, Satana, Nashik, Maharashtra, India*

[3]*Department of Pharmaceutics, Sandip Institute of Pharmaceutical Sciences, Nashik, Maharashtra – 422213, India*

[4]*Department of Pharmaceutics, Matoshri College of Pharmacy, Near Odhagaon, Eklahare, Nashik, Maharashtra – 422105, India*

[5]*Department of Pharmacology, MET's Institute of Pharmacy, Bhujbal Knowledge City (BKC), Affiliated to Savitribai Phule Pune University, Nashik, Maharashtra – 422003, India*

ABSTRACT

Metabolic syndrome (MS) is a multidimensional condition that is becoming more common around the world. Because of the established link between weight problems and metabolic syndrome, weight management strategies

The Metabolic Syndrome: Dietary Supplements and Food Ingredients. Raj K. Keservani, MPharm, Durgavati Yadav, PhD, Rajesh K. Kesharwani, PhD, Sippy Singh, DPhil & Kumar Sandeep, PhD (Eds.)

are crucial in preventing MS from spreading. Many epidemiological research offers proof linking nutritional styles to the occurrence of MS symptoms. As a significance of the epidemiology studies, dietary interference studies which analyzes the effects of supplementing diets with specific nutrients of attention on the indications of the MS have been showing. Indication has shown that routine interference encompassing changes in dietary consumption and physical activity leads to an enhanced metabolic profile both in the occurrence or deficiency of weight loss therefore highlighting the significance of a multi-faceted methodology in fighting MS. Nutritious treatment investigation is not focused exclusively on decreasing energy consumption and manipulating macronutrient intake but is examining the position of beneficial ingredients or bioactive materials of food. Such bioactive which mark weight maintenance and insulin sensitivity may have a possibly helpful effect on the indications of the MS. Nevertheless, the effectiveness of multiple useful vitamins wants to be additional distinct and obviously established. MS has become more common over the world, owing to increased consumption of energy-dense meals and less physical exercise. MS will increase the hazard of rising type 2 diabetes (T2D), cardiovascular diseases (CDVs), cancer, infertility, non-alcoholic fatty liver disease, dementia, etc. The main aim of aim of this chapter is to highlight the overview on the MS. We are confident this chapter will deliver valuable existing insights as well as critical regarding the MS which will guide future research.

11.1 INTRODUCTION

The term metabolic syndrome (Mets) has become first used withinside the National Cholesterol Education Program (NCEP) Adult Treatment Panel III (APP-III) to give an explanation for the co-occurrence of weight problems, dyslipidemia, hypertension, and atypical glucose metabolism. The link among metabolic issues and CVD threat factors, on the opposite hand, has been recognized for decades. Reaven coined the term "syndrome X" to characterize the link amongst insulin resistance, hypertension, type 2 diabetes (T2D), and cardiovascular disorder on the 1988 American Diabetes Association (ADA) Banting Conference. Other researchers have used the term "insulin resistance syndrome" to explain the clustering of cardiometabolic threat factors. Various groups have proposed distinctive standards to explain the connection of cardiovascular and metabolic illnesses. The International Diabetes Federation characterized the metabolic syndrome as signs and symptoms and physical or biochemical findings coexisting extra frequently than might be defined with the aid of using chance alone. The ADA mentioned the clustering of medical and laboratory functions in metabolic syndrome however wondered about the application of insulin resistance as a biomarker for cardiovascular risk. In 1998, the WHO released an operating definition for metabolic syndrome that specializes in insulin resistance, impaired glucose tolerance, or T2D, in addition to dyslipidemia (low HDL-C and excessive triglycerides), hypertension, and microalbuminuria. The European Group for the Study of Insulin Resistance utilized metabolic syndrome standards that have been corresponding to the ones utilized by the WHO, however, did now no longer encompass microalbuminuria. The NCEP ATP-III described metabolic syndrome primarily based totally on expanded waist circumference, lipids, blood pressure, and fasting glucose levels. The American Association of Clinical Endocrinologists (AACE) defines the metabolic syndrome because the prevalence of insulin resistance instead of diabetes. The IDF has proposed precise racial/ethnic cutoffs to account for populace variances; additionally, the new IDF metabolic syndrome criteria targeted on fasting plasma glucose concentration instead of insulin resistance. The variations in metabolic syndrome criteria amongst diverse firms are not going to be resolved. Future metabolic syndrome definitions may also want to mirror the contributions of adipokines, pro-inflammatory cytokines, and different humoral factors, given our present and growing information on the pathophysiology of weight problems and associated disorders. Factors connected to insulin resistance, diabetes, and cardiovascular illnesses. Cholesterol and blood pressure, and mortality prices because of diabetes or cardiovascular diseases (CDVs) (Keservani et al., 2010a, b). Metabolic syndrome-associated diseases have ended up extra not unusual place over the world, and those traits are impacted with the aid of using age, gender, race/ethnicity, loss of physical exercise, and different lifestyle factors. The occurrence of metabolic syndrome in young people additionally varies consistent with the definition, age, and populace beneath the study (Figure 11.1) (He et al., 2013; Edwardson et al., 2012; Simmons et al., 2010).

11.2 OBESITY

The specific molecular mechanisms which result in the pathogenesis and development of the MetS have not begun to be completely defined. Obesity, T2DM, and the MetS are complicated multi-factorial problems and not using a single derangement driving the epidemic. As there are various signs and symptoms of the MetS, so too are their numerous causes – all inextricably linked. Traditionally adipose tissue was thought of purely as an energy store – wherein excess nutrients were stored principally in the form of TAG. More recently, our knowledge of fatty tissue has expanded with the demonstration that fatty tissue is not simply an energy sink but a functioning endocrine organ. Correct adipose tissue function to allow for both these roles are a crucial aspect of whole-body glucose and fatty acid (energy) homeostasis. Obesity occurs because of a pathological increase in adipose tissue mass, either as the result of adipocyte hyperplasia or hypertrophy. There are significant and deleterious alterations in adipose tissue function that occur concurrently with this increase in fatty tissue mass, and these changes are crucial in the etiology of MetS. Obesity is a major risk factor for MetS, which has strong inheritance patterns and increases the likelihood of T2D and MetS (Jarouliya and Keservani, 2019). Obesity is not a single affliction. When adipocyte biology is perturbed, important systemic effects and alterations in adipose derived cytokines (or adipokines) are evident. Obesity impedes adiponectin secretion (Keller and Lemberg, 2003; Esser et al., 2014).

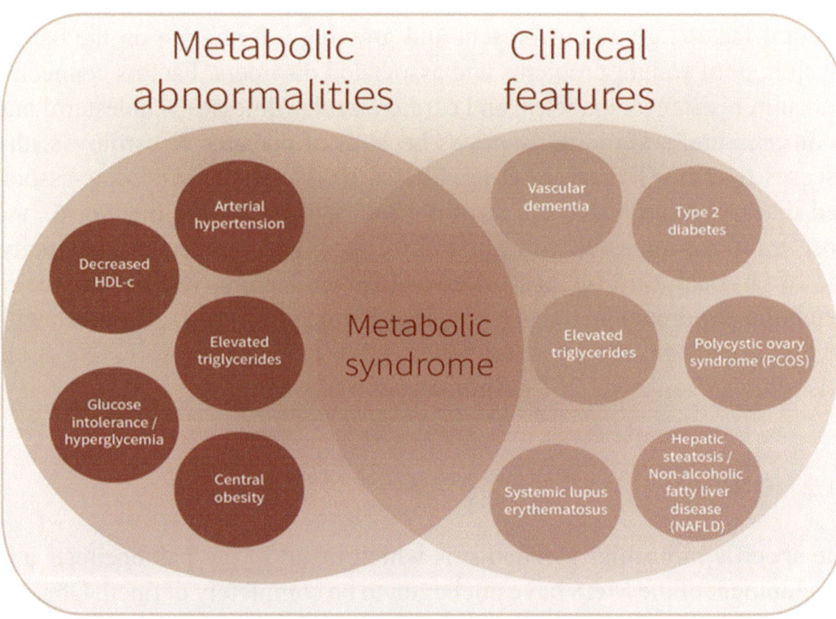

FIGURE 11.1 Metabolic syndrome abnormalities and clinical features.

On the other side, leptin secretion is seemingly unaltered – resistance to it signaling is associated with obesity and also with increased fatty acid release. Obesity is linked to an increase in free fatty acid (FFA) or non-esterified fatty acid (NEFA) release, according to recent research (NEFA). The conventional wisdom in this area of obesity is starting to be questioned. Whilst recent studies are exploring the possibility of normalized plasma NEFA levels in some obese patients, elevated NEFA in the peripheral tissues still exists. Insulin resistance in adipose and other peripheral tissues is promoted by immune cell infiltration and consequent chronic low-grade inflammation. The term "meta-inflammation" was coined to characterize the scenario of chronic low-grade inflammation linked to obesity and MetS. Increased levels of IL-1, IL-6, TNF-, and C-reactive protein (CRP), among other pro-inflammatory cytokines, are linked to macrophage infiltration of adipose tissue in chronic low-grade inflammation. There is also a decrease in the secretion of anti-inflammatory cytokines and so-called "protective" adipokines, e.g., adiponectin (Esser et al., 2014; Church, 2011).

11.2.1 INSULIN RESISTANCE

Insulin resistance reduces glucose uptake by peripheral tissues, including muscular and adipocytes, but gluconeogenesis in the liver proceeds unabated, resulting in hyperglycemia and, most likely, diabetes. Insulin resistance exists when a normal concentration of insulin elicits subnormal biological response. Many years prior to manifestation of the MetS or diabetes, insulin resistance is detectable in the body. It was during his Banting award lecture in the 1980's that Reaven put forward insulin resistance as the fundamental cause of what he described as "Syndrome X." However, Reaven did not recognize that obesity was a core component due to the fact that he identified insulin resistance in patients who were not obese.

Obesity and IR both cause and exacerbate each other. There is no conclusive proof that any one of these reasons is the sole cause of the Mets' failure (Romeo et al., 2012; Doyle et al., 2011).

11.2.2 PERIPHERAL TISSUES AND BIOMARKERS

In obesity, there's an enormous growth withinside the production and release of monoacylglycerols and diacylglycerols from adipocytes, main to lipotoxic consequences at peripheral tissues. FFA levels above a sure threshold disrupt insulin signaling and lift the chance of MetS and T2D. FFA do that with the aid of using inflicting insulin resistance on the muscle and liver and that they have the brought ability for unfavorable pancreatic Beta-cell function. Insulin resistance withinside the adipose ends in no suppression of hormone sensitive lipase (HSL) which in flip ends in increased lipolysis exacerbating the extended FFA situation. Lipotoxicity can end result now no longer most effective in insulin resistance on the peripheral tissues however additionally oxidative pressure and infection as demonstrated by raising FFA levels in wholesome individuals. Data from the UK Perspective take a look at turned into analyzed (Bremer and Jialal, 2013; Deng and Scherer, 2010).

11.3 UNDERLYING RISK FACTORS AND METABOLIC SYNDROME

The most common underlying risk factors for the syndrome appear to be abdominal fat and insulin resistance. Physical inactivity, age, and hormone imbalance are all connected disorders. Although an atherogenic diet (high in saturated fat and cholesterol) has been shown to increase the risk of cardiovascular disease in people who have the syndrome, it has not been proven to be an underlying risk factor. Insulin resistance, according to one opinion, is the fundamental cause of metabolic syndrome. Insulin resistance increases the likelihood of hyperglycemia in people with T2D without a doubt. Insulin resistance and compensatory hyperinsulinemia have been linked to a wide range of metabolic risk factors via many metabolic pathways. Some people who aren't overweight by traditional measures are insulin resistant and have high levels of metabolic risk indicators, according to a study. People with diabetic parents or one parent and a first- or second-degree relative, as well as many people of South Asian ancestry, are ideal examples. Despite the fact that insulin-resistant patients do not need to be clinically obese, their fat distribution is generally abnormal, with a lot of upper body fat. Insulin resistance is connected to excess body fat in the upper body. Extra upper body fat can accumulate intraperitoneally (visceral fat) or subcutaneously (subcutaneous fat). Excess visceral fats, consistent with many experts, is greater strongly related to insulin resistance than some other shape of adipose tissue; nevertheless, a few investigations have observed that extra subcutaneous belly (or truncal) fats are likewise connected to insulin resistance. Regardless of the way plenty of visceral fats and abdominal subcutaneous fats make a contribution to insulin resistance, a sample of belly (or upper-frame) weight problems correlates greater strongly with insulin resistance and the metabolic syndrome than a sample of lower-frame weight problems. In upper-body weight (BW) problems, the discharge of no esterified fatty acids from adipose tissue is substantially high, contributing to lipid accumulation in regions aside from adipose tissue. Insulin resistance and dyslipidemia appear like connected to ectopic lipid buildup in muscle and liver. Many experts trust that the growing occurrence of the metabolic syndrome is because of the growing burden of weight problems withinside the United States. The insulin resistance speculation has to be reconciled with this viewpoint. The root of the trouble will be abnormalities in adipose tissue metabolism. Obese people's adipose tissue is insulin resistant, elevating no esterified fatty acid levels, worsening insulin resistance in muscle, and converting hepatic metabolism; additionally, overweight people's adipose tissue has abnormalities withinside the manufacturing of numerous adipokines, which may also have an effect on insulin resistance and/or adjust ASCVD chance separately (Apridonidze et al., 2005; Roberts et al., 2013).

These changes have all been connected to metabolic risk factors. Metabolic syndrome can occur in adults with only a moderate level of total body fat, as previously stated. Many South Asians, for example, appear to be insulin resistant by nature, a condition exacerbated by modest abdominal obesity. Furthermore, insulin resistance varies greatly in the US population; those with more inherent insulin resistance can develop the metabolic syndrome Individuals with little or no underlying insulin resistance can develop the metabolic syndrome if they accumulate significant abdominal obesity, whereas those with little or no inherent insulin resistance can develop the metabolic syndrome if they collect significant abdominal obesity. These findings support the idea that body fat distribution, particularly excess abdominal fat, is important in the development of the condition. A condition of chronic, low-grade inflammation has recently been linked to this sickness. Some researchers believe that this form of inflammation causes or exacerbates the condition. Insulin resistance is hypothesized to be caused by inflammatory cytokines in both adipose tissue and muscle. Adipose tissue produces an overabundance of cytokines while releasing less adiponectin in the presence of obesity, and these processes appear to enhance the relationship between obesity and inflammation. Even in the absence of an increase in total body fat, insulin resistance is linked to low-grade inflammation. Finally, there is significant individual and ethnic variability in the clinical pattern of metabolic risk variables in obese/insulin-resistant people. Each metabolic risk factor's expression is expected to be influenced by its own genetic regulation, which determines the response to various environmental exposures. A range of polymorphisms in genes that impact lipoprotein metabolism, for example, have been linked to the exacerbation of dyslipidemia in obese patients. Similarly, when insulin resistance is paired with a genetic susceptibility to faulty insulin secretion, plasma glucose levels can rise to dangerously high levels. Despite the fact that the metabolic syndrome clearly predisposes to T2D, many cardiovascular disease researchers regard it as a complex risk factor for ASCVD. Several recent studies have connected the metabolic syndrome to an increased risk of cardiovascular disease, but the risk of cardiovascular disease skyrockets if T2D Mellitus develops. Finally, fatty liver, polycystic ovarian syndrome, cholesterol gallstones, sleep apnea, lipodystrophies, and HIV treatment with protease inhibitors have all been associated with insulin resistance and the metabolic syndrome. These connections are creating a lot of attention in a lot of different medical sectors (Figure 11.2) (Prasad et al., 2012; Panza et al., 2010; Sookoian and Pirola, 2011).

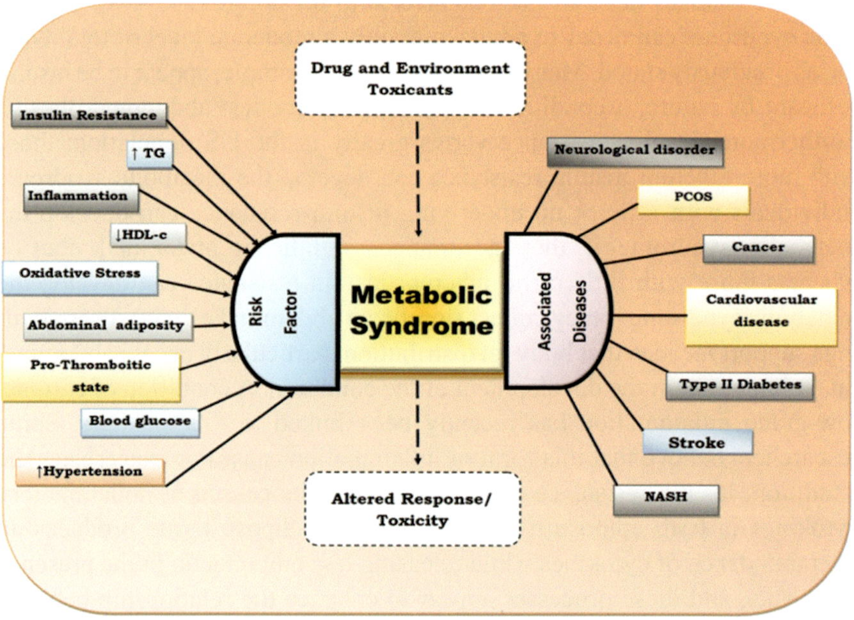

FIGURE 11.2 Metabolic syndrome with its associated risk factors and diseases at the intersection with drug toxicity.

11.4 TREATMENT OF THE METS

The MetS clinical management strategy's major goal is to reduce the risk of T2DM and CVD. Avoiding the MetS would be ideal because it would lower the risk of T2DM and CVD. This concept is known as "primordial prevention" a term coined in 1978 by Strasser and approved by the American Heart Association (AHA). Unfortunately, the number of people who manage to achieve primordial prevention is very low – less than 5% of middle-aged Americans have what is defined as ideal cardiovascular health as set out by the AHA. Primordial prevention is an ideal situation which when obtained is capable of conferring an additional 10 years of life to an individual. However, given that atherosclerosis has an early age of onset and also that childhood obesity is increasing worldwide, the likelihood is that primordial prevention percentages will not rise in the near future. With this in mind reversal of existing risk factors becomes of critical importance. Pharmacological agents to treat the symptoms of the MetS are aimed at reducing LDL-cholesterol, hypertension, and impaired glucose tolerance. The drug therapies aimed at reducing these metabolic risk factors include statins, fibrates, nicotinic acid, angiotensin-converting enzyme (ACE) inhibitors, metformin, thiazolidinediones or acarbose. The remainder of this review will focus on the role of diet in treating the MetS and its symptoms due to the principal role played by obesity in the etiology of the MetS. The manner of living is suitable. In fact, a change in lifestyle, such as increased physical activity and dietary changes, was found to be more helpful than metformin in lowering the risk of T2D in one study. "Let food be thy medicine, and medicine be thy food," Hippocrates, the Ancient Greek father of modern medicine, is said to have said, "Let food be thy medicine, and medicine be thy food." There are numerous studies both ongoing and concluded which are providing evidence for the role that diet plays in the development or prevention of the MetS (Eckel et al., 2010). The Framingham study is the primary source of information for markers of MetS. Framingham has demonstrated that diet quality is inextricably linked with obesity and therefore the MetS, particularly in women. Obesity has been linked to a diet high in saturated fatty acids and alcohol but poor in energy, carbohydrates, and minerals. Furthermore, women who consumed a high-energy diet consisting of protein, carbohydrates, and fat had a higher chance of becoming obese. Obesity, on the other hand, was inversely connected to a high calorie intake based on high fiber and vitamin E (Keservani et al., 2016). The Framingham study is a local study and fails to take into account dietary patterns based on global location – which is precisely what the interact study has attempted to do. One method which this study has made use of is the classification of diets into "Oriental," "Western" or "Prudent." Tofu, soy, and a variety of other sauces are all part of the Asian diet. The Western diet is rich in fried foods, salty snacks, eggs, and meat. The Prudent diet depends on large amounts of fruit and vegetables. Using this classification approach, they discovered that the Oriental diet had no link to acute myocardial infarction (AMI), whereas the Prudent diet, in addition to having no link to AMI, had a preventive effect at greater levels. The Western diet in comparison was associated with AMI. As it is the INTERHEART study which has highlighted a link between AMI and MetS similar to the levels in T2D the link between a Western diet and MetS prevalence is hard to ignore (Samson and Garber, 2014; Patti et al., 2018; Salas-Salvadó et al., 2014).

11.5 GOALS OF CLINICAL MANAGEMENT

Clinical treatment for people with mets is to lower their chances of developing clinical atherosclerosis. Even in people who have the metabolic syndrome, first-line treatment focuses on the primary risk factors: elevated LDL-C levels, hypertension, and diabetes. Another important goal is to avoid the development of T2D Mellitus in those with the metabolic syndrome. To lower the risk of ASCVD, people with diabetes who have been diagnosed must have their risk factors addressed more aggressively. The main goal of metabolic syndrome treatment is to create lifestyle changes that reduce the modifiable risk factors (obesity, physical inactivity, and a high-fat diet). All metabolic risk variables will be reduced if you adjust your lifestyle. Then, if the absolute risk is severe enough, pharmacological therapy may be considered as part of the treatment plan. LDL-C, blood pressure, and glucose elevations should be treated initially, and current treatment standards should be followed. Additionally, efforts should be made to assist cigarette smokers in quitting. These suggestions are based in part on NHLBI, AHA, and ADA guidelines for the management of certain risk factors. It's vital to keep in mind that people with ASCVD and/or diabetes might still have metabolic syndrome. The evidence-based basis for most of the recommendations has been provided in background materials for obesity, physical inactivity, lipids, high blood pressure, and diabetes. The goal of this statement is to show how to manage a multifaceted risk factor issue using an integrated approach (Grundy, 2016; Lackland and Voeks, 2014).

11.6 RISK ASSESSMENT

Many middle-aged patients with the metabolic syndrome are at higher absolute risk for ASCVD in the near future, according to a series of research (Han & Lean, et al., 2015) (e.g., 10-year risk). Furthermore, as previously stated, the long-term (lifetime) risk of ASCVD is increased even when the 10-year risk is not regarded significant, as in young adults who develop the syndrome, due to the high relative risk of ASCVD. A higher likelihood of developing T2D Mellitus early in life is an exacerbating factor that raises the lifetime risk of ASCVD. All persons diagnosed with metabolic syndrome deserve long-term therapy and follow-up in the therapeutic context in order to lower their lifetime risk of ASCVD. The primary goal is to alleviate the risk factors that are at the root of the problem. Individuals in this situation must be classified based on their absolute 10-year risk. Individuals with diabetes or any clinical indication of ASCVD are considered high-risk. Framingham risk scoring should be used to assess the 10-year risk of CAD in patients with metabolic syndrome who don't have ASCVD or diabetes (CHD) (Amiot, et al., 2016). In this assessment, individuals are divided into three risk categories depending on their 10-year CHD risk: high risk (10% to 20%), moderately high risk (10% to 20%), and minor to moderate risk (10% to 10%). As a result, recognizing metabolic syndrome is only one aspect of a comprehensive cardiovascular disease risk assessment. The metabolic syndrome by itself isn't a good way to predict CHD risk over the next 10 years. Although patients with the metabolic syndrome have a greater lifetime risk, they may not always have a high 10-year risk if they do not have diabetes. When determining 10-year risk, age, sex, smoking, and total cholesterol are all crucial factors to consider. Furthermore, unlike the risk variables included in Framingham scoring, metabolic syndrome risk factors are not evaluated for severity. According to the researchers, adding abdominal obesity, triglycerides, or fasting glucose to the Framingham 10-year risk algorithm provides little or no prognostic value for CHD (Cicero and Colletti, 2016).

11.7 MANAGEMENT OF UNDERLYING RISK FACTORS

Despite the fact that many people are genetically predisposed to the metabolic syndrome, it rarely manifests clinically in the absence of obesity and physical inactivity. As a result, medicines that target these underlying risk factors are the first line of defense. If you smoke cigarettes, which is another risk factor for ASCVD, you should make a concerted attempt to quit. The goal of modifying underlying risk factors is to avoid or delay the emergence of ASCVD, as well as to prevent T2D mellitus if it is not already present (Rochlani et al., 2017; Kaur, 2014).

11.7.1 ABDOMINAL OBESITY

Weight loss should be a top priority for people with abdominal obesity and metabolic syndrome. Reduced caloric consumption, increased physical activity, and the application of behavioral change concepts are the best ways to lose weight and keep it off. The first goal of weight loss is to reduce total BW by around 7% to 10% from baseline in 6 to 12 months. This will demand a 500–1,000 calorie daily caloric reduction. Increased physical activity aids in the reduction of caloric deficit. The severity of most, if not all, metabolic risk factors will be reduced if you lose the recommended amount of weight. Maintaining a lower weight is equally crucial, and this necessitates long-term monitoring and follow-up. Weight-loss medicines currently on the market have limited efficacy in the management of obesity. They may, however, be beneficial to certain patients. Bariatric surgery is becoming more common in the United States as a treatment for severe obesity. Obese people who are at a high risk of problems may benefit. However, weight-loss surgery is not without risk (Popa et al., 2016).

11.7.2 PHYSICAL INACTIVITY

Increased physical activity aids weight reduction, has favorable effects on metabolic chance factors, and, maximum significantly, lowers general ASCVD chance. The cutting-edge advice for the overall populace is half-hour of moderate-depth interest, consisting of brisk walking, on maximum, if now no longer all, days of the week; however, greater workout is beneficial. Going above and past cutting-edge requirements will therefore be particularly beneficial to men and women affected by the metabolic syndrome. Around 60 mins or greater of non-stop or intermittent cardio activity, ideally finished each day, can assist with weight reduction or maintenance.

It is recommended that you do 60 minutes of medium-intensity brisk walking, supplemented with other exercises. Multiple short (10–15-minute) stretches of action (strolling breaks at work, gardening, or housework), easy exercising equipment (e.g., treadmills), jogging, swimming, biking, golfing, group sports, and resistance training, in addition to averting not unusual place sedentary sports in amusement time, are all recommended (TV watching and pc games). Self-monitoring of physical activity can aid in sticking to an exercise program. The AHA suggests a clinical assessment of the risk of future ASCVD events before starting a new exercise program.

This provides an entire history of physical activity. For high-risk patients, physical exercise should be done under medical supervision (e.g., those with recent acute coronary syndromes or revascularization). In selected adults with cardiovascular disease, as well as those patients with symptoms or who are at high risk, the AHA recommends activity testing before vigorous exercise. Although this is a contentious topic, it is not necessary for all people starting a moderate-intensity, moderately progressive exercise programmed to undertake an exercise stress test (Han and Lean, 2011, 2015).

11.7.3 DIETS FOR ATHEROGENIC AND DIABETOGENIC PERSONS

A food plan low in saturated fats, trans fats, cholesterol, salt, and easy sugars is usually recommended similarly to weight reduction and calorie reduction. In addition, notwithstanding issues approximately mercury concentrations in unique species, better intakes of fruits, vegetables, and complete grains, in addition to fish eating, have to be encouraged. High carbohydrate consumption can irritate the metabolic syndrome's dyslipidemia. According to ATP-III (Amiot, et al., 2016), folks who are beginning a cholesterol-decreasing food plan have to devour 25 to 35% in their energy from overall fats. When the fats percent techniques 35%, keeping the minimum saturated fats consumption required to preserve a low LDL-C is difficult. Triglycerides can elevate and HDL-C levels can fall if the fats percent of the food plan is going beneath 25%; consequently, very-low-fats diets may also get worse atherogenic dyslipidemia (Klöting and Blüher, 2014). Some researchers favor fats intakes of 30% to 35% to keep away from growing atherogenic dyslipidemia in metabolic syndrome patients; others, though, are irritating approximately weight advantage from lengthy-time period intake of extra fats consumptions and like intakes of 25% to 30%. The challenge of whether or not enhancing the macronutrient content material of one's food regimen will assist one shed pounds has lengthy piqued human beings' hobby. For many years, a low-fat food regimen become endorsed because of the excessive caloric density of fats, which become a concept to boom the hazard of obesity. Recently, there was a surge of hobby withinside the capability that excessive-protein, low-carbohydrate diets can assist human beings shed pounds. The good judgment appears to be that fats and protein offer satiety that carbs do not. It's arguable if this effect of fats and protein on satiety makes the food regimen greater powerful for weight reduction. Furthermore, there may be a dearth of proof that excessive-fats/excessive-protein/

low-calorie diets may also hold a discounted frame weight over time. Those who observe a low-carbohydrate food regimen for 12 months lose no extra weight than folks that observe a regular weight loss strategy. Rich-fats diets aren't simplest excessive in saturated fats, however they're additionally low in fruits, vegetables, and entire grains, all of which might be important additives of modern-day diets. Excess protein will increase phosphorus load, which could purpose acidosis and worsen insulin resistance in people with persistent renal infection who've an appreciably decreased glomerular filtration rate; extra protein will increase phosphorus load, which could set off acidosis and get worse insulin resistance. Finally, specializing in macro-nutrient composition to inspire weight reduction misses the vital factors that affect frame weight (Amiot et al., 2016).

11.8 MANAGEMENT OF METABOLIC RISK FACTORS

In addition to lifestyle involvements expected at underlying risk factors, metabolic risk factors must be considered. If ASCVD or diabetes are present, or if the 10-year risk as calculated by Framingham risk factors is relatively high, risk factor medication therapy as described by current guidelines may be required (Grundy, 2016).

11.8.1 INCREASED BLOOD PRESSURE

The target blood pressure for antihypertensive therapy is 140/90 mm Hg when overt hypertension is present but there is no diabetes or chronic renal disease. In the case of diabetes or chronic renal illness, the blood pressure goal is 130/80 mm Hg. Patients through the metabolic syndrome should focus on lifestyle adjustments in addition to these clear treatment goals; the goals are to lower blood pressure as much as possible even in the deficiency of overt hypertension and to reap extra metabolic advantages from lifestyle changes. Weight loss, augmented physical movement, alcohol control, sodium decrease, and augmented eating of garden-fresh fruits and vegetables, as well as low-fat dairy products, as advocated by the DASH diet, can typically effectively moderate mild blood pressure rise (Rane et al., 2020; Rane and Keservani, 2020). Antihypertensive medicines are frequently required to prevent long-term harmful consequences such as myocardial infarction, stroke, and chronic kidney disease (CKD) if hyper-tension cannot be successfully controlled by lifestyle treatments. People with

T2D mellitus whose blood pressure is above target, as well as hypertensive patients with the metabolic disorder, may benefit from treatment. According to some experts, ACE inhibitors should be used as the first-line treatment for hypertension in individuals with the metabolic disorder, particularly if they have T2D or CKD. Inhibitors of the renin-angiotensin pathway, such as angiotensin-converting enzyme inhibitors (ACEIs) and angiotensin receptor blockers (ARBs), may decrease the threat of diabetes. ARBs may be used as a substitute to ACE inhibitors in persons with left ventricular failure who are unable to take ACE medications. The latter technique is still up for debate. Despite the fact that diuretics decrease the risk of cardiovascular events, the findings of an enormous clinical trial revealed that diuretics may raise the risk of T2D Mellitus progression in those with IFG or IGT. The potential advantage of low-dose diuretics in combined antihypertensive therapy, according to most hypertension researchers, outweighs their risk (Grundy, 2016, 2020).

11.8.2 ELEVATED FASTING GLUCOSE

Metabolic syndrome, which comprises both IFG and T2D Mellitus, is diagnosed by elevated fasting glucose (100 mg/dL). In metabolic syndrome individuals with IFG, weight loss, augmented physical activity, or mutually will postponement (or inhibit) the beginning of T2D Mellitus (or IGT if tested). In individuals with IFG or IGT, metformin, thiazolidinediones, and acarbose have all been proven to lower the risk of T2D. With the exception of an early trial with acarbose, there is no clinical trial evidence that oral hypoglycemic medications decrease the risk of cardiovascular events. Furthermore, neither metformin nor thiazolidinediones are suggested only for the inhibition of diabetes in this statement due to a lack of evidence of their cost-effectiveness and long-term safety. Clinical investigations have shown that treating dyslipidemia and hypertension reduces cardiovascular risk in persons with T2D Mellitus (Grundy, 2020; Kassi et al., 2011).

11.8.3 PROTHROMBOTIC STATE

In persons with the metabolic syndrome, elevated fibrinogen, plasminogen activator inhibitor-1, and other clotting factors are frequent. These aberrations, on the other hand, are not frequently discovered in clinical practice. The sole long-term option for primary prevention of their contribution to arterial thrombosis is low-dose aspirin or other antiplatelet medicines. If these drugs, particularly aspirin, are not contraindicated in patients with established ASCVD, they are recommended. Their effectiveness in people with T2D who don't have ASCVD hasn't been proven definitively in clinical trials, yet they're commonly advised in these people. Aspirin prophylaxis is an appealing treatment option for individuals with metabolic syndrome who are at moderately high risk for ASCVD events (Grundy, 2020; Kassi et al., 2011).

11.8.4 PROINFLAMMATORY STATE

Increased cytokines (such as tumor necrosis factor-_ and interleukin-6) and acute-phase reactants are prevalent in persons with the metabolic syndrome, indicating a pro-inflammatory condition (e.g., CRP, fibrinogen). CRP measurement is the easiest method for detecting a proinflammatory state in clinical practice. CRP values of 3 mg/L can be used to define such a situation in someone who has no other known reasons. If CRP is measured, an increased level indicates that lifestyle adjustments are required. CRP levels will be reduced as a result of the latter, notably weight loss, and the underlying inflammatory stimulation will probably be mitigated. There are no medications that work solely through this pathway to reduce cardiovascular risk. On the other hand, certain medications used to address other metabolic risk factors have been demonstrated to lower CRP levels (e.g., statins, nicotinic acid, fibrates, ACE inhibitors, and thiazolidinediones). These drugs, regardless of their other indications, cannot be recommended only for the goal of lowering a proinflammatory state at this time (Onat, 2011; Grundy, 2020).

11.9 FUTURE RESEARCH

This proclamation acknowledges that there are certain components of the metabolic syndrome that need to be researched further. Better methods for achieving and sustaining long-term weight loss and greater physical activity are the most important needs. Furthermore, creating new therapeutic options is difficult due to a lack of acceptance of the genetic and metabolic pathways that contribute to the syndrome's etiology. To better understand pathophysiology from the viewpoints of genetics, molecular biology, and cellular signaling, more fundamental and clinical research is required. The development of tools for determining short-term risk for SCVD and diabetes in individuals with the metabolic syndrome is urgently needed. Despite the fact that statins and other LDL-lowering medications diminish the risk of ASCVD, appropriate therapy for the remaining dyslipidemias is either unavailable or has nevertheless to be proved to reduce risk when combined with LDL-lowering pharmaceuticals. Insulin resistance is an appealing target for ASCVD prevention; however, clinical trials have yet to be conducted to prove ASCVD risk reduction by lowering insulin resistance per se. The apparent link between a proinflammatory states; the improvement of mutually ASCVD and diabetes warrants more research. Finally, a more thorough analysis of the cost efficacy of various medications, both alone and in combination therapy, is required (Simmons et al., 2010; Grundy, 2020).

KEYWORDS

- **C-reactive protein**
- **hormone-sensitive lipase**
- **metabolic syndrome**
- **non-esterified fatty acid**
- **nutrition**
- **World Health Organization**

REFERENCES

Amiot, M. J., Riva, C., & Vinet, A., (2016). Effects of dietary polyphenols on metabolic syndrome features in humans: A systematic review. *Obesity Reviews, 17*, 573–586.

Apridonidze, T., Essah, P. A., Iuorno, M. J., & Nestler, J. E., (2005). Prevalence and characteristics of the metabolic syndrome in women with polycystic ovary syndrome. *The Journal of Clinical Endocrinology & Metabolism, 90*, 1929–1935.

Bremer, A. A., & Jialal, I., (2013). Adipose tissue dysfunction in nascent metabolic syndrome. *Journal Of Obesity, 2013*, 393192.

Church, T., (2011). Exercise in obesity, metabolic syndrome, and diabetes. *Progress In Cardiovascular Diseases, 53*, 412–418.

Cicero, A. F. G., & Colletti, A., (2016). Role of phytochemicals in the management of metabolic syndrome. *Phytomedicine, 23*, 1134–1144.

Deng, Y., & Scherer, P. E., (2010). Adipokines as novel biomarkers and regulators of the metabolic syndrome. *Annals Of The New York Academy of Sciences, 1212*, E1–E19.

Doyle, S. L., Donohoe, C. L., Lysaght, J., & Reynolds, J. V., (2011). Visceral obesity, metabolic syndrome, insulin resistance and cancer. *Proceedings of the Nutrition Society, 71*, 181–189.

Eckel, R. H., Alberti, K., Grundy, S. M., & Zimmet, P. Z., (2010). The metabolic syndrome. *The Lancet, 375*, 181–183.

Edwardson, C. L., Gorely, T., Davies, M. J., Gray, L. J., Khunti, K., Wilmot, E. G., Yates, T., & Biddle, S. J. H., (2012). Association of sedentary behavior with metabolic syndrome: A meta-analysis. *Plos One, 7*, E34916.

Esser, N., Legrand-Poels, S., Piette, J., Scheen, A. J., & Paquot, N., (2014). Inflammation as a link between obesity, metabolic syndrome and type 2 diabetes. *Diabetes Research and Clinical Practice, 105*, 141–150.

Grundy, S. M., (2016). Metabolic syndrome update. *Trends in Cardiovascular Medicine, 26*, 364–373.

Grundy, S. M., (2020). Metabolic syndrome. In: Bonora, E., & Defronzo, R. A., (eds.), *Diabetes Complications, Comorbidities and Related Disorders.* Cham: Springer International Publishing.

Han, T. S., & Lean, M. E. J., (2011). Metabolic syndrome. *Medicine, 39*, 24–31.

Han, T. S., & Lean, M. E. J., (2015). Metabolic syndrome. *Medicine, 43*, 80–87.

He, S., Zheng, Y., Shu, Y., He, J., Wang, Y., & Chen, X., (2013). Hypertriglyceridemic waist might be an alternative to metabolic syndrome for predicting future diabetes mellitus. *Plos One, 8*, E73292.

Jarouliya, U., & Keservani, R. K., (2019). Pathways leading to child obesity. In: Bagchi, D., (ed.), *Global Perspectives on Childhood Obesity: Current Status, Consequences and Prevention* (2nd edn., pp. 137–146). Academic Press, Elsevier, Chapter 12. ISBN: 9780128128404.

Kassi, E., Pervanidou, P., Kaltsas, G., & Chrousos, G., (2011). Metabolic syndrome: Definitions and controversies. *BMC Medicine, 9*, 48.

Kaur, J., (2014). A comprehensive review on metabolic syndrome. *Cardiology Research and Practice, 2014*, 943162.

Keller, K. B., & Lemberg, L., (2003). Obesity and the metabolic syndrome. *American Journal of Critical Care, 12*, 167–170.

Keservani, R. K., Kesharwani, R. K., Sharma, A. K., Vyas, N., & Chadoker, A., (2010a). Nutritional supplements: An overview. *Int. J. of Curr. Pharm Rev. and Res., 1*(1), 59–75.

Keservani, R. K., Kesharwani, R. K., Vyas, N., Jain, S., Raghuvanshi, R., & Sharma, A. K., (2010a). Nutraceuticals and functional food as future food. *Der. Pharm. Lett, 2*(1), P106–116.

Keservani, R. K., Sharma, A. K., & Kesharwani, R. K., (2016). An overview and therapeutic applications of nutraceutical & functional foods. In: Keservani, R. K., Sharma, A. K., & Kesharwani, R. K., (eds.), *Recent Advances in Drug Delivery Technology* (pp. 160–201). IGI Global Publisher, USA, Ch 6. ISBN:9781522507543.

Klöting, N., & Blüher, M., (2014). Adipocyte dysfunction, inflammation and metabolic syndrome. *Reviews in Endocrine and Metabolic Disorders, 15*, 277–287.

Lackland, D. T., & Voeks, J. H., (2014). Metabolic syndrome and hypertension: Regular exercise as part of lifestyle management. *Current Hypertension Reports, 16*, 492.

Onat, A., (2011). Metabolic syndrome: Nature, therapeutic solutions and options. *Expert Opinion on Pharmacotherapy, 12*, 1887–1900.

Panza, F., Frisardi, V., Capurso, C., Imbimbo, B. P., Vendemiale, G., Santamato, A., D'onofrio, G., Seripa, D., et al., (2010). Metabolic syndrome and cognitive impairment: Current epidemiology and possible underlying mechanisms. *Journal of Alzheimer's Disease, 21*, 691–724.

Patti, A. M., Al-Rasadi, K., Giglio, R. V., Nikolic, D., Mannina, C., Castellino, G., Chianetta, R., et al., (2018). Natural approaches in metabolic syndrome management. *Archives of Medical Science: Ams, 14*, 422–441.

Popa, S., Moţa, M., Popa, A., Moţa, E., Serafinceanu, C., Guja, C., Catrinoiu, D., et al., (2016). Prevalence of overweight/obesity, abdominal obesity and metabolic syndrome and atypical cardiometabolic phenotypes in the adult Romanian population: Predator study. *Journal of Endocrinological Investigation, 39*, 1045–1053.

Prasad, H., Ryan, D. A., Celzo, M. F., & Stapleton, D., (2012). Metabolic syndrome: Definition and therapeutic implications. *Postgraduate Medicine, 124*, 21–30.

Rane, B. R., & Keservani, R. K., (2020). Introduction to nutraceutical & dietary supplements. In: Keservani, R. K., Sharma, A. K., & Kesharwani, R. K., (eds.), *Nutraceutical and Dietary Supplements* (pp. 35–55). Apple Academic Press, Taylor & Francis Group, Chapter 3. Isbn: 9781771888738.

Rane, B. R., Tadavi, S. A., & Keservani, R. K., (2020). Naturopathy. In: Sharma, A. K., Keservani, R. K., & Gautam, S. P., (eds.), *Herbal Product Development* (pp. 321–347). Apple Academic Press, Taylor & Francis Group, Chapter 12. ISBN: 9781771888776.

Roberts, C. K., Hevener, A. L., & Barnard, R. J., (2013). Metabolic syndrome and insulin resistance: Underlying causes and modification by exercise training. *Comprehensive Physiology, 3*, 1–58.

Rochlani, Y., Pothineni, N. V., Kovelamudi, S., & Mehta, J. L., (2017). Metabolic syndrome: Pathophysiology, management, and modulation by natural compounds. *Therapeutic Advances in Cardiovascular Disease, 11*, 215–225.

Romeo, G. R., Lee, J., & Shoelson, S. E., (2012). Metabolic syndrome, insulin resistance, and roles of inflammation – mechanisms and therapeutic targets. *Arteriosclerosis, Thrombosis, And Vascular Biology, 32*, 1771–1776.

Salas-Salvadó, J., Guasch-Ferré, M., Bulló, M., & Sabaté, J., (2014). Nuts in the prevention and treatment of metabolic syndrome. *The American Journal of Clinical Nutrition, 100*, 399s–407s.

Samson, S. L., & Garber, A. J., (2014). Metabolic syndrome. *Endocrinology And Metabolism Clinics of North America, 43*, 1–23.

Simmons, R. K., Alberti, K. G. M. M., Gale, E. A. M., Colagiuri, S., Tuomilehto, J., Qiao, Q., Ramachandran, A., et al., (2010). The metabolic syndrome: Useful concept or clinical tool? Report of a who expert consultation. *Diabetologia, 53*, 600–605.

Sookoian, S., & Pirola, C. J., (2011). Metabolic syndrome: From the genetics to the pathophysiology. *Current Hypertension Reports, 13*, 149–157.

CHAPTER 12

Beneficial Effect of Dietary Supplements on Metabolic-Induced Obesity

HEMANT P. SURYAWANSHI,[1] BHUSHAN R. RANE,[2] and RAJ K. KESERVANI[3]

[1]*PSGVPMS, College of Pharmacy, Shahada, Maharashtra – 425409, India*

[2]*Shri D. D. Vispute College of Pharmacy and Research Center, New Panvel, Maharashtra – 410206, India, Tel.: 9421534437, E-mail: rane7dec@gmail.com*

[3]*Faculty of B. Pharmacy, CSM Group of Institutions, Prayagraj, Uttar Pradesh – 212111, India*

ABSTRACT

Obesity is a multifaceted chronic disease having psychological, social, cultural, genetic, physiological, metabolic, and behavioral causes. Overweight and obesity are on the rise globally, in developing as well as developed nations. Obesity is fast escalating as a public health crisis, responsible for many complications. Obesity is primarily treated by calorie restriction and increased physical workouts. Dieting as part of a lifestyle change resulted in long-term weight loss, despite eventual weight rebound. When compared to continuous energy restriction, intermittent fasting provides no additional weight-loss benefits. There are a number of hypocaloric diets that work. Low-carbohydrate diets appear to be more helpful for weight loss in the short term than low-fat diets. However, all low-carbohydrate and low-fat diets are generally healthful. Sugary beverage consumption has also been linked to weight loss. Long-term weight loss maintenance with lifestyle changes has

The Metabolic Syndrome: Dietary Supplements and Food Ingredients. Raj K. Keservani, MPharm, Durgavati Yadav, PhD, Rajesh K. Kesharwani, PhD, Sippy Singh, DPhil & Kumar Sandeep, PhD (Eds.)

a low success rate, ranging from 2 to 20%. Evidence-based evaluations of effective weight-loss techniques are increasingly highlighting the significance of personal care, as well as realistic, health-outcome-focused goals and long-term lifestyle modifications such as increased physical activity and dietary alterations. There is no such thing as a "one-size-fits-all" diet that will suit everyone. Different dietary methods for increasing adherence are successful to varying degrees in different people. Many obese persons have tried and failed to lose weight on their own several times.

12.1 INTRODUCTION

Obesity is a complicated, chronic disease with complex psychological factors, social, cultural, genetic, physiological, metabolic, and behavioral reasons. Overweight and obesity are becoming increasingly common around the world, in both developing and developed countries (WHO, 2015; *Yazdi et al.,* 2015).

Excess body fat has accumulated to such a level in obesity that it can have a negative influence on health (WHO, 2015). When a person's body mass index (BMI), which is calculated by dividing a person's weight by the square of their height, exceeds 30 kg/m^2, they are considered obese. A range of 25 to 30 kg/m^2 is considered overweight (WHO, 2015). Some countries in Eastern Asia use lower values (*Kanazawa, 2005)*. Obesity has been linked to a number of diseases and ailments, most notably CVS disease, Type II diabetes mellitus, obstructive sleep apnea, some kinds of cancer, osteoarthritis, and asthma (*Haslam* et al., *2005*). As a result, we have seen that obesity reduces life expectancy.

While a high BMI is a risk factor, it has not been proven to be a direct cause of diseases caused by nutrition, exercise, or environmental variables (*Chiolero* et al., *2018*). Obesity raises the risk of depression, and depression is one of the leading causes of obesity (*Luppino* et al., *2010*).

Urbanization, genetic sensitivity, drugs, mental health issues, economic policy, exposure to endocrine disrupters, diet, exercise, automation, and the environment are all factors that contribute to obesity on an individual level (WHO, 2015; *Yazdi* et al., 2015; *Kassotis* et al., *2020; Bleich* et al., *2008*).

Preventing obesity calls for a complex approach, including community, family, and individual interventions (WHO, 2015; *Chiolero* et al., *2018*). The main treatments recommended by healthcare providers are diet and exercise (*Haslam* et al., *2005*). Reduced energy-dense foods, such as those heavy in fat or sugar, and increased dietary fiber consumption can improve diet quality (WHO, 2015). To suppress appetite or fat absorption, drugs are used in combination with a healthy diet (*Yanovski* et al., *2014*). A gastric balloon or surgery may be done if diet, exercise, and medication do not work to reduce the stomach's size or the length of the intestines, causing you to feel full sooner or limiting your ability to absorb nutrients from meals (*Colquitt* et al., *2014; Imaz* et al., *2008*).

Obesity is a prominent cause of death in both adults and children globally, with rising rates in both (WHO, 2015; Encyclopedia, 2015). In 2015, 600 million adults (12%) and 100 million children were obese in 195 countries (*Afshin* et al., *2017*). Obesity affects women more than males (WHO, 2015). Obesity is one of the 21st century's most important public health challenges (*Dibaise* et al., *2013*). In 2013, a number of medical organizations, including the American Medical Association and the American Heart Association (AHA), declared obesity to be a disease (*Jensen* et al., *2014; Pollack* et al., *2013; Weinstock* et al., *2013*).

Obesity is a condition in which extra body fat has accumulated to the point where it may be harmful to one's health (WHO, 2000). The waist-hip ratio and total cardiovascular risk factors are utilized to measure fat distribution, while the BMI is used to characterize it. BMI has a direct relationship with both percentage and total body fat (*Gray* et al., *1991*).

BMI is calculated by dividing a person's weight by their height squared, as follows:

$$\mathrm{BMI} = \frac{\mathrm{W}}{\mathrm{h}^2}$$

where; *W* is the weight; and *h* is the height; BMI is commonly measured in kilograms per square meter of height.

i. **Severe Obesity:** BMI $\geq$ 35 or 40 kg/m^2.
ii. **Morbid Obesity:** BMI of $\geq$ 35 kg/m^2 and or $\geq$ 40–44.9 kg/m^2.
iii. **Super Obesity:** BMI of $\geq$ 45 or 50 kg/m^2.

Because Asian populations have lower BMIs than Caucasians, several countries have redefined obesity. For example, Japan defines obesity as any BMI larger than 25 kg/m^2 (Kanazawa, 2005). In China, a BMI of more exceeding 28 kg/m^2 is used (*Bei-Fan* et al., *2002*).

12.2 MORTALITY

Obesity is one of the leading causes of death worldwide (Barness et al., 2007; Mokdad et al., 2004; Allison et al., 1999). According to several studies, mortality risk is lowest in nonsmokers with a BMI of 20–25 kg/m^2 (Aune et al., 2016; Whitlock et al., 2009; Di Angelantonio et al., 2016) and highest in current smokers with a BMI of 24–27 kg/m^2 (Aune et al., 2016; Whitlock et al., 2009; Di Angelantonio et al., 2016; Calle et al., 1999; Pischon et al., 2008). According to a 2013 review, grade 1 obesity, BMI 30–34.9 was not connected to an increased risk of death when compared to people of normal weight, and overweight, BMI 25–29.9 was linked to "lower" mortality than normal weight, BMI 18.5–24.9 (Flegal et al., 2013). The relationship between waist-to-hip and waist-to-height ratios and mortality is stronger (*Carmienke* et al., *2013*). Between 22–25 kg/m^2, the likelihood of harmful health impacts increases in Asians (WHO, 2004). Over a 16-year period, a BMI of more than 32 kg/m^2 was linked to a twofold mortality rate among women (Manson, 1995). Obesity is expected to kill 1,11,909 to 3,65,000 people in the United States per year (Haslam et al., 2005; Allison, 1999). While obesity is to blame for 1 million (7.7%) deaths in Europe (Tsigos et al., 2008; Fried et al., 2007). Obesity reduces life expectancy by an average of six to seven years (Haslam et al., 2005), a BMI of 30–35 kg/m^2 by two to four years, and extreme obesity (BMI 40 kg/m^2) by 10 years (Haslam et al., 2005; Whitlock et al., 2009).

12.3 MORBIDITY

Obesity puts people at risk for a wide range of physical and mental diseases. These comorbidities are most typically seen in metabolic syndrome (Haslam et al., 2005), a group of medical conditions that includes Type II diabetes, hypertension, hypercholesterolemia, with high triglyceride levels (Grundy et al., 2004).

There are two types of health consequences: those caused by a rise in the number of fat cells, such as osteoarthritis, obstructive sleep apnea and those brought on by a rise in the amount of fat cells in the body, such as such as diabetes, cancer, cardiovascular disease, and non-alcoholic fatty liver disease (Haslam et al., 2005; Bray et al., 2004). Increased body fat alters the body's insulin response, which can contribute to resistance of insulin. The rise in the level of fat triggers an inflammatory response (Shoelson et al., 2006, 2007).

12.4 COMPLICATIONS OF OBESITY

1. **Cardiovascular System:** Angina and myocardial infarction (Lu, 2014); congestive heart failure (Haslam et al., 2005; Aune et al., 2016); hypertension (Haslam et al., 2005); abnormal cholesterol levels (Haslam et al., 2005), deep vein thrombosis and pulmonary (Darvall et al., 2007).
2. **Skin:** Stretch marks, lymphedema, cellulitis (Yosipovitch et al., 2007), and intertrigo (Hahler et al., 2006).
3. **Endocrine System:** Diabetes mellitus (*Haslam* et al., *2005*).
4. **Reproductive System:** Polycystic ovarian syndrome, birth defects, menstrual disorders, infertility (*Haslam* et al., *2005*), complications during pregnancy (*Haslam* et al., *2005; Arendas* et al., *2008*), erectile dysfunction (*Corona* et al., *2014*).
5. **Nervous System:** Stroke, migraines (*Bigal* et al., *2008*), carpal tunnel syndrome (*Sharifi-Mollayousefi* et al., *2008*), loss of memory (*Beydoun* et al., *2008*), idiopathic intracranial hypertension (*Wall* et al., *2008*), sclerosis (*Munger* et al., *2009*).
6. **Gastroenterology:** Gastroesophageal reflux, fatty liver, cholelithiasis (*Dibaise* et al., *2013*).
7. **Cancer:** Colorectal, pancreatic, gallbladder, esophageal, kidney, Leukemia, malignant melanoma (*Basen* et al., *2011*), Hepatocellular carcinoma (*Dibaise* et al., *2013*).
8. **Respiratory System:** Hypoventilation syndrome, asthma (*Haslam* et al., *2005; Poulain* et al., *2006*), augmented complications during general anesthesia (*Haslam* et al., *2005*).
9. **Urinary System:** Urinary incontinence (*Hunskaar* et al., *2008*) renal failure (*Ejerblad* et al., *2006*).
10. **Others:** Gouty arthritis (*Aune* et al., *2014*), mobility (*Tukker* et al., *2009*), osteoarthritis (*Haslam* et al., *2005*), backache (*Molenaar* et al., *2008*).

12.5 CAUSES

The carbohydrate-insulin hypothesis of obesity argued that eating processed, high-glycemic-load carbohydrates causes hormonal changes in the body by promoting calorie accumulation in adipose tissue, intensify appetite, and reduce energy expenditure (Ludwig et al., 2018). According to the calorie is a calorie hypothesis, the majority of instances of obesity are caused by a combination of excessive food energy intake and a lack of physical activity (Lau et al., 2007). A small percentage of cases are caused by genetics, medical conditions, or psychiatric disease (Bleich et al., 2008). Excess food was identified as the key cause in a 2016 study (Bojanowska et al., 2016; Levitsky et al., 2019).

12.5.1 LIFESTYLE

There has been a significant movement toward less physically demanding jobs worldwide, and 30% of the world's population does not get adequate physical activity. This is owing to a rise in the use of mechanized transportation and a rise in the use of labor-saving technology in the house (Ness-Abramof et al., 2006). Physical activity levels in youngsters appear to be declining as a result of decreased walking and physical education (Salmon et al., 2007). In both children and adults, there is a relationship between television and mobile phone viewing time and the risk of obesity (Gortmaker et al., 1996; Vioque et al., 2000; Tucker et al., 1991).

12.5.2 GENETICS

Obesity, like many other medical diseases, is the consequence of a complex interaction of hereditary and environmental variables (Albuquerque et al., 2017). When there is enough food energy, polymorphisms in several genes influencing appetite and metabolism contributes to obesity. When a favorable situation presents itself, as of 2006, more than 41 of these genes in the human genome have been linked to the development of obesity (Poirier et al., 2006).

12.5.3 OTHER ILLNESSES

Obesity can be aggravated by a variety of physical and mental illnesses, with the pharmaceuticals used to treat them. Numerous uncommon genetic conditions, GH insufficiency are among the medical conditions that raise the risk of obesity (Rosén et al., 1993). Patients with psychiatric disorders have a higher risk of being overweight or obese than people who do not have psychiatric problems (Chiles et al., 2010).

12.5.4 SOCIAL FACTORS

There have been numerous theories presented to explain the relationship between BMI and socioeconomic level. The wealthy in industrialized countries are said to be able to buy more nutritious food, are subjected to more societal pressure to maintain their weight, and have more opportunities with higher physical fitness needs. The ability to purchase food, high energy expenditure from physical labor, and cultural values that promote a larger body size are all assumed to play a role in the observed trends in emerging nations (McLaren et al., 2007). Obesity appears to be linked to stress and a sense of low social status (Wilkinson et al., 2009; Bjorntorp et al., 2001; Goodman et al., 2003). Smoking has a substantial impact on a person's weight. Over the course of 10 years, those who stop smoking acquire an average of 4.4 kg for males and 5.0 kilograms for women (Flegal et al., 1995). In the developing world, urbanization is contributing to the rise in obesity rates. Early-life malnutrition is thought to play a part in the developing world's rising obesity rates (Caballero et al., 2001).

12.5.5 GUT BACTERIA

The research into the effects of infectious pathogens on metabolism is still in its early stages. The gut flora of lean and obese adults has been demonstrated to differ. There is evidence that gut flora has an impact on metabolic capacity. This apparent change is thought to confer a larger capability for energy harvesting, which contributes to obesity. Obesity has yet to be definitively determined as the direct cause or outcome of these variations (DiBaise et al., 2008).

12.5.6 USE OF ANTIBIOTICS

Antibiotic use in youngsters has been related to a higher risk of obesity later in life (Cox et al., 2015).

12.5.7 VIRUSES

A relationship between viruses and obesity has been discovered in humans and a range of animal species. It's unclear how much these connections have contributed to the obesity epidemic's rise (Falagas et al., 2006).

12.6 PREVENTION AND MANAGEMENT OF OBESITY WITH DIETARY

Obesity is primarily treated by losing weight through calorie restriction and physical activity (Jensen et al., 2014; Lau et al., 2007; US, 2017; Arnett et al., 2019). Dieting as part of a lifestyle change results in long-term weight loss, despite gradual weight recovery (Jensen et al., 2014). Intensive behavioral therapies that include both dietary and activity improvements are advised (Jensen, 2014; US, 2017; US Preventive, 2018). When compared to constant energy restriction, intermittent fasting provides no additional weight-loss benefits (Yannakoulia et al., 2019).

There are several hypocaloric diets that are effective (Jensen et al., 2014). In the near run, low carbohydrate diets appear to be more effective than low fat diets for weight loss (Services et al., 1997). However, in the long run, all low-carbohydrate and low-fat diets tend to be healthy (Services et al., 1997; Johnston et al., 2014). Reduced use of sugary beverages has also been connected to weight loss (Services et al., 1997). Long-term weight loss (Keservani et al., 2010a, b; Keservani and Sharma, 2014) maintenance with lifestyle changes has a low success rate, ranging from 2–20% (Wing et al., 2005).

12.6.1 DIETARY FACTORS RESPONSIBLE FOR WEIGHT LOSS

Evidence-based assessments of effective weight-loss strategies are increasingly emphasizing the necessity of tailored, multidisciplinary care, as well as realistic, health-outcome-focused goals and lasting lifestyle changes, such as increased physical activity and nutritional adjustments (Nihoei et al., 2000: Brownwell et al., 2005).

There is no one-size-fits-all diet that will work for everyone. Different dietary techniques for maximizing adherence are successful in different people to varying degrees. Many obese people have attempted to lose weight on their own numerous times.

12.6.1.1 DIETARY INTERVENTIONS

i. The amount of energy in various foods;
ii. Food intake (calories, lipids, carbohydrates, fiber, protein);
iii. Meal preparation and planning;

iv. Changes to the recipe;
v. Limit your consumption of high-energy but low-nutritional-value meals;
vi. Maintaining hydration and reducing alcohol intake;
vii. Examine the negative effects of obesity on one's health.

12.6.1.2 PHYSICAL ACTIVITY

Physical activity should be increased.

12.6.1.3 BEHAVIOR CHANGE

i. Examining food labels;
ii. Increasing the number of meals prepared at home;
iii. Mealtime measures to avoid overeating;
iv. Restaurant eating tactics;
v. Examining food labels;
vi. Emotional and eating awareness.

12.6.2 DIFFERENT WAYS TO REDUCE THE AMOUNT OF CALORIES CONSUMED IN ORDER TO CONTROL OBESITY

Obesity is caused by adipose tissue accumulating an excessive quantity of body fat. Losing one pound of fat necessitates an energy deficit of around 3,500 calories. To lose one to two pounds of body weight (BW) each week, a daily calorie reduction of 500 to 1,000 calories is recommended (Seagle et al., 2009).

12.6.2.1 CALORIC INTAKE SHOULD BE ADJUSTED

Calculated-deficit and fixed-calorie diets are the two primary techniques to setting a caloric amount for a weight-loss diet. The patient's current daily calorie needs are either measured or estimated on calculated deficit diets, and the shortfall is calculated by subtracting those needs from the total. Fixed-calorie diets feature a pre-determined total daily caloric intake prescription based on caloric levels that result in weight loss (Seagle et al., 2009). Fixed-calorie diets include low-calorie diets (LCD) and very-low-calorie diets (VLCD).

12.6.2.2 LOW-CALORIE DIETS (LCDS)

Diets that limit women's caloric intake to 1,200 to 1,500 calories and men's caloric intake to 1,500 to 1,800 calories, independent of macronutrient composition, force in losing weight if strictly followed (Freedman et al., 2001; Jensen et al., 2014). This is because most overweight and obese people will experience a caloric deficit as a result of their calorie intake. The National Institutes of Health recommends a low-calorie diet of 1,000 to 1,200 calories per day for women and 1,200 to 1,600 calories per day for males (Clinical Guidelines, 1998; Nihoei, 2000).

12.6.2.3 VERY-LOW-CALORIE DIETS (VLCDS)

VLCDs provide 800 calories or less per day, a total of 50 to 80 gms of protein, and 100% of the Reference Daily Intake (RDI) for vitamins and minerals per day to help people lose weight quickly while maintaining lean body mass (American Dietetic Association, 2009). VLCDs are a type of rigorous diet therapy that requires constant medical monitoring and should not be taken over a long period of time. They are designed to cause rapid and dramatic weight loss of 3–5 pounds each week, or 14 to 21 kg in a short period of time (11–14 weeks).

12.6.3 INTERMITTENT FASTING AND ALTERNATE DAY FASTING

Because it causes excessive lean tissue breakdown and ketosis, total fasting is not suggested for weight loss. Furthermore, compensatory decreases in resting metabolism and physical activity on total fasts are large and counterproductive, as they reduce energy output (Clinical Guidelines, 1998; Cummings et al., 2002).

12.6.3.1 COMPOSITION OF THE REDUCING DIET

Reduced-calorie diets should have a dietary composition that reduces the risk of nutritional deficiency and diet-related chronic illnesses. As a result, the diet should be sufficient in all nutrients to avoid deficiency, while also adhering to dietary requirements intended for healthiness (Clinical Guidelines, 1998; U.S. Department, 2012). The consumption of vegetables, legumes, fruits, whole grains, lean protein sources, and water should all be encouraged, with an emphasis on balance and moderation (Haven et al., 2006). Extreme dietary restriction or abnormally high intakes of any macronutrient or food should be limited to a brief period of time (Jensen et al., 2014).

12.6.4 MACRONUTRIENT DISTRIBUTION

The macronutrient composition of the diet does not appear to have a major impact on total weight reduction; reduced-calorie diets result in clinically relevant weight loss regardless of which macronutrients they favor (Sacks et al., 2009; Johnston et al., 2014). Low-carbohydrate diets, on the other hand, are ketogenic in the short term and may result in a greater loss of body water than body fat. When a diet is stopped or carbohydrate intake is increased, water weight is recovered. In this scenario, glycogen reserves, which contain water, are re-established. Regardless of the macronutrient distribution, any lowering diet that is maintained over time and is hypocaloric will result in a reduction of body fat – regardless of the macronutrient distribution (Wadden et al., 2006).

For the most part, moderate fat, balanced macronutrient reduction diets are nutritionally appropriate. Very low-fat diets are deficient in vitamin E, B12, calcium, iron, and zinc. High-fat, low-carbohydrate diets are nutritionally deficient, and several nutrients must be supplemented to make them adequate (Adam et al., 2006; Foster et al., 2005; Schoeller et al., 2005).

12.6.4.1 PROTEIN

Dietary protein required during weight loss. Protein has a Recommended Dietary Allowance of 0.8 grams per kg per day, but most people consume about 1.2 grams per kg per day, or about 15% of their overall caloric consumption. The World Health Organization (WHO) recommends that dietary protein account for 10 to 15% of total energy consumption for adults who are in energy balance and have a stable weight (Singh et al., 2017).

12.6.4.2 FAT

Even on low-calorie diets, the human body requires tiny amounts of essential fatty acids (e.g., 3–6 grams) (linoleic or arachidonic acid). Fat is also required to transport the fat-soluble vitamins A, D, E, and K. As a result, a fat-free diet is not recommended. However, because fat is a high-calorie food, it is frequently reduced on calorie-restricted diets in order to reduce energy consumption while increasing bulk (Singh et al., 2017).

12.6.4.2.1 Moderate to Low-Fat Balanced Diets

The AHA weight-loss diet standards should be followed in terms of dietary fat, saturated fat, trans fat, polyunsaturated fat, monounsaturated fat, and cholesterol levels. In other cases, lower quantities may be suitable, but they more than meet the needs while also sustaining cardiovascular health (Clinical Guidelines, 1998).

Weight loss diets that are moderate to low in fat (20 to 30% of calories) are referred to as "balanced deficit" diets because they maintain a decent balance of macronutrients (U.S. Department, 2010, 2012).

12.6.4.2.2 Very Low-Fat Reducing Diets

Diets with very low-fat content, such as the Pritikin Diet (Pritikin et al., 1980), and the Ornish Diet (Ornish et al., 1993), have been promoted not only for weight loss but also for improved cardiovascular risk profiles. The Ornish Diet is part of a program that combines nonsmoking, exercise, and behavior change. It's low in fat and saturated fat (13% of calories), rich in carbohydrate (81% of calories), and high in fiber (38 grams).

12.6.4.2.3 Weight Reduction by Using High Fat Diet

Diets that are high in fat and low in carbohydrate (less than 200 grams per day) are common. Obese people who adopt high fat, low carbohydrate diets for themselves have been shown to consume fewer calories and lose weight (Freedman et al., 2001).

12.6.4.3 CARBOHYDRATES

To feed their central nervous system, red blood cells, and other glucose-dependent tissues, most people need at least 50 grams of carbohydrates each day. Gluconeogenesis is likely to occur if carbohydrate consumption falls below this level. Because fatty acids can't be used as a carbon source for gluconeogenesis, amino acids are required to keep blood glucose levels stable and to fuel glucose-dependent functions in tissues. For diets with more than 800 calories per day, at least 100 grams of carbohydrate should be provided, preferably within the acceptable macronutrient distribution ranges (AMDR) of 45 to 65% of total energy intake. Both hypocaloric diets high in sugars (mono and disaccharides) and equi-caloric diets high in starches (digestible polysaccharides) show similar weight loss effects under experimental conditions (Surwit et al., 1997; Saris et al., 2000).

12.6.4.3.1 Low Carbohydrate Diets

Ketogenic diets have a carbohydrate content of fewer than 100 grams per day, and especially less than 50 grams per day. Ketosis can be a problem on a variety of popular low-carbohydrate diets (Atkins et al., 2002).

12.6.4.3.2 Diets with Low Glycemic Index (GI)

GI was originally developed to aid in the treatment of diabetes, but it has subsequently gained popularity as a weight-loss tool. The GI is a measurement of the blood glucose response to a certain quantity of carbohydrate (typically 50 grams) in a known diet evaluated to the similar quantity of carbohydrate in a control item (usually white bread) (Wolever et al., 1991). In more detail, the GI is a grading system for carbohydrate-containing foods based on their postprandial blood glucose responses to standardized carbohydrate servings. Whole grains, legumes, fruits, vegetables, and low-GI whole foods will help you meet your fiber goals while also assisting in weight loss. Obese people may benefit from a well-balanced, hypocaloric, low-glycemic-index diet. In conclusion, GI-based diets may provide certain patients with short- and long-term weight loss benefits.

12.6.4.3.3 Paleo Diet

Despite the fact that no large-scale scientific study has looked into the Paleo Diet in depth, notably for the goals of weight loss, it has gained substantial appeal among consumers in recent years. This dietary approach implies that people eat foodstuffs that would have been found in Paleolithic humans' diets (Lindeberg et al., 2012). As a result, all processed meals, and beverages, as well as grains, dairy, and legumes, are prohibited. Animal protein, fish, poultry products, vegetables, fruit, and kernel are all permitted foods (not peanuts). This diet has the advantage of being high in nutrients from vegetables and fruits (Rizzoli et al., 2014; Turner et al., 2013; Venn et al., 2004).

12.6.5 WATER

To avoid dehydration, it's vital to drink plenty of water while on a weight-loss diet, especially if it's ketogenic, low in calories, or done in hot weather or with physical activity. Losses of body water are followed by losses of body glycogen and protein. It is important to stress the consumption of water (low calorie or calorie-free fluids) (Kleiner et al., 1999). Increased physical activity raises water requirements, not just because of sweat losses, but also because of increased water losses owing to breathing (Wildman et al., 2004).

12.6.6 ELECTROLYTES

Electrolyte balance is maintained under normal conditions on a well-balanced diet that is not too energy-restrictive. Electrolytes may typically be supplied with normal diets if an individual is losing too many electrolytes owing to excessive sweating or urine loss (Wildman, 2004). Electrolyte values are especially important on VLCD since hypokalemia has been linked to cardiac arrhythmias in the past (Tsai et al., 2005).

12.6.7 VITAMINS AND MINERALS

The importance of vitamin and mineral supplementation during weight loss and maintenance cannot be overstated. Even on reduced diets for all other nutrients, the levels of nutrients prescribed in the Recommended Dietary Allowance (RDA) for an individual's age and sex must be reached. Potassium, copper, magnesium, Vitamin E, Vitamin B6, folic acid, iron, and calcium are all essential vitamins, minerals, and electrolytes that are more likely to be lacking in a low-calorie diet. In diets of less than 1,200 calories per day, vitamin, and mineral supplements in levels that approximate the necessary dietary allowances are likely to be required. Due to their high demands for iron, calcium, and folic acid, women in reproductive age groups may require iron, calcium, and folic acid supplements beyond 1,200 calories per day, but most other nutritional requirements can be met by a well-balanced diet that follows the Dietary Guidelines for Americans (U.S. Department, 2010). As a result, foods that are rich in micronutrient density but low in caloric density are particularly crucial to include in a calorie-restricted diet. Fruits, vegetables, lentils, and barely processed whole grains are among them.

12.6.7.1 CALCIUM SUPPLEMENTATION

Some studies have suggested that calcium supplements and/or dairy product supplements play a major function in controlling obesity in the past. However, not all evidence supports this concept and some studies have revealed that calcium or dairy consumption does not aid in weight loss, nor does calcium supplementation prevent weight gain (Lanou et al., 2008; Yanovski et al., 2009; Holecki et al., 2008; Wagner et al., 2007).

12.7 DIETARY SUPPLEMENTS AND WEIGHT LOSS

Currently, dietary supplements are used by more than half of the adult population. Despite the fact that only about a quarter of the supplements eaten were recommended by a doctor or healthcare provider, the majority of consumers claimed they took them to improve their health (Table 12.1) (Bailey et al., 2013).

TABLE 12.1 General Constituent Use to Reduce Weight Obtain from Diets (IFICF, 2004)

Ingredient	Mechanism
Bitter orange	Mild appetite suppressant; increased energy expenditure and lipolysis
Caffeine	Increased thermogenesis and fat oxidation due to central nervous system stimulation
Calcium	Breakdown of lipids and increase accumulation of fat but its absorption is reduced.
Chitosan	In the digestive tract, dietary fat binds.
Chromium	Reduced food intake, hunger, and fat cravings; increased lean muscle mass; fat loss promotion
Coleus forskohlii	Lipolysis is increase, and appetite is decreased.

TABLE 12.1 *(Continued)*

Ingredient	Mechanism
Ephedra	Central nervous system stimulation; enhanced thermogenesis; decreased appetite.
Garcinia cambogia	Lipogenesis was inhibited, and food intake was reduced.
Glucomannan	Extended stomach emptying time; enhanced sensations of fullness and satiety.
Green coffee bean extract (*C. arabica, C. canephora, C. robusta*)	Fat accumulation was slowed, and glucose metabolism changes.
Green tea and its extract	Reduced lipogenesis and fat absorption.
Guar gum	Intestinal bulking agent; delayed gastric emptying; heightened satiety symptoms.
Hoodia	Reduced food intake due to suppressed appetite.
Pyruvate	Lipolysis and energy expenditure have both increased.
Raspberry ketone	Change lipid metabolism.

12.8 FIBER

Dietary fiber is chemically similar to carbohydrate in most forms, but it is practically non-caloric since the human body lacks the enzymes needed to break the glycosidic bonds in the fiber. Increased dietary fiber consumption appears to lower weight gain in some short-term experimental and cross-sectional studies. Fiber, on the other hand, is ineffective as a weight-loss help (Iqbal, 2006). To aid laxation, fiber should be incorporated in lowering diets at a rate of 25 to 38 grams per day. Although there is no conclusive evidence, both soluble and insoluble dietary fiber may help to decrease hunger and sustain satiety (Melanson et al., 2004; Howarth et al., 2001; Raben et al., 1995). On a low-fiber diet, eating five or more servings of fruits and vegetables per day, as well as eating enough whole grain breads and cereals, will help you meet your soluble and insoluble fiber goals. Water requirements rise as dietary fiber intake rises; thus, fluid intake should be high as well (Slavin et al., 2003).

12.9 BEVERAGES

At present, typical daily intake of added sugars in the diet is expected to be around 15.8% of total energy, with calorically sweetened liquids including soft drinks, fruit juices, and other sweetened beverages accounting for 47% of total added sugars in the diet (Dennis et al., 2009; Guthrie et al., 2000).

When compared to water or energy-free beverages, the use of energy-containing beverages tends to increase total energy intake from meals (Dennis et al., 2009). The Beverage Guidance Panel advises that caloric, sweetened liquids with no nutritional value be limited to eight ounces per day for the average individual. Drinking calorie-dense, nutrient-dense beverages must be limited (The Beverage Guidance Panel, 2006).

While there are health benefits to drinking energy and nutrient-dense beverages like milk and 100% fruit juice, to fulfill weight reduction or maintenance objectives, the increased energy given by these beverages must be compensated by an increase in energy expenditure or a decrease in other areas of energy consumption (The Beverage Guidance Panel, 2006).

12.10 ALCOHOL

Alcohol contains around 7 calories per gram, implying that it contains more energy per unit of weight. Non-nutritive energy, sometimes known as "empty calories," is provided by alcoholic beverages. When protein, carbohydrate, and fat are consumed at the same time as alcohol, their oxidation is slowed because alcohol is oxidized first, and the other macronutrients are balanced due to alcohol's fat-sparing effect. This could lead to more fat storage. Chronic alcohol intake can result in fatty liver, dyslipidemia, and weight gain.

Alcohol may boost hunger, according to certain studies (Poppitt et al., 1996; Tremblay et al., 1995). Alcohol is at the bottom of the satiating efficiency hierarchy of metabolic fuels ingested by humans (Melanson et al., 2004). As a result of these issues, alcohol is commonly forbidden on weight-loss programs. Alcohol has high calorie content and may bypass satiety pathways, making it difficult to control your eating (Dennis et al., 2009).

12.11 LOW AND NO CALORIE SWEETENERS

These substances are referred to as low and no-calorie sweeteners, non-nutritive sweeteners, sugar alternatives, reduced-calorie sweeteners, and artificial sweeteners. They are used to sweeten foods and beverages without adding considerable amounts of sugar or calories to the final product. Low-calorie sweeteners appeal to us because they meet our intrinsic desire for sweetness while avoiding the calories associated with sugar consumption. They may aid in the preparation of low-calorie diets.

Low-calorie sweeteners, on their own, have no evidence of helping people lose weight; they utilize in combination with a hypocaloric diet (Phelan et al., 2009; Anderson et al., 2012). Water, non-caloric liquids, and low-calorie sweetened beverages, are a better choice for weight maintenance than high-calorie beverages including soft drinks, sweet teas, other sugary drinks, and alcohol.

12.11.1 LOW CALORIE NON-NUTRITIVE SWEETENERS APPROVED BY FDA

In the United States, seven non-nutritive sweeteners have been approved by the FDA for use in meals and beverages. The majority of them are controlled as food additives, such as:

1. **Saccharin:** It is a calorie-free, non-nutritive sweetener that is not digested by the body. It has a longer history on the market than any other low-calorie sweetener. Its sweetness ranges from 300 to 500 times that of sugar and is heat stable (International Food Information Council Foundation, 2009).
2. **Acesulfame K (Ace-K):** It does not result in a net increase in calories or potassium in humans because it remains unchanged when removed from the body. It is 200 times sweeter than sugar. It can withstand high temperatures, making it suitable for cooking and baking (Calorie Control Council, 2011).
3. **Sucralose:** Although sucralose is made from sugar, it is not absorbed or metabolized by human stomach enzymes as a carbohydrate. It is calorie-free (International Food Information Council Foundation, 2014). It's 600 times sweeter than sugar and is extremely heat resistant.
4. **Aspartame:** It is a sweetener manufactured from aspartic acid and phenylalanine, two amino acids (Magnuson et al., 2007). It's sold to customers as a tabletop sweetener and as an ingredient. It has a sweetness of 160–220 times that of sugar. Because of its strong sweetness, only a small quantity is required to sweeten a food, and the energy provided by the compounds is therefore minimal.

5. **Neotame:** It is a low-calorie sweetener that is nonnutritive and was approved by the FDA in 2002. It can be used as an ingredient as well as a flavor enhancer to change the flavor of other meals and beverages (International Food Information Council Foundation, 2009). Because Neotame has a sweetness of 7,000–13,000 times that of sugar, it only takes a little quantity to sweeten.
6. **Advantame:** This component is a non-nutritive, low-calorie sweetener that was approved in 2014. It is mostly used as a food ingredient and is not available over-the-counter. In 2014, it was approved for usage in meals and beverages. It has a structure that is comparable to aspartame (Calorie Control Council, 2014).
7. **Stevia:** Stevia leaves, often known as "sweet leaf," have long been utilized in Asian and South American cultures as a non-caloric sweetener and traditional medicine (Goyal et al., 2010).

12.11.2 SWEETENERS HAVING LOW CALORIE

1. **Polyols:** Sugar alcohols (polyols) are nutritive sweeteners since they contribute calories to the diet. However, compared to table sugar, it has fewer calories per gram (Calorie Control Council, 2014; International Food Information Council Foundation, 2004). The sweetness of most polyols is around half that of sugar. They can be used to replace sugar in products for sweetness and volume, allowing them to be branded as "sugar-free."
2. **Erythritol:** Natural sources include pears, melons, grapes, and mushrooms. Erythritol has fewer calories per gram than the other polyols (Sandrou et al., 2000; Livesey et al., 2001).
3. **Sorbitol:** Sugar-free candies, chewing gums, baked goods, frozen desserts, and toothpaste all include sorbitol, a non-cariogenic sugar replacement (Sandrou et al., 2000). As a result, it may be advantageous to people who are obese or diabetic.
4. **Mannitol:** Fresh mushrooms, brown algae, tree bark, and the majority of fruits and vegetables all naturally contain mannitol (Mortensen et al., 2006).
5. **Xylitol:** Birch, raspberry, plum, and corn are all sources of xylitol. It is typically found in sugar-free gums, cough drops, mints, and other oral health products (Magnuson et al., 2007).

6. **Maltitol:** Maltitol is commercially delivered from the starch of cereals like corn, wheat and potatoes (Magnuson et al., 2007).
7. **Lactitol:** It is found in baked goods, hard, and soft sweets, frozen dairy desserts, and chocolate (IFICF, 2004).
8. **Isomalt:** It can be found in sweets, toffees, wafers, etc., in most European countries. Around 90% of isomalt is fermented further in the colon, making it a possible prebiotic (Livesey et al., 2003).
9. **Hydrogenated Starch Hydrolysate (HSH):** They are mostly employed as bulking agents and food additives in commercially made "sugarless" food products to offer sugariness, quantity, and feel. HSH has a glycemic index similar to maltitol, and in the colon, around 40% of it gets digested (Livesey et al., 2001). As a result, it may be appropriate for diabetics and obese people.

12.12 CONCLUSION

1. None of the FDA-approved low-calorie sweeteners on the market in the United States are required in individual diet. However, some customers love the sweet taste of low-calorie sweeteners, whether nutritive or non-nutritive, on the basis of safety or health, there is no reason to prohibit their use.
2. All FDA-approved sweeteners are safe for use by the general public, pregnant women (excluding phenylketonurics), and children (except phenylketonurics) (ADI) in doses below the acceptable daily intake. Aspartame is safe for pregnant women and their growing children, according to the American Academy of Pediatrics (save for those with PKU). In terms of taste, mouth feel, and utility in food processing and preparation, none of the low-calorie nutritive or non-nutritive sweeteners are perfect substitutes for sugar, yet they are acceptable to many individuals.
3. The majority of low-calorie sweeteners, both nutritive and non-nutritive, are not responsible for tooth decay. Xylitol having anti-cariogenic properties, promoting tooth enamel remineralization.
4. The most well-documented negative effects of low-calorie sweeteners are the laxative effects of poorly absorbed nutritional low-calorie sweeteners (sugar alcohols) when ingested in high doses. With very heavy use of aspartame by people with phenylketonuria, there is a possible risk (PKU).

5. If low-calorie sweeteners are substituted for caloric sweeteners in the diet and calorie intakes are lowered, low-calorie sweeteners may help with weight loss. Otherwise, they don't seem to make much of an impact. It's unlikely that they'll help you lose weight if you don't combine them with healthy food, physical activity, and behavioral changes.
6. There is currently no scientific evidence to suggest that people seeking to lose weight should avoid low-calorie sweeteners.
7. If patients or other users are concerned about excessive intakes of nutritive or non-nutritive low-calorie sweeteners, the best advice is to mix and combine low-calorie sweeteners to maintain doses of any one component low and below the ADI, or to avoid them entirely.

KEYWORDS

- **acceptable macronutrient distribution ranges**
- **body mass index**
- **carbohydrate**
- **dieting**
- ***Glycemic index***
- **hypocaloric diets**
- **obesity**
- **weight loss**

REFERENCES

Adam-Perrot, A., Clifton, P., & Brouns, F., (2006). Low-carbohydrate diets: Nutritional and physiological aspects. *Obes. Rev., 7*(1), 49–58.

Afshin, A., Forouzanfar, M. H., Reitsma, M. B., Sur, P., & Estep, K., (2017). Health effects of overweight and obesity in 195 countries over 25 years. *N. Engl. J. Med., 377*(1), 13–27.

Albuquerque, D., Nóbrega, C., Manco, L., & Padez, C., (2017). The contribution of genetics and environment to obesity. *Br. Med. Bull., 123*(1), 159–173.

Allison, D. B., Fontaine, K. R., Manson, J. E., Stevens, J., & VanItallie, T. B., (1999). Annual deaths attributable to obesity in the United States. *JAMA, 282*(16), 1530–1538.

American Dietetic Association, (2009). In: Nonas, C. & Foster G., (ed.), Managing Obesity: A Clinical Guide (2nd edn.). American Dietetic Association.

Anderson, G. H., Foreyt, J., Sigman-Grant, M., & Allison, D. B., (2012). The use of low-calorie sweeteners by adults: Impact on weight management. *J. Nutr., 142*(6), 1163S–1169S.

Arendas, K., Qiu, Q., & Gruslin, A., (2008). Obesity in pregnancy: Pre-conceptional to postpartum consequences. *J. Obstet. Gynaecol., 30*(6), 477–488.

Arnett, D. K., Blumenthal, R. S., Albert, et. al., (2019). ACC/AHA guideline on the primary prevention of cardiovascular disease. *Circulation, 140*(11), e596–e646.

Atkins, R. C., (2002). *Dr. Atkins' Diet Revolution*. New York: Collins.

Aune, D., Norat, T., & Vatten, L. J., (2014). Body mass index and the risk of gout: A systematic review and dose-response meta-analysis of prospective studies. *Eur. J. Nutr., 53*(8), 1591–1601.

Aune, D., Sen, A., Norat, T., Janszky, I., Romundstad, P., Tonstad, S., & Vatten, L. J., (2016). Body mass index, abdominal fatness, and heart failure incidence and mortality: A systematic review and dose-response meta-analysis of prospective studies. *Circulation, 133*(7), 639–649.

Aune, D., Sen, A., Prasad, M., Norat, T., Janszky, I., Tonstad, S., Romundstad, P., & Vatten, L. J., (2016). BMI and all-cause mortality: Systematic review and non-linear dose-response meta-analysis of 230 cohort studies with 3.74 million deaths among 30.3 million participants. *BMJ, 353*, i2156.

Bailey, R. L., Gahche, J. J., Miller, P. E., Thomas, P. R., & Dwyer, J. T., (2013). Why US adults use dietary supplements. *JAMA Intern. Med., 173*(5), 355–361.

Barness, L. A., Opitz, J. M., & Gilbert-Barness, E., (2007). Obesity: Genetic, molecular, and environmental aspects. *American Journal of Medical Genetics. Part A. 143A*(24), 3016–3034.

Basen-Engquist, K., & Chang, M., (2011). Obesity and cancer risk: Recent review and evidence. *Curr. Oncol. Rep., 13*(1), 71–76.

Bei-Fan, Z., (2002). Predictive values of body mass index and waist circumference for risk factors of certain related diseases in Chinese adults: Study on optimal cut-off points of body mass index and waist circumference in Chinese adults. *Asia Pac. J. Clin. Nutr., 11*, (Suppl 8), S685–693.

Beydoun, M. A., Beydoun, H. A., & Wang, Y., (2008). Obesity and central obesity as risk factors for incident dementia and its subtypes: A systematic review and meta-analysis. *Obesity Reviews (Meta-analysis)., 9*(3), 204–218.

Bigal, M. E., & Lipton, R. B., (2008). Obesity and chronic daily headache. *Curr. Pain Headache Rep., 12*(1), 56–61.

Björntorp, P., (2001). Do stress reactions cause abdominal obesity and comorbidities? *Obes. Rev., 2*(2), 73–86.

Bleich, S., Cutler, D., Murray, C., & Adams, A., (2008). Why is the developed world obese? *Annu. Rev. Public Health., 29*, 273–295.

Bojanowska, E., & Ciosek, J., (2016). Can we selectively reduce appetite for energy-dense foods? An overview of pharmacological strategies for modification of food preference behavior. *Curr. Neuropharmacol., 14*(2), 118–142.

Bray, G. A., (2004). Medical consequences of obesity. *J. Clin. Endocrinol. Metab., 89*(6), 2583–2589.

Brownwell, K. D., & Fairburn, C. G., (2005). *Eating Disorders and Obesity: A Comprehensive Handbook*. Guilford Press: New York.

Caballero, B., (2001). Introduction. Symposium: Obesity in developing countries: Biological and ecological factors. *J. Nutr., 131*(3), 866S–870S.

Calle, E. E., Thun, M. J., Petrelli, J. M., Rodriguez, C., & Heath, C. W., (1999). Body-mass index and mortality in a prospective cohort of U.S. adults. *N. Engl. J. Med., 341*(15), 1097–1105.

Calorie Control Council (CCC), (2011). *Sweeteners and Lite.* https://caloriecontrol.org/sweetener-facts/sugar-substitutes/sucralose/page/33/ (accessed on 25 October 2022).

Calorie Control Council (CCC), (2014). *Aspartame.* Available at: https://caloriecontrol.org/sweetener-facts/sugar-substitutes/aspartame/ (accessed on 25 October 2022).

Carmienke, S., Freitag, M. H., Pischon, T., Schlattmann, P., Fankhaenel, T., Goebel, H., & Gensichen, J., (2013). General and abdominal obesity parameters and their combination in relation to mortality: A systematic review and meta-regression analysis. *Eur. J. Clin. Nutr., 67*(6), 573–585.

CCC. The Calorie Control Council, (2014). *Polyols.* Available at: http://www.caloriecontrol.org/sweeteners-and-lite/polyol (accessed on 02 August 2022).

Chiles, C., & Van-Wattum, P. J., (2010). Psychiatric aspects of the obesity crisis. *Psychiatr. Times, 27*(4), 47–51.

Chiolero, A., (2018). Why causality, and not prediction, should guide obesity prevention policy. *Lancet Public Health, 3*(10), e461, e462.

Clinical Guidelines on the Identification, Evaluation, and Treatment of Overweight and Obesity in Adults: The Evidence Report. (1998). National Institutes of Health. *Obes Res. 6* Suppl 2, 51S–209S. Erratum in: *Obes Res* 1998 Nov; *6*(6), 464. PMID: 9813653.

Colquitt, J. L., Pickett, K., Loveman, E., & Frampton, G. K., (2014). Surgery for weight loss in adults. *Cochrane Database Syst Rev., 8*(8), CD003641.

Corona, G., Rastrelli, G., Filippi, S., Vignozzi, L., Mannucci, E., & Maggi, M., (2014). Erectile dysfunction and central obesity: An Italian perspective. *Asian J. Androl., 16*(4), 581–591.

Cox, L. M., & Blaser, M. J., (2015). Antibiotics in early life and obesity. *Nat. Rev. Endocrinol., 11*(3), 182–190.

Cummings, S., Parham, E. S., & Strain, G. W., (2002). Position of the American Dietetic Association: Weight management. *J. Am. Diet. Assoc., 102*(8), 1145–1155.

Darvall, K. A., Sam, R. C., Silverman, S. H., Bradbury, A. W., & Adam, D. J., (2007). Obesity and thrombosis. *Eur. J. Vasc. Endovasc. Surg., 33*(2), 223–233.

Dennis, E., Flack, K., & Davy, B., (2009). Beverage consumption and adult weight management: A review. *Eat. Behav., 10*(4), 237–246.

Di Angelantonio, E., Bhupathiraju, S., Wormser, D., Gao, P., Kaptoge, S., & Berrington De, G. A., (2016). Body-mass index and all-cause mortality: Individual-participant-data meta-analysis of 239 prospective studies in four continents. *Lancet, 388*(10046), 776–786.

Dibaise, J. K., & Foxx-Orenstein, A. E., (2013). Role of the gastroenterologist in managing obesity. *Expert Rev. Gastroenterol. Hepatol., 7*(5), 439–451.

DiBaise, J. K., Zhang, H., Crowell, M. D., Krajmalnik-Brown, R., Decker, G. A., & Rittmann, B. E., (2008). Gut microbiota and its possible relationship with obesity. *Mayo Clin. Proc., 83*(4), 460–469.

Ejerblad, E., Fored, C. M., Lindblad, P., Fryzek, J., McLaughlin, J. K., & Nyrén, O., (2006). Obesity and risk for chronic renal failure. *Journal of the American Society of Nephrology, 17*(6), 1695–1702.

Falagas, M. E., & Kompoti, M., (2006). Obesity and infection. *Lancet, 6*(7), 438–446.

Flegal, K. M., Kit, B. K., Orpana, H., & Graubard, B. I., (2013). Association of all-cause mortality with overweight and obesity using standard body mass index categories: A systematic review and meta-analysis. *JAMA, 309*(1), 71–82.

Flegal, K. M., Troiano, R. P., Pamuk, E. R., Kuczmarski, R. J., & Campbell, S. M., (1995). The influence of smoking cessation on the prevalence of overweight in the United States. *N Engl. J. Med., 333*(18), 1165–1170.

Foster, G. D., & Makris, A. P., (2005). Low-carbohydrate diets in the treatment of obesity. *Nutr. Today, 40*(4).

Freedman, M. R., King, J., & Kennedy, E., (2001). Popular diets: A scientific review. *Obes. Res., 9*(Suppl 1), 1S–40S.

Fried, M., Hainer, V., Basdevant, A., Buchwald, H., Deitel, M., Finer, N., Greve, J. W., et al., (2007). Bariatric Scientific Collaborative Group Expert Panel. Interdisciplinary European guidelines for surgery for severe (morbid) obesity. *Obes Surg. 17*(2), 260–270.

Friedman, H., (2015). *Encyclopedia of Mental Health* (2nd edn., p. 158). Academic Press; CA, USA.

Goodman, E., Adler, N. E., Daniels, S. R., Morrison, J. A., Slap, G. B., & Dolan, L. M., (2003). Impact of objective and subjective social status on obesity in a biracial cohort of adolescents. *Obes. Res., 11*(8), 1018–1026.

Gortmaker, S. L., Must, A., Sobol, A. M., Peterson, K., Colditz, G. A., & Dietz, W. H., (1996). Television viewing as a cause of increasing obesity among children in the United States, 1986–1990. *Arch. Pediatr. Adolesc. Med., 150*(4), 356–362.

Goyal, S. K., Samsher, & Goyal, R. K., (2010). Stevia (*Stevia rebaudiana*) a bio-sweetener: A review. *Int. J. Food Sci. Nutr., 61*(1), 1–10.

Gray, D. S., & Fujioka, K., (1991). Use of relative weight and body mass index for the determination of adiposity. *J. Clin. Epidemiol., 44*(6), 545–550.

Grundy, S. M., (2004). Obesity, metabolic syndrome, and cardiovascular disease. *J. Clin. Endocrinol. Metab., 89*(6), 2595–2600.

Guthrie, J., & Morton, J., (2000). Food sources of added sweeteners in the diets of Americans. *J. Am. Diet. Assoc., 100*(1), 43–51.

Hahler, B., (2006). An overview of dermatological conditions commonly associated with the obese patient. *Ostomy Wound Manag., 52*(6), 34–36, 38, 40. Passim.

Haslam, D. W., & James, W. P., (2005). Obesity. *Lancet, 366*(9492), 1197–1209.

Haven, J., & Britten, P., (2006). My pyramid-the complete guide. *Nutr. Today, 41*(6).

Holecki, M., Zahorska-Markiewicz, B., Wiecek, A., Mizia-Stec, K., Nieszporek, T., & Zak-Golab, A., (2008). Influence of calcium and vitamin D supplementation on weight and fat loss in obese women. *Obesity Facts: J. Obes., 1*(5), 274–279.

Howarth, N. C., Saltzman, E., & Roberts, S. B., (2001). Dietary fiber and weight regulation. *Nutr. Rev., 59*(5), 129–139.

Hunskaar, S., (2008). A systematic review of overweight and obesity as risk factors and targets for clinical intervention for urinary incontinence in women. *Neurourol. Urodyn., 27*(8), 749–757.

Imaz, I., Martínez-Cervell, C., García-Alvarez, E. E., Sendra-Gutiérrez, J. M., & González-Enríquez, J., (2008). Safety and effectiveness of the intragastric balloon for obesity. A meta-analysis. *Obes. Surg., 18*(7), 841–846.

International Food Information Council Foundation, (2004). *Sugar Alcohols Fact Sheet.* Available from: https://ingredients.net.au/images/IFIC%20Sugar%20Alcohols.pdf (accessed on 25 October 2022).

International Food Information Council Foundation, (2009). *Low-Calorie Sweeteners and Health.* Available from: https://foodinsight.org/2009-food-health-survey-consumer-attitudes-toward-food-nutrition-and-health/ (accessed on 25 October 2022).

International Food Information Council Foundation, (2014). *Facts About Low-Calorie Sweeteners.* from https://foodinsight.org/wp-content/uploads/2020/06/IFIC-Food-and-Health-Survey-2020.pdf (accessed on 25 October 2022).

Iqbal, S. I., Helge, J. W., & Heitmann, B. L., (2006). Do energy density and dietary fiber influence subsequent 5-year weight changes in adult men and women? *Obes., 14*(1), 106–114.

Jensen, M. D., (2014). Expert panel report: Guidelines (2013) for the management of overweight and obesity in adults. *Obesity, 22*, S41–S410.

Jensen, M. D., Ryan, D. H., Apovian, C. M., Ard, J. D., & Comuzzie. (2014). American College of cardiology/American heart association task force on practice, guidelines. *Obes. Soc., Circulation, 129*(25 Suppl 2), S102–138.

Johnston, B. C., Kanters, S., Bandayrel, K., Wu, P., Naji, F., Siemieniuk, R. A., Ball, G. D., et al., (2014). Comparison of weight loss among named diet programs in overweight and obese adults: A meta-analysis. *JAMA, 312*(9), 923–933.

Kanazawa, M., Yoshiike, N., Osaka, T., Numba, Y., Zimmet, P., & Inoue, S., (2005). Criteria and classification of obesity in Japan and Asia-Oceania. Nutrition and fitness: Obesity, the metabolic syndrome, cardiovascular disease, and cancer. *World Rev. Nutr. Diet., 94*, 1–12.

Kassotis, C. D., Vandenberg, L. N., Demeneix, B. A., Porta, M. S., & Remy, T. L., (2020). Endocrine-disrupting chemicals: Economic, regulatory, and policy implications. *Lancet Diabetes Endocrinol., 8*(8), 719–730.

Keservani, R. K., & Sharma, A. K., (2014). Flavonoids: Emerging trends and potential health benefits. *J. Chin. Pharm. Sci, 23*(12), 815–822.

Keservani, R. K., Kesharwani, R. K., Sharma, A. K., Vyas, N., & Chadoker, A., (2010b). Nutritional supplements: An overview. *Int. J. of Curr. Pharm. Rev. and Res., 1*(1), 59–75.

Keservani, R. K., Kesharwani, R. K., Vyas, N., Jain, S., Raghuvanshi, R., & Sharma, A. K., (2010a). Nutraceuticals and functional food as future food. *Der. Pharm. Lett., 2*(1), 106–116.

Kleiner, S. M., (1999). Water: An essential but overlooked nutrient. *J Am Diet Assoc., 99*(2), 200–206.

Lanou, A., & Barnard, N., (2008). Dairy and weight loss hypothesis: An evaluation of the clinical trials. *Nutr. Rev., 66*(5), 272–279.

Lau, D. C., Douketis, J. D., Morrison, K. M., Hramiak, I. M., Sharma, A. M., & Ur, E., (2007). Canadian clinical practice guidelines on the management and prevention of obesity in adults and children [summary]. *CMAJ, 176*(8): S1–13.

Levitsky, D. A., Sewall, A., Zhong, Y., Barre, L., Shoen, S., Agaronnik, N., LeClair, J. L., et al., (2019). Quantifying the imprecision of energy intake of humans to compensate for imposed energetic errors: A challenge to the physiological control of human food intake. *Appetite., 133*, 337–343.

Lindeberg, S., (2012). Paleolithic diets as a model for prevention and treatment of Western disease. *Am J Hum Biol., 24*(2), 110–115.

Livesey, G., (2001). Tolerance of low-digestible carbohydrates: A general view. *Br. J. Nutr., 85*(1), S7–16.

Livesey, G., (2003). Health potential of polyols as sugar replacers, with emphasis on low glycaemic properties. *Nutr. Res. Rev., 16*(2), 163–191.

Lu, Y., Hajifathalian, K., Ezzati, M., Woodward, M., Rimm, E. B., & Danaei, G., (2014). Metabolic mediators of the effects of body-mass index, overweight, and obesity on coronary heart disease and stroke: A pooled analysis of 97 prospective cohorts with 1·8 million participants. *Lancet, 383*(9921), 970–983.

Ludwig, D. S., & Ebbeling, C. B., (2018). The carbohydrate-insulin model of obesity: Beyond calories in, calories out. *JAMA, 178*(8), 1098.

Luppino, F. S., De Wit, L. M., Bouvy, P. F., Stijnen, T., Cuijpers, P., Penninx, B. W., & Zitman, F. G., (2010). Overweight, obesity, and depression: A systematic review and meta-analysis of longitudinal studies. *Arch. Gen. Psychiatry, 67*(3), 220–229.

Magnuson, B. A., Burdock, G. A., Doull, J., Kroes, R. M., Marsh, G. M., & Pariza, M. W., (2007). Aspartame: A safety evaluation based on current use levels, regulations, and toxicological and epidemiological studies. *Crit. Rev. Toxicol., 37*(8), 629–727.

Manson, J. E., Willett, W. C., Stampfer, M. J., Colditz, G. A., Hunter, D. J., Hankinson, S. E., Hennekens, C. H., & Speizer, F. E., (1995). Body weight and mortality among women. *New Eng. J. Med., 333*(11), 677–685.

McLaren, L., (2007). Socioeconomic status and obesity. *Epidemiol. Rev., 29*, 29–48.

Melanson, K. J., (2004). CE test: Food intake regulation in body weight management: A primer. *Nutr. Today, 39*(5).

Mokdad, A. H., Marks, J. S., Stroup, D. F., & Gerberding, J. L., (2004). Actual causes of death in the United States, 2000. *JAMA, 291*(10), 1238–1245.

Molenaar, E. A., Numans, M. E., Van, A. E. J., & Grobbee, D. E., (2008). Considerable comorbidity in overweight adults: Results from the Utrecht Health Project. *Nederlands Tijdschrift voor Geneeskunde* (English abstract) (in Dutch), *152*(45), 2457–2463.

Mortensen, A., (2006). Sweeteners permitted in the European Union: Safety aspects. *Food & Nutr Res., 50*(3), 104–116.

Munger, K. L., Chitnis, T., & Ascherio, A., (2009). Body size and risk of MS in two cohorts of US women. *Neurology, 73*(19), 1543–1550.

National Institutes of Health Obesity Education Initiative, (2000). *The Practical Guide Identification, Evaluation, and Treatment of Overweight and Obesity in Adults*. 00-4084.

National Institutes of Health. Office of Dietary Supplements, (2014). *Dietary Supplement for Weight Loss: Fact Sheet for Health Professionals*. http://ods.od.nih.gov/factsheets/WeightLoss-HealthProfessional (accessed on 02 August 2022).

Ness-Abramof, R., & Apovian, C. M., (2006). Diet modification for treatment and prevention of obesity. *Endocrine, 29*(1), 5–9.

Ornish, D., (1993). Diet and fitness changes reverse coronary artery disease. *World Rev. Nutr. Diet., 77*, 38–48.

Phelan, S., Lang, W., Jordan, D., & Wing, R., (2009). Use of artificial sweeteners and fat-modified foods in weight loss maintainers and always-normal weight individuals. *Int. J. Obes., 33*(10), 1183–1190.

Pischon, T., Boeing, H., Hoffmann, K., Bergmann, M., & Schulze, M., (2008). General and abdominal adiposity and risk of death in Europe. *N. Engl. J. Med., 359*(20), 2105–2120.

Poirier, P., Giles, T. D., Bray, G. A., Hong, Y., Stern, J. S., Pi-Sunyer, F. X., & Eckel, R. H., (2006). Obesity and cardiovascular disease: Pathophysiology, evaluation, and effect of weight loss. *Arterioscler. Thromb. Vasc. Biol., 26*(5), 968–976.

Pollack, A. A. M. A., (2013). *Recognizes Obesity as a Disease*. The New York Times.

Poppitt, S. D., Eckhardt, J. W., McGonagle, J., Murgatroyd, P. R., & Prentice, A. M., (1996). Short-term effects of alcohol consumption on appetite and energy intake. *Physiol. Behav., 60*(4), 1063–1070.

Poulain, M., Doucet, M., Major, G. C., Drapeau, V., Sériès, F., Boulet, L. P., Tremblay, A., & Maltais, F., (2006). The effect of obesity on chronic respiratory diseases: Pathophysiology and therapeutic strategies. *CMAJ., 174*(9), 1293–1299.

Pritikin, M., & McGrady, P. M., (1980). *The Pritikin Program for Diet and Exercise*. Bantam Books New York.

Raben, A., Jensen, N. D., Marckmann, P., Sandstrom, B., & Astrup, A., (1995). Spontaneous weight loss during 11 weeks' ad libitum intake of a low fat/high fiber diet in young, normal weight subjects. *Int. J. Obes. Relat. Metab. Disord., 19*(12), 916–923.

Rizzoli, R., (2014). Dairy products, yogurts, and bone health. *Am. J. Clin. Nutr., 99*(5 Suppl), 1256S–1262S.

Robert F. Kushner & Daniel H. Bessesen (2007). *Treatment of the Obese Patient*. Humana Press, Totowa.

Rosén, T., Bosaeus, I., Tölli, J., Lindstedt, G., & Bengtsson, B. A., (1993). Increased body fat mass and decreased extracellular fluid volume in adults with growth hormone deficiency. *Clin. Endocrinol., 38*(1), 63–71.

Sacks, F., Bray, G., & Carey, V., (2009). Comparison of weight-loss diets with different compositions of fat, protein, and carbohydrates. *N Engl J Med., 360*(9), 859–873.

Salmon, J., & Timperio, A., (2007). Prevalence, trends and environmental influences on child and youth physical activity. *Pediatric Fitness. J. Sports Sci. Med., 50*, 183–199.

Sandrou, D. K., & Arvanitoyannis, I. S., (2000). Low-fat/calorie foods: Current state and perspectives. *Crit. Rev. Food Sci. Nutr., 40*(5), 427–447.

Saris, W. H., Astrup, A., & Prentice, A. M., (2000). Randomized controlled trial of changes in dietary carbohydrate/fat ratio and simple vs complex carbohydrates on body weight and blood lipids: The carmen study. The carbohydrate ratio management in European national diets. *Int. J. Obes. Relat. Metab. Disord., 24*(10), 1310–1318.

Schoeller, D. A., & Buchholz, A. C., (2005). Energetics of obesity and weight control: Does diet composition matter? *J. Am. Diet. Assoc., 105*(5 Suppl 1), S24–S28.

Seagle, H. M., Strain, G. W., Makris, A., & Reeves, R. S., (2009). Position of the American dietetic association: Weight management. *J. Am. Diet. Assoc., 109*(2), 330–346.

Services, (1987). Swedish agency for health technology assessment and assessment of social services (SBU) dietary treatment of obesity. *Annals of the New York Academy of Sciences, 499*(1), 250–263.

Sharifi-Mollayousefi, A., Yazdchi-Marandi, M., Ayramlou, H., Heidari, P., Salavati, A., Zarrintan, S., & Sharifi-Mollayousefi, A., (2008). Assessment of body mass index and hand anthropometric measurements as independent risk factors for carpal tunnel syndrome. *Folia Morphol., 67*(1), 36–42.

Shoelson, S. E., Herrero, L., & Naaz, A., (2007). Obesity, inflammation, and insulin resistance. *Gastroenterology, 132*(6), 2169–2180.

Shoelson, S. E., Lee, J., & Goldfine, A. B., (2006). Inflammation and insulin resistance. *J. Clin. Invest, 116*(7), 1793–1801.

Singh, P., Kesharwani, R. K., & Keservani, R. K., (2017). Protein, Carbohydrate and fats: Energy metabolism. In: Bagchi, D., (ed.), *Sustained Energy for Enhanced Human Function and Activity*. Elsevier. Ch 6. ISBN: 9780128054130.

Slavin, J., (2003). Why whole grains are protective: Biological mechanisms. *Proc. Nutr. Soc., 62*(1), 129–134.

Surwit, R. S., Feinglos, M. N., & McCaskill, C. C., (1997). Metabolic and behavioral effects of a high-sucrose diet during weight loss. *Am. J. Clin. Nutr, 65*(4), 908–915.

Sweeting, H. N., (2007). Measurement and definitions of obesity in childhood and adolescence: A field guide for the uninitiated. *Nutr. J., 6*(1), 32.

The Beverage Guidance Panel, (2006). *Beverage Intake in the United States: Noncalorically Sweetened Beverages.* The University of North Carolina at Chapel Hill.

Tremblay, A., Wouters, E., Wenker, M., St-Pierre, S., Bouchard, C., & Despres, J. P., (1995). Alcohol and a high-fat diet: A combination favoring overfeeding. *Am. J. Clin. Nutr., 62*(3), 639–644.

Tsai, A. G., & Wadden, T. A., (2005). Systematic review: An evaluation of major commercial weight loss programs in the United States. *Ann Intern Med., 142*(1), 56–66.

Tsigos, C., Hainer, V., Basdevant, A., Finer, N., Fried, M., Mathus-Vliegen, E., Micic, D., et al., (2008). Management of obesity in adults: European clinical practice guidelines. *Obes. Facts., 1*(2), 106–116.

Tucker, L. A., & Bagwell, M., (1991). Television viewing and obesity in adult females. *Am. J. Public Health, 81*(7), 908–911.

Tukker, A., Visscher, T. L., & Picavet, H. S., (2009). Overweight and health problems of the lower extremities: Osteoarthritis, pain and disability. *Public Health Nutr., 12*(3), 359–368.

Turner, B. L., & Thompson, A. L., (2013). Beyond the paleolithic prescription: Incorporating diversity and flexibility in the study of human diet evolution. *Nutr. Rev., 71*(8), 501–510.

U.S. Department of Health and Human Services and U.S. Department of Agriculture, (2010). *Dietary Guidelines for Americans* (7th edn.) U.S. Government Printing Office, Washington, DC.

U.S. Department of Health and Human Services, (2012). *Very Low-Calorie Diets.* Version current.

US Department of Health and Human Services. *2015–2020 Dietary Guidelines for Americans* – health.gov. 2017, health.gov. Skyhorse Publishing Inc.

US Preventive Services Task Force, (2018). Behavioral weight loss interventions to prevent obesity-related morbidity and mortality in adults US preventive services task force recommendation statement. *JAMA, 320*(11), 1163–1171.

Venn, B. J., & Mann, J. I., (2004). Cereal grains, legumes and diabetes. *Eur J Clin Nutr. 58*(11), 1443–1461.

Vioque, J., Torres, A., & Quiles, J., (2000). Time spent watching television, sleep duration and obesity in adults living in Valencia, Spain. *Int. J. Obes. Relat. Metab. Disord., 24*(12), 1683–1688.

Wadden, T. A., Bryne, K., & Krauthamer-Ewing, S., (2006). Obesity: Management. In: Shils, M., Shike, M., Olsen, J., & Ross, C., (eds.), *Modern Nutrition in Health and Disease* (pp. 1029–1042). Baltimore: Lippincott, Williams, and Wilkons.

Wagner, G., Kindrick, S., Hertzler, S., & DiSilvestro, R., (2007). Effects of various forms of calcium on body weight and bone turnover markers in women participating in a weight loss program. *J. Am. Coll. Nutr., 26*(5), 456–461.

Wall, M., (2008). Idiopathic intracranial hypertension (pseudotumor cerebri). *Curr. Neurol. Neurosci. Rep., 8*(2), 87–93.

Weinstock, M., (2013). *The Facts About Obesity*. H & HN.

Whitlock, G., Lewington, S., Sherliker, P., Clarke, R., Emberson, J., Halsey, J., Qizilbash, N., et al., (2009). Body-mass index and cause-specific mortality in 900 000 adults: Collaborative analyses of 57 prospective studies. *Lancet, 373*(9669), 1083–1096.

WHO Expert Consultation, (2004). Appropriate body-mass index for Asian populations and its implications for policy and intervention strategies. *Lancet, 363*(9403), 157–163.

WHO. (2015). *Obesity and Overweight Fact Sheet N°311*. https://www.who.int/news-room/fact-sheets/detail/obesity-and-overweight (accessed on 25 October 2022).

Wildman, R., & Miller, B. S., (2004). *Sports and Fitness Nutrition*. Thomson/Wadsworth, Australia.

Wilkinson, R., & Pickett, K., (2009). *The Spirit Level: Why More Equal Societies Almost Always Do Better* (pp. 91–101). London: Allen Lane.
Wing, R. R., & Phelan, S., (2005). Long-term weight loss maintenance. *Am. J. Clin. Nutr., 82*(1 Suppl), 222S–225S.
Wolever, T. M., Jenkins, D. J., Jenkins, A. L., & Josse, R. G., (1991). The glycemic index: Methodology and clinical implications. *Am. J. Clin. Nutr., 54*(5), 846–854.
World Health Organization, (2000). *Evaluation of Certain Food Additives and Contaminants, 896*, 1–128.
Yannakoulia, M., Poulimeneas, D., Mamalaki, E., & Anastasiou, C. A., (2019). Dietary modifications for weight loss and weight loss maintenance. *Metab., 92*, 153–162.
Yanovski, J., Parikh, S., & Yanoff, L., (2009). Effects of calcium supplementation on body weight and adiposity in overweight and obese adults: A randomized trial. *Ann. Intern. Med., 150*(12), 821–829.
Yanovski, S. Z., & Yanovski, J. A., (2014). Long-term drug treatment for obesity: A systematic and clinical review. *JAMA, 311*(1), 74–86.
Yazdi, F. T., Clee, S. M., & Meyre, D., (2015). Obesity genetics in mouse and human: Back and forth, and back again., *Peer J., 3*, e856.
Yosipovitch, G., DeVore, A., & Dawn, A., (2007). Obesity and the skin: Skin physiology and skin manifestations of obesity. *J. Am. Acad. Dermatol., 56*(6), 901–916, quiz 917–920.

CHAPTER 13

Nutraceuticals in the Prevention and Management of Type II Diabetes Mellitus

URMILA JAROULIYA

School of Studies in Biochemistry, Jiwaji University, Gwalior, Madhya Pradesh – 474011, India, E-mail: urmi_gudia@rediffmail.com

ABSTRACT

Nutritional dietary supplements, vitamins, and herbs are essential additives of nutraceuticals, which make them instrumental in maintaining health, act towards numerous illnesses, and therefore promote the high quality of existence. Diabetes mellitus (DM) is considered one of them. It's far a multi-factorial metabolic disease meditated through excessive blood sugar/glucose levels. In spite of remarkable improvements in drug discovery to treat this disorder, enormous efforts are on for finding an entire method from herbal compounds. Sufferers struggling with prediabetes and diabetes have notably elevated chance for cardiovascular diseases (CDVs) and other complications. In present conditions, control of hyperglycemia consists of pharmacological medications, physical activity, alternate in lifestyles and food plan behavior. Now ever days dietary supplements have become to be attractive desire to prevent or treat hyperglycemia, especially for subjects with moderate hyperglycemia. Diets rich in complete grains, fruits, green veggies, legumes, nuts, and decrease in delicate grains, red/processed meats, and sugar-sweetened drinks have showed huge results in lowering diabetes risk, enhancing glycemic index and blood lipid profile in patients with diabetes. In lots of cases, a right eating regimen and elevated physical activity are powerful in the prevention and remedy of insulin resistance and pre-diabetes. This chapter focuses on functional herbal foods, diverse

The Metabolic Syndrome: Dietary Supplements and Food Ingredients. Raj K. Keservani, MPharm, Durgavati Yadav, PhD, Rajesh K. Kesharwani, PhD, Sippy Singh, DPhil & Kumar Sandeep, PhD (Eds.)

micronutrients and physical activity which commonly allows in preventing and managing the DM and its related complications.

13.1 INTRODUCTION

Diabetes mellitus (DM) is a persistent degenerative disorder with growing morbidity and mortality, characterized by hyperglycemia and frequently related to hyperlipidemia and oxidative stress. Diabetes is the metabolic ailment, that could have an effect on almost each organ system within the body and is typically regarded as a clinical syndrome with variable phenotypic expression in preference to a single disease with a particular etiology (Jarouliya et al., 2012; American Diabetes Association (ADA), 2013).

There are three foremost types of DM: type 1, type 2, and gestational diabetes (taking place during pregnancy), which have similar signs, symptoms, and results, but distinctive reasons and population distributions. In the long run, all forms are because of incorrect functioning of beta cells of the pancreas, not able to provide enough insulin to prevent hyperglycemia. Type 1 or insulin dependent DM, is generally because of autoimmune destruction of the pancreatic beta cells, which produce insulin. Type 2 or non-insulin dependent DM or adult-onset diabetes, is characterized with the aid of tissue wide insulin resistance, which means cell does not reply to insulin, normally takes place in fats, liver, and muscle cells. Gestational diabetes is just like type 2 diabetes (T2D), in that it involves insulin resistance because of predisposal of being pregnancy hormones (William and John, 2001). Type 2 may be resulting from a variety of factors which includes environmental (diet and way of life), metabolic, and genetic. There is a lot of proof records that counseled that 9 of 10 instances of T2DM are due to modifications in way of life factors (Hu et al., 2001).

It leads to long-term pathogenic situations along with micro-vascular and macro-vascular complications, neuropathy, retinopathy, nephropathy, and a consequent reduction in quality of lifestyles and a growth in the rate of mortality. A few of the multiple threat elements underling the incidence and development of T2D Mellitus, diet is the principle changeable factor (Nimesh and Ashwlayan, 2018; Yeung et al., 2018).

13.2 PATHOPHYSIOLOGY FOR TYPE 2 DIABETES (T2D) MELLITUS

In ordinary state, Insulin secretion is often induced by a reaction to excessive glucose concentrations. While circulating glucose levels increase, cells absorb glucose specifically via the glucose transporter 2 (GLUT2), a solute carrier protein that still works as a glucose sensor for β-cells. As soon as glucose enters, glucose catabolism is activated, growing the intracellular ATP/ADP ratio, which induces the closing of ATP-dependent potassium channels within the plasma membrane. This results in membrane depolarization and opening of the voltage dependent Ca^{2+} channels, allowing Ca^{2+} to go into the cell. The upward thrust inside the intracellular Ca^{2+} concentration triggers the priming and fusion of the secretory insulin-containing granules to the plasma membrane, resulting in insulin release (Fu et al., 2013; Boland et al., 2017).

Insulin resistance is a crucial characteristic of T2D mellitus. Within the mechanisms of insulin resistance, adipose tissue, skeletal muscle, and liver are primary organs concerned in glucose metabolism and, consequently, play vital roles in insulin resistance or disorder of β-cells. This is probably due to a more complicated network of interactions among the surroundings and distinct molecular pathways intimated in molecular biology (Christensen and Gannon, 2019). The latest evidence recommended that inflammatory stress, oxidative or metabolic stress results in a lack of beta cells physiology and accordingly affects the insulin functioning. Numerous molecular mechanisms are postulated for insulin resistance, including an extra of FFAs and hyperglycemia result in β-cell disorder through inducing ER stress via the activation of the apoptotic unfolded protein response (UPR) pathways (Yamamoto et al., 2019). In other mechanism it showed that an increase in the intracellular concentration of fatty acid metabolites turns on a serine kinase cascade, which results in defects in insulin signaling downstream to the insulin receptor. Many molecules involved inside the intracellular processing of the signal provided via insulin, the insulin receptor substrate-2, the protein kinase B and the transcription factor FoxO1. Information has provided sturdy proof that dysfunction of these proteins' outcomes in insulin resistance. Currently, research has found out that phosphoinositide dependent kinase 1-independent phosphorylation of protein kinase C causes a reduction in insulin receptor gene expression (Saini, 2010). Research has proven that insulin resistance is also related to mitochondrial disorder, studies are still taking place to find the role of mitochondrial disorder within the improvement of insulin resistance and etiology of type 2. Moreover, it's been recommended that mitochondrial dysfunction triggers activation of numerous serine kinases, and diminishes the insulin signal transduction.

13.3 DIETARY FACTORS FOR THE PREVENTION OF TYPE 2 DIABETES (T2D) MELLITUS

Economic growth and environmental modifications have led to intense modifications in food manufacturing, processing, and distribution systems and have improved the supply of unhealthful meals. With dietary habit changes, women, and men around the world have experienced extra body weight (BW) gain in large part because of modifications in nutritional styles and reduced physical interest related to elevated diabetes prevalence and mortality (Ezzati and Riboli, 2010). Extra adiposity meditated by way of higher body mass index (BMI) is the most powerful threat factor for diabetes, and Asians generally tend to increase diabetes at a much reduced BMI than the ones of European ancestry (Hu, 2008). The chance of diabetes rises as an excess growth in body fat, beginning from BMI or waist circumference.

Research has provided proof that individuals (who consumes numerous food items) and food organizations can also play a role in diabetes prevention. Complete grain intake has been continuously related to decrease diabetes risk (Aune et al., 2013). Contrarily, extra consumption of white rice, a processed grain, related to elevated diabetes risk (Hu et al., 2012), particularly amongst Asian populations with white rice as a staple meal and a major supply of calories. Frequent intake of red meats, specifically processed red meats together with Bacon, sausages, and hot dogs, become strongly related to higher diabetes threat.

In an analysis through Wu et al. (2012), they determined that fish and/or seafood intake was not considerably related to diabetes risk. Rather, a difference in the route of the association among fish/seafood intake and diabetes danger was determined between geographical areas. Higher fish/seafood intake was related to higher diabetes risk in North America and Europe; however, it was related to decrease threat in Asia (Wu et al., 2012).

Consumption of fruits and vegetables is no longer related to diabetes risk, but consumption of green leafy vegetables was associated with reduced risk (Carter et al., 2010; Cooper et al., 2012). Similarly, intake of particular entire fruits which includes blueberries, grapes, and apples were extensively related to decrease diabetes risk (Muraki et al., 2013).

Eating a more quantity of dairy products has been related to reasonably decrease diabetes risk (Tong et al., 2011) and the advantages of yogurt appear to be more constant than different kinds of dairy products. Intake of nuts that are high in PUFA and monounsaturated fatty acids (MUFA), might also have useful outcomes on diabetes prevention. More nut intake, particularly walnuts, was related to decrease diabetes risk (Pan et al., 2013; Keservani et al., 2010a, b).

Researchers experimentally confirmed that these dietary styles favoring fruits, green vegetables, whole grains, honey as a sugar alternative (Agarwal et al., 2007) and legumes on the rate of red meats, refined grains, are useful for diabetes prevention (Imamura et al., 2009; Liese et al., 2009).

13.4 FUNCTIONAL HERBAL FOOD INGREDIENTS USED FOR TYPE II DIABETES

Scientific observations have proven that nutrients perform an essential function in the prevention of chronic diseases like diabetes, as a maximum of them may be associated with diet. Functional food most effectively taken into consideration not only for living but additionally as a supply of mental and physical well-being, presenting the prevention and reduction of danger factors for numerous illnesses or improving certain physiological features (López-Varela et al., 2002; Rane et al., 2020).

The natural dietary supplements discussed in this chapter have been used for human intake for hundreds of years, and their safety in humans has been nicely established. With appropriate protection profile and comfort, these dietary supplements having attractive attention to the ordinary pharmacological treatment options in diabetic patients, especially for those whose blood glucose levels are not properly managed through the therapies alone (Lee and Dugoua, 2011; Davì et al., 2010; Najm and Lie, 2010; Bartlett and Eperjesi, 2008).

13.4.1 MOMORDICA CHARANTIA

Bitter gourd (Momordica charantia) is a famous vegetable that's historically recognized to have health useful outcomes mainly on diabetes. It's far a reasonably-priced vegetable which allows the economically susceptible people to manage their blood glucose levels. It consists of bioactive chemical compounds, vitamins, minerals, and antioxidants. Analyzes of bioactive compounds (Charantin, Polypeptide-p, Vicine, and others) have proven that bitter gourd is rich in nutrients and phytochemicals (triterpene, proteid, steroid, alkaloid, inorganic, lipid, and phenolic compounds) which might be accountable for the anti-diabetic effects. Consumption of bitter gourd will increase insulin secretion from the pancreas, decreases intestinal glucose uptake, and elevates uptake and usage of glucose in peripheral tissues (Sandra et al., 2014) (Figure 13.1(a)).

13.4.2 ALLIUM SATIVUM (GARLIC)

Allium sativum L. (garlic), has a completely long folks records of use, to prevent and deal with a numerous range of health issues. More than 400 bioactive compounds were recognized from garlic. It has been discovered to be powerful in reducing serum glucose levels by way of diverse proposed mechanism. Compounds such as allicin, allyl propyl disulfide and S-allyl cysteine sulfoxide, an isolated product from garlic, may additionally make contributions to its useful impact in diabetes via the prevention of the liver's inactivation of insulin.

A few scientific observations have postulated that garlic may also act as an antidiabetic agent through increasing either the pancreatic secretion of insulin from the β cells or release of bound insulin. Consequently, the hypoglycemic action of garlic may probably be because of an increase in secretion of insulin from β cells of the pancreas, release of certain insulin or enhancement of insulin sensitivity (Perera and Li, 2012) (Figure 13.1(b)).

13.4.3 EMBLICA OFFICINALIS (AMLA)

Amla forms an essential part of the Indian system of medicine and is effective in the prevention of DM, hyperlipidemia, oxidative stress, and other diseases. It's far the richest source of vitamin C and consists of gallic acid, ellagic acid and flavonoids (Keservani and Sharma, 2014). Amla prevents or delays absorption of glucose by means of the inhibition of carbohydrate hydrolyzing enzymes, α-amylase, and α-glucosidase, within the digestive organs. It may also act by stimulating the insulin release from the beta cells of the pancreas in type 2 diabetics. Therefore, it's far having hypoglycemic effects and might be used for managing hyperglycemia (Patell, 2011) (Figure 13.1(c)).

13.4.4 FENUGREEK

Fenugreek, additionally called by its scientific name *Trigonella foenum-graecum* L., *leguminosae*. Fenugreek has many health-promoting properties, while its miles taken as vegetables, food supplements or in medicinal treatments. Fenugreek's pharmacological consequences are attributed to a variety of bioactive compounds consisting of polyphenols, steroids, lipids, alkaloids, saponins, flavonoids, hydrocarbons, carbohydrates, galactomannan fiber, and amino acids. Because of its antioxidant activity, it performs a substantial function in hepatoprotection and blood glucose reduction (Naicker et al., 2016) (Figure 13.1(d)).

13.4.5 CINNAMON

Cinnamon is commonly used as a spice internationally. The dry bark of cinnamon trees is rich in botanical source of polyphenolics and has been used to enhance fashionable health and deal with lots of ailment conditions including DM (Altschuler et al., 2007; Blevins et al., 2007). Cinnamaldehyde turns on PPAR gamma to upregulate insulin and thereby exhibit a glucose-decreasing impact. It additionally activates AMPK pathway to inhibit gluconeogenesis. By means of these two mechanisms, it is clear that cinnamaldehyde acts similarly to thiazolidinediones and alpha-glucosidase inhibitors (Sarian et al., 2017; Hosni et al., 2017) (Figure 13.1(e)).

13.4.6 MURRAYA KOENIGII (L.) (NIGELLA SATIVA)

Murraya koenigii L., famous in medical usage. The leaves of *Murraya koenigii* are also used as an herb in Indian medicinal drug. The seeds of Nigella sativa, additionally referred to as black seeds are been used for hundreds of years as a natural treatment for numerous illnesses. It is having antihyperglycemic, antioxidants along with anti-hypertensive, antihyperlipidemic, and anti-microbial activity. The chemical components are nigellicine, nigellidine, thymol, linoleic, oleic, and so forth (Yimer et al., 2019). The bioactive constituent for diabetes is thymoquinone (1.8%–48%). The plant enables in insulin secretion by means of restoring the activity of enzymes involved in glucose metabolism which includes glucose-6-phosphatase and fructose-1,6-phosphatase, and fructose-1,6-bisphosphatase of gluconeogenesis. It also acts like SGLT-2 inhibitors and thiazolidinediones (Gholamnezhad et al., 2016) (Figure 13.1(f)).

13.4.7 GURMAR (GYMNEMA SYLVESTRE)

Gymnema sylvestre (*G. sylvestre*) popularly referred to as "gurmar," this means that "destroyer of sugar" in Hindi (Yeh et al., 2003). The phytochemicals answerable for antidiabetic activity consists of triterpene saponins called gymnemic acids, gymnemasaponins, and a polypeptide, gurmarin. The herb reveals a vast variety of therapeutic effects as a powerful natural treatment for diabetes, besides being used for arthritis, diuretic, anemia, osteoporosis, hypercholesterolemia, bronchial asthma, constipation, microbial infections, indigestion, and anti-inflammatory. It is well known herb in the remedy of diabetes because it indicates effective results on blood sugar homeostasis, controls sugar cravings, and promotes regeneration of pancreas (Tiwari et al., 2014; Leach, 2007; Khramov et al., 2008; Sunil et al., 2018) (Figure 13.1(g)).

13.4.8 EUGENIA JAMBOLANA (JAMUN)

Eugenia jambolana, Family: Myrtaceae), normally called black plum or Jamun is a plant local to India. The presence of bioactive compounds which include alkaloids, tannins, phenols, lipids, flavonoids in its leaves, barks, fruits, stems, and roots contributes to rich supply for nutrients and medicinal drug. Because of the presence of these bioactive compounds, it possesses antioxidant, antimicrobial, anti-inflammatory, hepatoprotective, and antidiabetic properties. Eugenia jambolana does have very well pronounced hypoglycemic, or blood-sugar reducing action (Achrekar et al., 1991) (Figure 13.1(h)).

13.4.9 AZADIRACHTA INDICA (NEEM)

Neem (Azadirachta indica; family: Meliaceae), a tropical evergreen tree, is broadly disbursed in Asia. Neem leaves and seeds are utilized in Ayurveda as a household treatment. Distinctive bioactive compounds were remoted from neem leaves, which includes nimbin, azadirachtin, nimbidiol, quercetin, nimbidin, and so on. Lots of *vitro* and *in vivo* research have mentioned hypoglycemic activity (Sanni et al., 2019; Satyanarayana et al., 2015; Perez et al., 2014; Saidu et al., 2017). It slows down the action of epinephrine on glucose metabolism, arising usage of peripheral glucose and displaying hypoglycemic activity without modifying the serum cortisol concentration (Dubey et al., 2019; Ponnusamy et al., 2015) (Figure 13.1(i)).

13.4.10 OLIVE OIL

Olive oil, as meals consists specially of fatty acids and bioactive compound, has been related to the prevention and management of many continual illnesses which includes T2D. Even though it became also the principle component of the Mediterranean food regimen. Numerous bioactive compounds including monounsaturated fatty acids, and bio-phenols which include hydroxytyrosol and oleuropein extracted from olive oil. They had been associated with preventing inflammation and cytokine-triggered oxidative damage, reducing blood glucose level, lowering carbohydrate absorption, increasing insulin sensitivity and associated gene expression. Numerous different antidiabetic mechanisms were mentioned, in addition to it is also used as a potential Immunomodulatory (Keservani et al., 2015), antiproliferative, and antioxidative agent (Alkhatib et al., 2018) (Figure 13.1(j)).

13.4.11 MICROALGAE

Microalgae are photosynthetic, filamentous organisms that occur in fresh and saltwater (Ramos-Romero et al., 2021). Arthrospira (also called Spirulina), Chlorella, and Tetraselmis are presently used for human nutrition (Neyrinck et al., 2017). They are rich in proteins (approx. 70%) (Koyande et al., 2019; Batista et al., 2013) and consists of numerous bioactive additives which includes dietary fiber, polyphenols, carotenoids, phycobiliproteins, polysaccharides, vitamins, minerals, sterols, and omega-3 polyunsaturated fatty acids (PUFAs) along with eicosapentaenoic acid (EPA, 20:5 n–3) and docosahexaenoic acid (DHA, 22:6 n–3) (Archana et al., 2008). Presence of these bioactive components makes it powerful towards hepatotoxicity, inflammation, enhancing immunity, CDVs, microbial infection, rheumatic arthritis, oxidative stress, DM, and other illnesses. Microalgae are one of the ingredients of the future, as they are the source of high-quality protein and consist of an array of bioactive molecules with healing capability to act towards the current epidemics of obesity, malnutrition, DM, and associated complications (Jarouliya et al., 2015; Wang et al., 2015; Gómez-Zavaglia et al., 2019) (Figure 13.2).

FIGURE 13.1 (a) *Momordica charantia*; (b) *Allium sativum*; (c) *Emblica officinalis*; (d) fenugreek; (e) cinnamon; (f) *Murraya koenigii*; (g) gurmar; (h) *Eugenia jambolana*; (i) *Azadirachta indica*; (j) olive oil.

FIGURE 13.2 Microalgae in powder and tablet form.
Source: Indiatimes.com.

13.5 MICRONUTRIENTS IN MANAGING TYPE 2 DIABETES (T2D)

Micronutrients consist of vitamins, minerals, and other bioactive compounds that our body needs in small amounts in order to accomplish certain functions and helps in regulating macromolecules metabolism.

13.5.1 ALPHA LIPOIC ACID

One of the most promising herbal remedies for diabetes is alpha-lipoic acid (ALA). Animal products like red meat (pork) and organ meats are excellent resources of alpha-lipoic acid, however plant meals like broccoli, tomatoes, spinach, and Brussels sprouts additionally include it. The researchers concluded that a 3-week remedy using 600 mg/day of the component rich in ALA decreased the chief symptoms of diabetic neuropathy – a complication of diabetes that influences the nerves (Packer et al., 2001; Baldi et al., 2013). ALA is a strong anti-oxidants and has been proven to remarkably diminish the signs and symptoms of neuropathy. It has also been proven to enhance insulin sensitivity via recruiting glucose transporter-4 to plasma membranes inflicting uptake of glucose. New research has additionally recommended that ALA stimulates glucose disposal in patients with T2D (Starface et al., 2002).

13.5.2 L-CARNITINE

L-carnitine, a natural vitamin-like compound, is provided via dietary assets (e.g., meat, dairy products), and is synthesized commonly within the liver and kidneys from lysine and methionine. Research have proven that L-carnitine promotes insulin sensitivity and has lipid-reducing actions. L-carnitine plays some of essential intracellular and metabolic functions, which include fatty acid transport across the inner mitochondrial membrane into the matrix for beta-oxidation, detoxing of potentially toxic metabolites, regulation of the mitochondrial acyl-Co A/CoA ratio, and stabilization of cell membranes (Dayananadan et al., 2001; Nimesh and Ashwalayan, 2018). Two feasible mechanisms are probably concerned in the metabolic effect of L-carnitine: the primary is a management of acetyl and acyl cellular trafficking for correct meeting for the energy demand; the second action is for the synthesis of key glycolytic and gluconeogenic enzymes.

13.5.3 OMEGA-3 FATTY ACID

The body can't make omega-3 fatty acids, so ought to get them from nutritional resources. The great resources are fatty fish (together with salmon, mackerel, sardines, herring, and tuna), flaxseed, night time primrose, borage, walnuts, and wheat germ (American Heart Association (AHA), 2016). Omega-3 fatty acids have insulin-sensitizing results via elevated production and secretion of adipocytokines which includes adiponectin and leptin (Perez-Matute et al., 2007; Oh et al., 2010; Gonzalez-Periz et al., 2009; Serhan et al., 2008). Through modulation of transcription factors, omega-3 fatty acid may also enhance fatty acid oxidation and decrease de novo lipogenesis. These outcomes show a reduction in hepatic fat accumulation and maintain hepatic insulin sensitivity (Tanaka et al., 2010; Lewis et al., 2017).

13.5.4 CHROMIUM

Chromium is a trace mineral that the body often desires in small doses, and have a useful role within the regulation of insulin action and its impact on lipids, proteins, and carbohydrates metabolism. Chromium is found in many foods, particularly in liver, Brewer's yeast, American cheese, wheat germ, vegetables such as carrots, potatoes, broccoli, and spinach, and is likewise found in alfalfa, brown sugar, molasses, dried beans, nuts, seeds, mushrooms, and animal fat. The binding of chromium stabilizes the active conformation of insulin receptor tyrosine kinase, there by way of facilitating the action of insulin (Cefalu et al., 2002).

13.5.5 MAGNESIUM

Magnesium is an important cofactor located in lots of herbal foods like nuts, whole grains, green leafy vegetables, and legumes in addition to fortified foods (O'Connell, 2001). Intracellular Mg performs a key function in regulating insulin action, insulin-mediated glucose uptake, and vascular tone. Mg deficit has been proposed as a probable underlying common mechanism of the "insulin resistance" because of defect in defective tyrosine-kinase activity.

13.5.6 VITAMIN D AND CALCIUM

Vitamin D/calcium may be consumed via certain meals, which include egg yolk, saltwater fish, animal liver, cheese, nuts, and vitamin D-fortified milk products and cereals. It's far required for the proper functioning of the nerves, muscle tissue, and immune system. It is able to play a useful role in glucose tolerance via its effects on insulin secretion and insulin sensitivity (Mitri et al., 2011). Multiple mechanism had given for the outcomes of vitamin D. On T2D, research have observed that the pancreas consists of receptors for the active vitamin D metabolite known as 1,25-dihydroxyvitamin D, which is required for the synthesis and secretion of insulin through the pancreatic beta cells (Medline Plus, 2019).

13.5.7 VITAMIN H (BIOTIN)

Vitamin H, more usually referred to as biotin, is a part of the B complex group of vitamins. All B vitamins assist the body to convert food (carbohydrates) into fuel (glucose), that is used to supply energy. Biotin levels have an effect on blood sugar levels and have a tendency to be decrease in humans with T2D. A mixture supplement that consists of chromium picolinate and biotin can also lessen healthcare prices via managing blood sugar level. Biotin supplementation has been proven to enhance the overall performance of insulin, the hormone that performs an essential role in metabolizing blood sugar. The dietary supplements also can increase the activity of an enzyme, glucokinase, which the liver uses early in the process of metabolizing blood sugar (Will and Byers, 1996).

13.5.8 VITAMIN C

Vitamin C is a water-soluble antioxidant that has been investigated therapeutically in human beings with T2D. Vitamin C supplementation improves glycemic control, blood lipids, blood pressure, and oxidative stress in human beings with T2D. It enables the prevention or reduction of cataract and nerve problems that usually arise in humans with diabetes. It inhibits protein glycosylation, that's an essential pathway in the development of the long-term complications related to diabetes (Mason et al., 2021; Keservani et al., 2016).

13.6 PHYSICAL ACTIVITY

A sedentary way of life is every other threat element for T2DM. Physical activity have significant effect in the prevention and treatment of T2D along with recommended diet. The recommended type of activity is aerobic exercise, which seems to be the healthiest and safest for diabetic patients. The effort of adequate intensity – low or moderate – burns sufficient calorie which promotes weight loss. Also, glucose control is improved, tissue insulin sensitivity is increased, and HbA1c levels in the blood is decreased. The recommended duration of exercise is about 30 minutes. Short-term, but more frequent physical activity is more beneficial than long-term training performed less often. In light of the studies done by researchers, 30-minute aerobic exercise with low or moderate intensity seems to be the best choice for obese people. It should be repeated as often as possible, preferably every day for effectiveness and healthy life (Klimek et al., 2019).

13.7 CONCLUSION

Nutraceuticals, commonly known as food supplements obtained from natural sources have been studied for DM for centuries. *Gymnema Sylvestre, Momordica charantia, Nigella sativa, Allium sativum, Emblica officinalis, Murraya koenigii, Eugenia Jambolana, Azadirachta indica and*

Cinnamomum cassia have good mechanism and efficacy in preventing DM supported by good quality clinical trials and previous animal studies (Keservani et al., 2016). Anti-hyperglycemic effects of these drugs are attributed to their ability to increase insulin secretion, decrease insulin resistance, and reduce the intestinal absorption of glucose. Few nutraceuticals are used as a preventive intervention in clinical studies.

Micronutrients (vitamins and minerals) are a subset of the larger organization of acknowledged dietary important nutrients for humans' health which functions in preserving the metabolism and physiological processes of the human body.

Microalgae (unicellular algae) will represent one of the fundamental foods of the future for nutritional and environmental reasons. They're resources of high-quality protein and bioactive molecules with the ability to manage obesity and diabetes in present situations. Combining these nutraceuticals food supplements along with physical activity could potentially reduce and prevent the adverse effects associated with DM. Thus, physicians can rely on these nutraceuticals as complementary therapeutics, in managing the T2D mellitus.

KEYWORDS

- **body mass index**
- **glucose transporter 2**
- **hyperglycemia**
- **micronutrients**
- **monounsaturated fatty acids**
- **nutraceutical-functional foods**
- **physical activity**
- **type 2 (non-insulin-dependent) diabetes mellitus**

REFERENCES

Achrekar, S., Kaklij, G. S., Pote, M. S., & Kelkar, S. M., (1991). Hypoglycemic activity of *Eugenia jambolana* and *Ficus bengalenesis*: Mechanism of action. *In Vivo, 5*, 143–147.

Agrawal, O. P., Pachauri, A., Yadav, H., Urmila, J., Goswamy, H. M., Chapperwal, A., Bisen, P. S., & Prasad, G. B., (2007). Subjects with impaired glucose tolerance exhibit a high degree of tolerance to honey. *J. Med. Food., 10*, 473–478.

Ahmad, A., Catherine, T., & Jaakko, T., (2018). Olive oil nutraceuticals in the prevention and management of diabetes: From molecules to lifestyle. *Int. J. Mol. Sci., 19*, 2024.

Altschuler, J. A., Casella, S. J., MacKenzie, T. A., & Curtis, K. M., (2007). The effect of cinnamon on A1C among adolescents with type 1 diabetes. *Diabetes Care, 30*, 813–816.

American Diabetes Association, (2013). Diagnosis and classification of diabetes mellitus. *Diabetes Care, 36*, S67–S74.

American Heart Association, (2016). *Fish and Omega-3 Fatty Acids*. Retrieved from: http://www.heart.org (accessed on 25 October 2022).

Archana, K., Anish, Z., Urmila, J., Pratiksha, B., Prasad, G. B. K. S., & Bisen, P. S., (2008). Spirulina in health care management. *Curr. Pharm. Biotechnol., 9*, 400–405.

Aune, D., Norat, T., Romundstad, P., & Vatten, L. J., (2013). Whole grain and refined grain consumption and the risk of type 2 diabetes: A systematic review and dose-response meta-analysis of cohort studies. *Eur. J. Epidemiol., 28*, 845–858.

Baldi, A., Choudhary, N., & Kumar, S., (2013). Nutraceuticals as therapeutic agents for holistic treatment of diabetes. *Int. J. Green Pharm., 7*, 278–287.

Bartlett, H. E., & Eperjesi, F., (2008). Nutritional supplementation for type 2 diabetes: A systematic review. *Ophthalmic Physiol. Opt., 28*, 503–523.

Batista, A. P., Gouveia, L., Bandarra, N. M., Franco, J. M., & Raymundo, A., (2013). Comparison of microalgal biomass profiles as novel functional ingredient for food products. *Algal Res*., 2, 164–173.

Blevins, S. M., Leyva, M. J., Brown, J., Wright, J., Scofield, R. H., & Aston, C. E., (2007). Effect of cinnamon on glucose and lipid levels in non-insulin-dependent type 2 diabetes. *Diabetes Care, 30*, 2236, 2237.

Boland, B. B., Rhodes, C. J., & Grimsby, J. S., (2017). The dynamic plasticity of insulin production in beta-cells. *Mol. Metab., 6*, 958–973.

Carter, P., Gray, L. J., Troughton, J., Khunti, K., & Davies, M. J., (2010). Fruit and vegetable intake and incidence of type 2 diabetes mellitus: Systematic review and meta-analysis. *BMJ., 341*, 4229.

Cefalu, W. T., Wang, Z. Q., Zhang, X. H., Baldor, L. C., & Russell, J. C., (2002). Oral Chromium pico-Linate improves carbohydrate and lipid metabolism and enhances skeletal muscle Glut-4 translocation in obese, hyperinsulinemic (JRC-LA Corpulent) rats. *J. Nutr., 132*, 1107–1114.

Christensen, A. A., & Gannon, M., (2019). The beta cell in type 2 diabetes. *Curr. Diabetes Rep., 19*, 81.

Cooper, A. J., Forouhi, N. G., Ye, Z., Buijsse, B., Arriola, L., Balkau, B., Barricarte, A., Beulens, J. W., Boeing, H., et al., (2012). Fruit and vegetable intake and type 2 diabetes: EPIC-interact prospective study and meta-analysis. *Eur. J. Clin. Nutr., 66*, 1082–1092.

Davì, G., Santilli, F., & Patrono, C., (2010). Nutraceuticals in diabetes and metabolic syndrome. *Cardiovasc. Ther., 28*, 216–226.

Dayanandan, A., Kumar, P., & Panneerselvam, C., (2001). Protective role of L-carnitine on liver and heart lipid peroxidation in atherosclerotic rats. *J. Nutr. Biochem., 12*, 254–257.

Dubey, R., Patil, K., Dantu, S. C., et al., (2019). Azadirachtin inhibits amyloid formation, disaggregates pre-formed fibrils and protects pancreatic βcells from human islet amyloid polypeptide/amylin-induced cytotoxicity. *Biochem. J., 476*, 889–907.

Ezzati, M., & Riboli, E., (2013). Behavioral and dietary risk factors for non-communicable diseases. *N. Engl. J. Med., 369*, 954–964.

Fu, Z., Gilbert, E. R., & Liu, D., (2013). Regulation of insulin synthesis and secretion and pancreatic Beta-cell dysfunction in diabetes. *Curr. Diabetes Rev., 9*, 25–53.

Gholamnezhad, Z., Havakhah, S., & Boskabady, M. H., (2016). Preclinical and clinical effects of *Nigella sativa* and its constituent, thymoquinone: A review. *J. Ethnopharmacol., 190*, 372–386.

Gómez-Zavaglia, A., Prieto, M. A., Jiménez-López, C., Mejuto, J. C., & Simal-Gandara, J., (2019). The potential of seaweeds as a source of functional ingredients of prebiotic and antioxidant value. *Antioxidants, 8*, 406.

Gonzalez-Periz, A., Horrillo, R., Ferre, N., et al., (2009). Obesity-induced insulin resistance and hepatic steatosis are alleviated by omega-3fattyacids: A role for resolvins and protectins. *FASEB J., 23*, 1946-57.

Habicht, S. D., Ludwig, C., Yang, R. Y., & Krawinkel, M. B., (2014). *Momordica charantia* and type 2 diabetes: From *in vitro* to human studies. *Curr. Diabetes Rev., 10*, 48–60.

Hosni, A. A., Abdel-Moneim, A. A., Abdel-Reheim, E. S., Mohamed, S. M., & Helmy, H., (2017). Cinnamaldehyde potentially attenuates gestational hyperglycemia in rats through modulation of PPARγ, proinflammatory cytokines and oxidative stress. *Biomed. Pharmacother., 88*, 52–60.

Hu, E. A., Pan, A., Malik, V., & Sun, Q., (2012). White rice consumption and risk of type 2 diabetes: Meta-analysis and systematic review. *BMJ, 344*, e1454.

Hu, F. B., (2008). Obesity epidemiology. *Metabolic Consequences of Obesity* (pp. 149–173). New York, New York: Oxford University Press.

Hu, F. B., Manson, J. E., Stampfer, M. J., Colditz, G., Liu, S., Solomon, C. G., & Willett, W. C., (2001). Diet, lifestyle, and the risk of type 2 diabetes mellitus in women. *N. Engl. J., Med., 345*, 790–797.

Imamura, F., Lichtenstein, A. H., Dallal, G. E., Meigs, J. B., & Jacques, P. F., (2009). Generalizability of dietary patterns associated with incidence of type 2 diabetes mellitus. *Am. J. Clin. Nutr., 90*, 1075–1083.

Jarouliya, U., Zacharia, A., Keservani, R. K., & Prasad, G. B. K. S., (2015). Spirulina maxima and its effect on antioxidant activity in fructose induced oxidative stress with histopathological observations. *Eur. Pharm. J., 62*, 13–19.

Jarouliya, U., Zacharia, J. A., Kumar, P., Bisen, P. S., & Prasad, G. B., (2012). Alleviation of metabolic abnormalities induced by excessive fructose administration in Wistar rats by Spirulina maxima. *Indian J. Med. Res., 135*, 22–28.

Keservani, R. K., & Sharma, A. K., (2014). Flavonoids: Emerging trends and potential health benefits. *J. Chin. Pharm. Sci., 23*(12), 815–822.

Keservani, R. K., Kesharwani, R. K., Sharma, A. K., & Jarouliya, U., (2015). Dietary supplements, nutraceutical and functional foods in immune response (immunomodulators). In: Bagchi, D., Preuss, H. G., & Swaroop, A., (ed.), *Nutraceutical and Functional Foods in Human Life and Disease Prevention* (pp. 343–358). Boca Raton, Fl. Taylor and Francis, Ch 20. ISBN: 9781482237214.

Keservani, R. K., Kesharwani, R. K., Sharma, A. K., Vyas, N., & Chadoker, A., (2010b). Nutritional supplements: An overview. *Int J. of Curr. Pharm. Rev. and Res., 1*(1), 59–75.

Keservani, R. K., Kesharwani, R. K., Vyas, N., Jain, S., Raghuvanshi, R., & Sharma, A. K., (2010a). Nutraceuticals and functional food as future food. *Der Pharm. Lett, 2*(1), 106–116.

Keservani, R. K., Sharma, A. K., & Kesharwani, R. K., (2016). An overview and therapeutic applications of nutraceutical & functional foods. In: Keservani, R. K., Sharma, A. K., &

Kesharwani, R. K., (eds.), *Recent Advances in Drug Delivery Technology* (pp. 160–201). IGI Global Publisher, USA, Ch 6. ISBN:9781522507543.

Khramov, V. A., Spasov, A. A., & Samokhina, M. P., (2008). Chemical composition of dry extracts of *Gymnema sylvestre* leaves. *Pharm. Chem. J.*, 42, 29–31.

Klimek, M., Knap, J., Reda, M., & Masternak, M., (2019). Physical activity in prevention and treatment of type 2 diabetes mellitus. *J. Educ. Health. Sport.,* 9, 1175–1181.

Koyande, A. K., Chew, K. W., Rambabu, K., Tao, Y., Chu, D. T., & Show, P. L., (2019). Microalgae: A potential alternative to health supplementation for humans. *Food Sci. Hum. Wellness, 8*, 16–24.

Leach, M. J., (2007). *Gymnema sylvestre* for diabetes mellitus: A systematic review. *J. Altern. Complement. Med., 13*, 977–983.

Lee, T., & Dugoua, J. J., (2011). Nutritional supplements and their effect on glucose control. *Curr. Diab. Rep., 11*, 142–148.

Lewis, E. J., Perkins, B. A., Lovblom, L. E., Bazinet, R. P., Wolever, T. M., & Bril, V., (2017). Omega-3 supplementation on neuropathy in type 1 diabetes. *Neurology, 88*, 2294–2301.

Liese, A. D., Weis, K. E., Schulz, M., & Tooze, J. A., (2009). Food intake patterns associated with incident type 2 diabetes: The insulin resistance atherosclerosis study. *Diabetes Care, 32*, 263–268.

López-Varela, S., Gonzαlez-Gross, M., & Marcos, A., (2002). Functional foods and the immune system: A review. *Eur. J. Clin. Nutr., 56*, S29.

Mason, S. A., Keske, M. A., & Wadley, G. D., (2021). Effects of vitamin C supplementation on glycemic control and cardiovascular risk factors in people with type 2 diabetes: A GRADE-assessed systematic review and meta-analysis of randomized controlled trials. *Diabetes Care, 44*, 618–630.

Medline Plus, (2019). *Vitamin D*. https://medlineplus.gov/vitamind.html (accessed on 02 August 2022).

Mitri, J., Dawson-Hughes, B., Hu, F. B., & Pittas, A. G., (2011). Effects of vitamin D and calcium supplementation on pancreatic β cell function, insulin sensitivity, and glycemia in adults at high risk of diabetes: The calcium and vitamin D for diabetes mellitus (CaDDM) randomized controlled trial. *Am. J. Clin. Nutr., 94*, 486–494.

Muraki, I., Imamura, F., Manson, J. E., Hu, F. B., Willett, W. C., Van, D. R. M., & Sun, Q., (2013). Fruit consumption and risk of type 2 diabetes: Results from three prospective longitudinal cohort studies. *BMJ*, 347.

Naicker, N., Nagiah, S., Phulukdaree, A., & Chuturgoon, A., (2016). *Trigonella foenum-graecum* seed extract, 4-hydroxyisoleucine, and metformin stimulate proximal insulin signaling and increase expression of glycogenic enzymes and GLUT2 in HepG2 cells. *Metab. Syndr. Relat. Disord., 14*, 114–120.

Najm, W., & Lie, D., (2010). Herbals used for diabetes, obesity, and metabolic syndrome. *Prim. Care, 37*, 237–254.

Neyrinck, A. M., Taminiau, B., Walgrave, H., Daube, G., Cani, P. D., Bindels, L. B., & Delzenne, N. M., (2017). Spirulina protects against hepatic inflammation in aging: An effect related to the modulation of the gut microbiota? *Nutrients, 9*, 633.

Nimesh, S., & Ashwlayan, V. D., (2018). Nutraceuticals in the management of diabetes mellitus. *Pharm. Pharmacol. Int. J., 6*, 114–120.

O'Connell, B. S., (2001). Select vitamins and minerals in the management of diabetes. *Diabetes Spectr., 14*, 133.

Oh, D. Y., Talukdar, S., Bae, E. J., Imamura, T., Morinaga, H., Fan, W., Li, P., Lu, W. J., Watkins, S. M., (2010). GPR120 is an omega-3 fatty acid receptor mediating potent anti-inflammatory and insulin-sensitizing effects. *Cell, 3*(142), 687–698.

Packer, L., Kraemer, K., & Rimbach, G., (2001). Molecular aspects of alpha lipoic acid in the prevention of diabetes complications. *Nutrition, 17*, 888–895.

Pan, A., Sun, Q., Manson, J. E., Willett, W. C., & Hu, F. B., (2013). Walnut consumption is associated with lower risk of type 2 diabetes in women. *J. Nutr., 143*, 512.

Patell, V. M., (2011). *Screening Method (Metabolite Grid) for Therapeutic Extracts and Molecules for Diabetes.* US. 20110159118.

Pe´rez-Matute, P., Pe´rez-Echarri, N., Martínez, J. A., & Marti, A., (2007). Eicosapentaenoic acid actions on adiposity and insulin resistance in control and high-fat-fed rats: Role Of apoptosis, adiponectin and tumor necrosis factor-alpha. *Br. J. Nutr., 97*, 389–398.

Perera, P. K., & Li, Y., (2012). Functional herbal food ingredients used in type 2 diabetes mellitus. *Pharmacogn. Rev., 6*, 37–45.

Perez, G. R. M., & De Jesus, M. O. M., (2014). Beneficial effect of *Azadirachta indica* on advanced glycation end-product in streptozotocin-diabetic rat. *Pharm. Biol., 52*, 1435–1444.

Ponnusamy, S., Haldar, S., Mulani, F., Zinjarde, S., Thulasiram, H., & Ravikumar, A., (2015). Gedunin and azadiradione: Human pancreatic alpha-amylase inhibiting limonoids from neem (*Azadirachta indica*) as anti-diabetic agents. *PLoS One, 10*, e0140113.

Ramos-Romero, S., Torrella, J. R., Pagès, T., Viscor, G., & Torres, J. L., (2021). Edible microalgae and their bioactive compounds in the prevention and treatment of metabolic alterations. *Nutrients, 13*, 563.

Rane, B. R., Bharath, M. S., Patil, R. R., Keservani, R. K., & Jain, A. S., (2020). Novel approaches in nutraceuticals. In: Kesharwani, R. K., Keservani, R. K., & Sharma, A. K., (eds.), *Enhancing the Therapeutic Efficacy of Herbal Formulations Through Novel Drug Delivery Systems* (pp. 241–266). IGI Global International Publisher, Pennsylvania, USA, Chapter 11. ISBN13: 9781799844532.

Saidu, Y., Muhammad, S. A., Abbas, A. Y., Onu, A., Tsado, I. M., & Muhammad, L., (2017). *In vitro* screening for protein tyrosine phosphatase 1B and dipeptidyl peptidase IV inhibitors from selected Nigerian medicinal plants. *J. Intercult. Ethnopharmacol., 22*(6), 154–157.

Saini, V., (2010). Molecular mechanisms of insulin resistance in type 2 diabetes mellitus. *World J. Diabetes, 15*(1), 68–75.

Sanni, O., Erukainure, O. L., Chukwuma, C. I., Koorbanally, N. A., Ibeji, C. U., & Islam, M. S., (2019). *Azadirachta indica* inhibits key enzyme linked to type 2 diabetes *in vitro,* abates oxidative hepatic injury and enhances muscle glucose uptake ex vivo. *Biomed. Pharmacother., 109*, 734–774.

Sarian, M. N., Ahmed, Q. U., Mat So'ad, S. Z., Alhassan, A. M., Murugesu, S., Perumal, V., Syed, M. S. N. A., et al., (2017). Antioxidant and antidiabetic effects of flavonoids: A structure-activity relationship based study. *Biomed. Res. Int.*, 8386065.

Satyanarayana, K., Sravanthi, K., Shaker, I., & Ponnulakshmi, R., (2015). Molecular approach to identify antidiabetic potential of *Azadirachta indica*. *J. Ayurveda Integr. Med., 6*, 165–174.

Serhan, C. N., Chiang, N., & Van, D. T. E., (2008). Resolving inflammation: Dual anti-inflammatory and pro-resolution lipid mediators. *Nat. Rev. Immunol., 8*, 349–361.

Sharon, Y., Jane, S., & Nissa, M., (2018). Nutritional supplements for the prevention of diabetes mellitus and its complications. *J. Nutr. Intermed. Metab., 14*, 16–21.

Straface, E., Rivabene, R., Masella, R., Santulli, M., Paganelli, R., & Malorni, W., (2002). Structural changes of the erythrocyte as a marker of noninsulin-dependent diabetes: Protective effects of N-acetylcysteine. *Biochem. Biophys. Res. Commun., 290*, 1393–1398.

Sunil Mahajan, Pratibha Chauhan, Meerambika Mishra, Dhananjay Yadav, Urmila Jarouliya, & GBKS Prasad (2018). Pterocarpus Marsupium and Gymnema Sylvestre Powder in Alleviation of Biochemical Abnormalities Associated with Human Type 2 Diabetes Mellitus. *J. Complement. Med. Alt. Healthcare., 7*(4), 555718.

Tanaka, N., Zhang, X., Sugiyama, E., Kono, H., Horiuchi, A., Nakajima, T., Kanbe, H., et al., (2010). Eicosapentaenoic acid improves hepatic steatosis independent of PPARα activation through inhibition of SREBP-1maturationinmice.*Biochem.Pharmacol., 80*, 1601–1612.

Tiwari, P., Mishra, B. N., & Sangwan, N. S., (2014). Phytochemical and pharmacological properties of *Gymnema sylvestre*: An important medicinal plant. *Biomed. Res. Int.,* 830285.

Tong, X., Dong, J. Y., Wu, Z. W., Li, W., & Qin, L. Q., (2011). Dairy consumption and risk of type 2 diabetes mellitus: A meta-analysis of cohort studies. *Eur. J. Clin. Nutr., 65*, 1027–1031.

Wang, S. Y., Chang, X. Y., Zhao, S., Zhang, R. N., Pan, Y. L., & Zhou, X. R., (2015). Effect of spirulina polysaccharide on blood glucose and antioxidant activity in diabetic rats. *Zhiye Yu Jiankang, 31*, 3229–3235.

Will, J. C., & Byers, T., (1996). Does diabetes mellitus increase the requirement for biotin? *Nutr. Rev.,* 54, 193–202.

Williams, G., & John, C. P., (2001). *Text Hand book of Diabetes* (2nd edn.). Hoboken: Blackwell Science.

Wu, J. H., Micha, R., Imamura, F., Pan, A., Biggs, M. L., Ajaz, O., Djousse, L., et al., (2012). Omega-3 fatty acids and incident type 2 diabetes: A systematic review and meta-analysis. *Br. J. Nutr.,* 2, S214–227.

Yamamoto, W. R., Bone, R. N., Sohn, P., Syed, F., Reissaus, C. A., Mosley, A. L., Wijeratne, A. B., True, J. D., Tong, X., Kono, T., et al., (2019). Endoplasmic reticulum stress alters ryanodine receptor function in the murine pancreatic beta cell. *J. Biol. Chem., 294*, 168–181.

Yeh, G. Y., Eisenberg, D. M., Kaptchuk, T. J., & Phillips, R. S., (2003). Systematic review of herbs and dietary supplements for glycemic control in diabetes. *Diabetes Care, 26*, 1277–1294.

Yimer, E. M., Tuem, K. B., Karim, A., Ur-Rehman, N., & Anwar, F., (2019). *Nigella sativa* L, (black cumin): A promising natural remedy for a wide range of illnesses. *Evid. Based Complement. Alternat. Med.*, 1528635.

CHAPTER 14

The Effect of Medicinal Plants on the Metabolic Syndrome

BUI THANH TUNG, DUONG VAN THANH, and
NGUYEN PHUONG THANH

Department of Pharmacology and Clinical Pharmacy, VNU University of Medicine and Pharmacy, Vietnam National University, Ha Noi, 144 Xuan Thuy, Cau Giay, Ha Noi, Vietnam, E-mail: tungasia82@gmail.com (B. T. Tung)

ABSTRACT

Metabolic syndrome (MS) is a group of metabolic abnormalities that occur together and raise the risk of cardiovascular disease, type 2 diabetes (T2D), and stroke. Despite significant advancements in prevention and treatment, the majority of patients still experience serious adverse effects. On the other hand, due to modernized lifestyles and bad eating habits, the present prevalence of metabolic illnesses is increasing. Plants are an excellent source of chemicals for traditional medicine researchers looking for new treatments. Because medicinal plants often include bioactive mixes of varied active compounds that might support the movement of one another or have a synergistic effect, they can be utilized to treat or predict metabolic illnesses, providing them with a particular advantage over a single chemical entity. More than 1,200 different plants have been recorded for the treatment of diabetes in traditional medicine, with many of them potentially targeting other risk factors linked with MS, such as hypercholesterolemia and hypertension (HTN). The purpose of this review is to present particular phytochemical, preclinical, and clinical concerns of plant species related to MS using examples of commonly recognized and partially defined plant species.

The Metabolic Syndrome: Dietary Supplements and Food Ingredients. Raj K. Keservani, MPharm, Durgavati Yadav, PhD, Rajesh K. Kesharwani, PhD, Sippy Singh, DPhil & Kumar Sandeep, PhD (Eds.)

14.1 INTRODUCTION

Metabolic syndrome is one of the most frequent health issues all over the world. This disease is caused mostly by a sedentary lifestyle, high levels of subjective stress, and a poor diet. Life is busier, and people often turn to fast foods, and unfortunately, these foods contain a lot of sugar and fat. The accumulation of these compounds in the body causes metabolic disorders, increased oxidative stress, and free radical activity that causes health problems evidenced by a dizzying increase in cases of diabetes, cardiovascular disease, or obesity (Zimmet et al., 2001). Based on the diagnostic signs and guidelines of medical organizations, Ritthikrai Thaweecharoen has compiled the table diagnostic manual of Metabolic Syndrome as shown in Table 14.1 (Thaweecharoen et al., 2019). Besides, the treatment of metabolic disorders is also very complicated and long, accompanied by many side effects of the drug. Therefore, the treatment of diseases with herbs is getting more and more attention, especially the combination of modern medicine and traditional medicine.

TABLE 14.1 Diagnostic Guidelines of the Metabolic Syndrome

	WHO (1999)	**NCEP-ATP III (2001)**	**NCEP-R (2004)**	**IDF (2005)**	**AACE**
Obesity	WHR > 0.90 (male) > 0.85 (female) or BMI > 30 kg/m^2	WC ≥ 102 cm (male) ≥ 88 cm (female)	WC ≥ 102 cm (male) ≥ 88 cm (female)	[Requirement] WC ≥ 94 cm (male) ≥ 80 cm (female)	Overweight/ obesity BMI ≥ 25 kg/m^2
Serum triglycerides	≥ 150 mg/dl	≥ 150 mg/dl	≥ 150 mg/dl or medication	≥ 150 mg/dl or medication	≥ 150 mg/dl
Serum HDL Cholesterol	< 35 mg/dl (male) < 39 mg/dl (female)	< 40 mg/dl (male) < 50 mg/dl (female)	< 40 mg/dl (male) < 50 mg/dl (female) or medication	< 40 mg/dl (male) < 50 mg/dl (female) or medication	< 40 mg/dl (male) < 50 mg/dl (female)
Blood pressure	≥ 140/90 mmHg	≥ 130/85 mmHg or medication	≥ 130/85 mmHg or medication	≥ 130/85 mmHg or medication	≥ 130/85 mmHg or medication
Fasting plasma glucose	[Requirement] FPG ≥ 110 mg/dl	≥ 100 mg/dl	≥ 100 mg/dl	≥ 100 mg/dl or previously diagnosed T2DM	110–126 mg/dl

TABLE 14.1 *(Continued)*

	WHO (1999)	NCEP-ATP III (2001)	NCEP-R (2004)	IDF (2005)	AACE
Other risk factors	Urinary albumin excretion rate ≥ 20 μg/min or albumin/ creatinine ratio ≥ 30 mg/g	–	–	–	Family history of T2DM, HTN, or CVD. Polycystic ovary syndrome, sedentary life style, advancing age and ethnic groups having high risk for DM or CVD physician's judgment.
Diagnosis	Impaired FPG + any two criteria	Any 3 criteria	Any 3 criteria	WC + any 2 criteria	

There are many species of plants that have been studied and given positive results in the treatment and prevention of metabolic syndrome. We chose plants that are near to human everyday life to offer in this overview among hundreds of medicinal plants that have been investigated for the prevention of MS. These are foods that are used as regular seasonings, such as ginger, garlic, and celery, or eaten as vegetables, such as okra and bitter melon. There are also grapes – daily fruit or aloe vera, green tea – famous items in the beauty field of women (Keservani et al., 2010a, b; Keservani and Sharma, 2014).

14.2 MEDICINAL PLANTS IN METABOLIC SYNDROME

14.2.1 *CINNAMOMUM CASSIA (L.)*

Cinnamomum cassia also called Chinese cinnamon is a tropical aromatic evergreen tree of the Lauraceae family, commonly used as a traditional medicine in Asia to treat bellyaches, dysmenorrhea, arthritis, etc. (Zhang et al., 2019). Nowadays, *C. cassia* is a traditional spice, widely used around the world. Besides, scientists have conducted many studies to understand the impact of cinnamon on metabolic syndrome. Table 14.2 summarizes some of the studies.

TABLE 14.2 Effect of *Cinnamomum cassia* (L.) on Disease Risk Associated with MS

Study	Part of Plant/ Constituent	Result	Effect	References
In vivo (long Evans rat with standard balanced diet)	Cinnamon powder from cinnamon bark (cinnamaldehyde)	↓TC, LDL-C, TG levels.	Lipid-lowering	Rahman et al. (2013)
In vivo (non-fasted Wistar rats with standard pellet diet and tap water, injured by halothane)	*C. cassia* bark 85.7 mg/kg and extract of *C. cassia* 5.96 mg/kg	↓BG	Anti-diabetic	Verspohl et al. (2005)
In vitro (INS-1 cells)	85.7 mg/kg (bark) and 5.96 mg/kg (extract)	↑Insulin release	Anti-diabetic	Verspohl et al. (2005)
Clinical trial with healthy people (compare dietary and lifestyle changes alone to the one with *C. cassia*	1,000 mg/capsule/ person/day	↓BG levels	Anti-diabetic	Hoehn et al. (2012)
In vivo (Sprague-Dawley rats injured by STZ 60 mg/kg)	*C. cassia* nanoparticles (CcAgNPs) 5 and 10 mg/kg; and extract of *C. cassia* 200 mg/kg	↓FBG significantly ↓Levels of urea, creatinine Bowman's space and capillary tufts are normal	Anti-diabetic Improve kidney function in diabetes	Kouame et al. (2019)

14.2.2 *SMALLANTHUS SONCHIFOLIUS* (YACON)

Yacon (*Smallanthus sonchifolius*) belongs to the family Asteraceae, native to the Andean regions of South America (Lachman et al., 2003). It is a perennial herbaceous plant, 2–2.5 m high. Yacon roots come in a variety of shapes and sizes, containing starch, usually consuming raw and tasting sweet like apple (Ojansivu et al., 2011). In addition, yacon can be grown in many different climates, altitudes, and soils. Therefore, many parts of the world have grown yacon, such as Argentina, Bolivia, Brazil, Italy, Japan, Korea, Peru, US, New Zealand, etc.

In recent years, there have been many studies about the use of yacon. Scientists have announced that yacon is a promising plant as a functional food treating MS, such as diabetes, obesity, cardiovascular diseases (CDVs), and stroke (Choque et al., 2013). Habib et al. roots of yacon contain high concentration fructooligosaccharides (FOS), which was beneficial for the treatment of diabetes. In this study, the rats in the yacon flour treatment group showed a decrease in food intake, abdominal, and epididymal adipose tissue, TG, VLDL; besides, they also increased in the stimulation of insulin production. B. Genta's study showed that the methanol, butanol, and chloroform extracts can effective hypoglycemic activity at minimum doses of 50, 10, 20 mg/kg, respectively. In particular, the Wistar rats were induced type 2 diabetes (T2D) with STZ and then were treated with the leaf extract for 8 weeks; the results showed that these extracts lowered BG level, increased plasma insulin level and decreased post-Grandia; glucose (Genta et al., 2010). In addition, Valentová et al. demonstrated the *in vitro* anti-oxidative effect of yacon leaf extract, based on radical scavenging and anti-lipoperoxidation reactions (Valentová et al., 2005). Table 14.3 summarizes some studies about the *Smallanthus sonchifolius* on MS.

TABLE 14.3 Effect of *Smallanthus sonchifolius* on Disease Risk Associated with MS

Study	Part of plant/ Constituent	Result	Effect	References
In vivo (diabetic Wistar rats)	Yacon flour with 340 mg/kg and 6,800 mg/kg	↓Food intake ↓Abdominal and epididymal adipose tissue ↓TG, VLDL ↑Insulin release	Lipid-lowering Anti-diabetic ↓Stress oxidative	Habib et al. (2011)
In vivo (Wistar rats injured by STZ)	Extract of *S. sonchifolius* leaf	Hypoglycemic ↑Plasma insulin level ↓Post prandial glucose	Anti-diabetic	Genta et al. (2010)

TABLE 14.3 *(Continued)*

Study	Part of plant/ Constituent	Result	Effect	References
In vitro (radical scavenging and anti-lipoperoxidation)	Phenolic total extract of yacon leaf	Slow down the formation of free radicals Inhibited xanthine oxidase, t-Butyl hydroperoxide Inhibited stress oxidative	Prevention of diseases related to oxidative stress	Valentová et al. (2005)

14.2.3 *HELIANTHUS ANNUUS* (SUNFLOWER)

Helianthus annuus or sunflower is a species of the Asteraceae family grown around the world offering a variety of nutritional and medicinal benefits. The common sunflower seed contains a multitude of nutritious components including protein, unsaturated fats, fiber, vitamins (E), selenium, copper, zinc, folate, iron, and more (Alagawany et al., 2015). A famous commercial product from sunflower is cooking oil, which is beneficial for health, especially cardiovascular. Table 14.4 summarizes some studies on MS of *Helianthus annuus (Sunflower).*

TABLE 14.4 Effect of *Helianthus annuus* (Sunflower) on Disease Risk Associated with MS

Study	Part of Plant/ Constituent	Result	Effect	References
In vitro	Extract of sunflower sprout	1 mg/mL concentration of extract inhibited 83.29% AGE (advanced glycation end products)	Anti-diabetic	Sun et al. (2012)
In vivo (both sexes of bred Wister strain rats induced diabetes by alloxan monohydrate)	Extract of *H. annuus* seed (dose 150 mg/kg)	↓Cholesterol, TG, LDL-C ↓Significantly glucose levels	Hypoglycemic Anti-diabetic	Cd et al. (2013)

TABLE 14.4 *(Continued)*

Study	Part of Plant/ Constituent	Result	Effect	References
In vitro	Sunflower protein	Inhibited angiotensin I converting enzyme	Antihypertensive	Megías et al. (2004)
In vitro (scavenging of DPPH free radical, inhibition of lipid peroxidation in rat liver, cumene hydroperoxide or CCl_4 reduced form of NADPH)	Extract of *H. annuus*	DPPH radical scavenging ↓Iron Inhibited beta-carotene oxidization	Antioxidant	Xi et al. (2008)

The effect of *H. annuus* on metabolic syndrome has been mentioned in multiple studies. Zheng et al. demonstrated that the extract of *H. annuus* sprout inhibited 83.29% AGE, which is directly related to the development of diabetes (Sun et al., 2012). The anti-diabetic benefits of sunflower seed extract were also studied in normal, Luka' study used both sexes of bred Wister strain rats, they were induced diabetes by alloxan monohydrate 150 mg/kg. After 14 days of treatment with 150 mg/kg seed extract of sunflower, cholesterol, triglyceride, LDL-C was reduced, especially BG levels were significantly declined compared to the diabetic control, with 152.70±87.85 mg/dL, 495.50±00.98 mg/dL respectively (Cd et al., 2013). Hydrolyzed protein from sunflower oil has an antihypertensive effect by inhibiting angiotensin I converting enzyme (Megías et al., 2004).

14.2.4 PRUNUS DULCIS

Prunus dulcis or almonds belongs to the Rosaceae family. It has long been known as a source of essential nutrients including lipid, vitamin E, polyphenols (Bolling, 2017). Recently, almonds are in demand as a healthy food. Bento et al. conducted a 6-week clinical trial, volunteers took 20 g of almonds daily for 6 weeks. The results showed that almonds reduced TC, LDL-C, non-HDL-C but had no effect on HDL-C, TG, Apo A, Apo B, and homocysteine (Bento et al., 2014). In addition, Liu et al. proven that 58 g almonds per day during 20 weeks reduced levels of TC, HDL-C, LDL-C, TG, VLDL (Liu et al., 2018). A list of several Prunus dulcis research is shown in Table 14.5.

TABLE 14.5 Effect of *Prunus dulcis* on Disease Risk Associated with MS

Study	Part of Plant/ Constituent	Result	Effect	References
Randomized, controlled, crossover	20 g Baru almonds/ day during 6 weeks	↓TC, non-HDL-C, LDL-C	Reduced bad lipid	Bento et al. (2014)
Randomized, controlled	58 g almonds/day during 20 weeks	↓Levels of TC, HDL-C, LDL-C, TG, VLDL		Liu et al. (2018)

14.2.5 TITHONIA DIVERSIFOLIA

Tithonia diversifolia or Mexican sunflower is a shrub-like perennial plant belonging to the Asteraceae family. It is widely grown in north and central America, Africa, Australia, and Asia (Varnham, 2006). *T. diversifolia* was used as a traditional medicine to treat abscesses, hematomas, muscular cramps, and malaria (Frei et al., 1998; Ajaiyeoba et al., 2006). Nowadays, some effects of *T. diversifolia* on metabolic syndrome have been discovered, especially T2D. Studies on the anti-diabetic effects of *Tithonia diversifolia* are presented in Table 14.6.

TABLE 14.6 Effect of *Tithonia diversifolia* on Disease Risk Associated with MS

Study	Part of Plant/ Constituent	Result	Effect	References
In vivo (diabetic mice)	Ethanol extract dose 1,00,500, 1,500 mg/kg during 3 weeks	↓BG, blood insulin levels ↓Insulin resistance	Anti-diabetic	Miura et al. (2002)
In silico and *in vitro* (isolated compounds and computer-based modeling program)	3 sesquiterpene lactone isolated from *T. diversifolia*	Against PPARγ/α	Anti-diabetic	Lin (2012)

TABLE 14.6 *(Continued)*

Study	Part of Plant/ Constituent	Result	Effect	References
In vivo (ICR adult male mice induced by alloxan	Aqueous fraction dose 100, 200, 300 mg/kg	↓BG levels ↑Insulin levels Dose-dependent reduction of GLUT2 mRNA expression	Anti-diabetic	Chunudom et al. (2020)
In vivo (male Wistar rats)	Saponin isolated dose 20, 40, 60, 80, 100 mg/kg during 21 days	↓TG, cholesterol, HDL-C levels significantly	↓The risk of metabolic disease including arteriosclerosis, diabetes, and obesity	Chunudom et al. (2020)

14.2.6 LACTUCA INDICA

Lactuca indica belongs to the family Compositae and is distributed throughout Asia. It has been used as a folk medicine for the treatment of intestinal disorders (Choi et al., 2016). Phytochemical constituents of *L. indica* were investigated including constituents that treat diabetes. Hou et al. announced that *L. indica* contained latucain C and lactucaside compounds, which showed significant anti-diabetic activity (Hou et al., 2003). Besides, Choi et al. proven eight phenolic compounds inside *L. indica* with antioxidant and alpha-glucosidase inhibitory effects (Choi et al., 2016).

14.2.7 ZINGIBER OFFICINALE

Ginger (*Zingiber officinale Roscoe*, Zingiberaceae) is a commonly used spice or medicinal herb in the treatment of various ailments across the world. Ginger is well-known for its high concentration of phytochemicals and biologically active compounds like phenolics and flavonoids. Some studies on the effects of ginger are summarized in Table 14.7.

TABLE 14.7 Effect of *Zingiber officinale* on Disease Risk Associated with MS

Study	Part of Plant/ Constituent	Results	Effect	References
In vivo (HFD-induced obese rats)	6-Gingerol (75 mg/kg) for 30 days	• ↓BW, glucose, insulin • ↓The activity and mRNA level of HMG-CoA reductase • ↑The activities and mRNA levels of CPT-1, LCAT, and LDL. • ↓The activities of FAS, PPARγ, and SREBP1 • ↓The expression levels of inflammatory markers (TNFα- and IL-6)	Anti-obesity	Brahma Naidu et al. (2016)
Double-blinded, placebo-controlled clinical trial, 70 T2DM	1,600 mg ginger daily for 12 weeks	• ↓Fasting plasma glucose, HbA1C, insulin, HOMA, TG, TC, CRP, and PGE_2	Anti-diabetic	Arablou et al. (2014)
A parallel, randomized, single-blind placebo-controlled clinical trial (204 Patients with T2DM)	3 g ginger with three glasses of black tea for 8 weeks	• ↓Intercellular adhesion molecule-1 concentrations • ↓SBP	Antihypertensive	Azimi et al. (2016)
A randomized, double-blind, placebo-controlled clinical trial (80 eligible obese women (aged 18–45 years))	2 g ginger rhizomes powder for 12 weeks	• ↓BW, BMI, WC, HC, appetite score	Anti-obesity	Ebrahimzadeh et al. (2015)

Much research has been conducted on the impact of ginger on metabolic illness. An example is an *in vivo* research on HFD-induced obesity rats with a dosage of 6-Gingerol (75 mg/kg) for 30 days with anti-obesity benefits (Brahma et al., 2016). Another randomized, double-blind, placebo-controlled experiment (80 suitable women (aged 18–45 years) conducted by Ebrahimzadeh et al. demonstrated anti-obesity effects (Ebrahimzadeh et al., 2015). There are also human studies. Arable et al. conducted a double-blind, placebo-controlled clinical experiment with 70 T2DM and dosage of 1,600 mg ginger daily for 12 weeks, which resulted in reduced diabetes-related indicators: Fasting plasma glucose, HbA1c, insulin, HOMA, TG, TC, CRP, and PGE2 (Arablou et al., 2014). Not only that, but Azimi et al. in 2016 reported an effect on Intercellular adhesion molecule-1 concentrations and SBP by combining 3 g of ginger with three cups of black tea for eight weeks. This demonstrates ginger's antihypertensive impact. In T2DM patients (this is a parallel, randomized, single-blind, placebo-controlled clinical study including 204 T2DM patients) (Azimi et al., 2016). From the examples above, it is clear that ginger is widely used in folk medicine to treat MS symptoms such as obesity, diabetes, and high BP.

14.2.8 ALLIUM SATIVUM (GARLIC)

Garlic, like ginger, is a staple in every kitchen. Garlic has been used medicinally for many years. Its anti-diabetic, antihypertensive, and lipid-lowering effects are recognized. The most current human research has been added to the earlier animal experiments. Garlic has long been known to reduce BP (SBP, DBP); it also lowers cholesterol and LDL while raising HDL (Durak et al., 2004; Choudhary et al., 2018). Table 14.8 summarizes the findings of two recent human trials on the use of garlic for therapy.

TABLE 14.8 Effect of *Allium sativum* (Garlic) on Disease Risk Associated with MS

Study	Part of Plant/ Constituent	Results	Effect	References
Randomized controlled trial (40 patients with T2DM)	Daily dose 100 mg/kg crushed raw garlic 2 times per day for 4 weeks	• ↓WC, SBP, DBP, FBG, TG • ↑HDL-C • ↓BMI levels, but not significant	Anti-obesity Antihypertensive Anti-diabetic	Choudhary et al. (2018)

TABLE 14.8 *(Continued)*

Study	Part of Plant/ Constituent	Results	Effect	References
Randomized controlled trial (23 hyperlipidemia patients (of them, 13 patients were hypertensive)	Garlic extract for 4 months	• ↑HDL • ↓LDL and cholesterol • ↓SBP, DBP in hypertensive patients • ↓MDA	Antihypertensive Hypolipidemic	Durak et al. (2004)

14.2.9 GINKO BILOBA

Ginkgo biloba (G. Biloba), often known as ginkgo, provides substantial benefits to traditional medicine, including weight reduction effects as well as anti-diabetic, antihypertensive, and antipyretic characteristics. The main constituents of *Ginkgo biloba* are terpene lactones (bilobalide and ginkgolides A, B, and C) and flavone glycosides (isorhamnetin, quercetin, and kaempferol). There has been a lot of studies done on the effects of ginkgo on obesity, high BP, dyslipidemia, and hyperglycemia. Table 14.9 details some of the *G. biloba* 's impacts.

TABLE 14.9 Effect of *Ginko bilobaon* Disease Risk Associated with MS

Study	Part of Plant/ Constituent	Results	Effect	References
A double-blind, randomized, placebo-controlled trial Human, T2DM patients.	120 mg/day, p.o for 90 day	• ↓HbA1c • ↓FBG • ↓Plasma insulin levels • ↓BMI • ↓WC • ↓VAI	Anti-diabetic	Aziz et al. (2018)
Human patients	A variety of doses	• ↓TG • ↓TC • ↓LDL-C • ↑HDL-C	Hypolipidemic	Fan et al. (2018)

TABLE 14.9 *(Continued)*

Study	Part of Plant/ Constituent	Results	Effect	References
In vivo, rats	100 mg/kg/ day, p.o for 12 weeks	• ↓BP • ↑RNA eNOS expression • ↑Protein expressions of iNOS, TNF-α, IL-6 and IL-1β in the kidney • ↓MDA • ↓Nitrite levels • ↓GSH • ↓Cholesterol	Antihypertensive	Abdel-Zaher et al. (2017)
In vivo, rats	500 mg/kg, p.o for 14 days	• ↓food/energy intake • ↓BW • ↑Adipo R1 and IL-10 gene expressions • ↑IR and Akt phosphorylation • ↓NF-κB p65 phosphorylation • ↓TNF-α levels	Anti-obesity	Hirata et al. (2015)

Ginkgo biloba dosage varies a lot, but it's generally between 100 and 500 mg. The duration of most experimental designs is around three months. *In vivo* trials to lower BP, BW, food/energy intake, cholesterol, or increase RNA eNOS expression, protein expressions of iNOS, TNF-α, IL-6, and IL-1 in the kidney, as well as adipo R1 and IL-10 gene expressions, IR, and Akt phosphorylation were used to establish safety and efficacy. Furthermore, *G. Biloba* tri has been found to lower MDA, nitrite levels, NF-B p65 phosphorylation, and TNF-levels, all of which are useful in decreasing BP and preventing obesity (Hirata et al., 2015; Abdel-Zaher et al., 2017). Not only that, but a study published in 2018 by Aziz et al. revealed that the plant has an anti-diabetic impact. The findings achieved in T2DM patients after 90 days of testing with a dosage of 120 mg/day are extremely positive. Blood glucose-related markers such as

HbA1c, FBG, and plasma insulin levels all decreased. Not only that, but these individuals had lower BMI, WC, and VAI (Aziz et al., 2018). Simultaneously, Fan Yu et al. proved *G. biloba's* lipid-reducing efficacy by human indicators of decreasing TG, TC, LDL-C, and raising HDL-C (Fan et al., 2018).

14.2.10 MOMORDICA CHARANTIA (BITTER MELON)

Bitter melon is a widespread species that originated in Asia and is a member of the Cucurbitaceae family. Bitter melon includes phytochemicals and substances such as triterpenoid, saponins steroid, insulin-like peptides, alkaloids, and so on. It has traditionally been used to treat hypoglycemia. Table 14.10 summarizes three common instances of three categories of fundamental study utilizing bitter melon.

TABLE 14.10 Effect of *Momordica charantia* on Disease Risk Associated with MS

Study	Part of Plant/ Constituent	Results	Effect	References
A randomized, double-blinded, placebo-controlled, clinical trial (24 patients)	2,000 mg/day, 3 months	• ↓A1c, 2 h glucose, glucose AUC • ↑Insulin AUC, first phase and total insulin secretion. • BW, BF%, BMI, WC	Anti-diabetic	Cortez-Navarrete et al. (2018)
In vivo (HDF induced fat rats)	0.5 g/kg/day, 1.0 g/kg/day P extracts, or 0.2, 1.0 g/kg/day G extracts	• ↑Insulin sensitivity • ↓BW and visceral FM. • ↓plasma glucose, TG, and TC • ↑FFA • ↑mRNA expression of leptin, PPAR-γ, PPAR-α • ↓Expression of resistin.	Antidiabetic Anti-obesity	Shih et al. (2008)
In vitro (HIV-1-protease inhibitor-treated human hepatoma cells, HepG2)	Bitter melon juice (BMJ)	• ↓TG and TC • ↑HDL-C • Inhibits apoC-III* mRNA expression	Lipid-lowering	Nerurkar et al. (2006)

*ApoC-III: inhibits the interactions between apoE and LDL receptor (LDLR) or with the LDL receptor-related protein (LRP), involved in the hepatic uptake of remnant particles.

Many experts have investigated the impact of bitter melon on metabolic disorders in a variety of methods. Nerurkar et al. conducted an *in vitro* case study. They employed BMJ on human liver tumor cells that had been treated with the HIV-1 protease inhibitor HepG2. The findings revealed lipid-lowering effects (through a reduction in TG, TC, and increase in HDL-C; and inhibition of apoC-III mRNA expression) (Nerurkar et al., 2006). Not only that, but *in vivo*, the typical scientist, Chun-Ching Shih et al. have shown the effects on metabolism. They have demonstrated anti-diabetes (insulin sensitivity, plasma glucose lowering); anti-obesity effects (loss of body mass, fatty acids without visceral fat mass) and increased mRNA expression of leptin, PPAR-γ, PPAR-αor decreased expression of resistin using different dosages different and bitter melon extracts in HDF-induced fatty rats (Shih et al., 2008). In particular, in 2018, scientist Cortez-Navarrete conducted 3-month research in 24 patients with a daily dosage of 2,000 mg. This was a randomized, double-blind, placebo-controlled clinical trial. The trial's findings were beneficial in the fight against diabetes (Cortez-Navarrete et al., 2018).

14.2.11 VITIS VINIFERA

Grapes or Vitis species, which are members of the Vitaceae family, are grown in many parts of the world. Grapes, particularly the seeds, are considered to be high in polyphenols, powerful antioxidants that have been shown to lower the risk of metabolic syndrome-related conditions. Several studies on the effects of grapes on MS are included in Table 14.11.

TABLE 14.11 Effect of *Vitis viniferaon* Disease Risk Associated with MS

Study	Part of Plant/ Constituent	Results	Effect	References
In vivo (Aged male rats)	500 mg/kg BW/ day, GSPE (8 days)	• ↓Food intake • ↑Energy expenditure • ↓BW	Anti-obesity	Serrano et al. (2016)
In vivo (Diet-induced obese male rats)	25 mg/kg BW/ day, 3 weeks	• ↓Plasma glucose and insulin levels	Anti-diabetic	Pascual-Serrano et al. (2017)

TABLE 14.11 *(Continued)*

Study	Part of Plant/ Constituent	Results	Effect	References
In vivo (DOCA HTN-induced SD rats)	150, 240, 384 mg/kg BW/day, GSPE 4 weeks	• ↓SBP, reversion of morphological hypertrophy of heart • ↑In aortic rings vasodilatation	Antihypertensive	Huang et al. (2015)
A randomized (double-blind, placebo-controlled clinical trial, 24 patients with diagnosis of metabolic syndrome)	500 mg; 90 days (three times per day), transresveratrol	• ↓BG and insulin levels, as well as BW, BMI, FM, WC, insulin area under the curve, and total insulin secretion.	Anti-diabetic	Méndez-del Villar et al. (2014)
A single-blind study (61 healthy subjects, >30 and <70 years old)	200 or 400 mg/kg/day, Proanthocyanidin rich grape seed extract	• ↓LDL oxidization	Lipid-lowering	Sano et al. (2007)

Grapes have been utilized in research to reduce food and energy consumption for the body while also reducing BW, BMI, fat mass, and WC (Méndez-del Villar et al., 2014; Serrano et al., 2016). Grapes also lower systolic BP, reverse morphological hypertrophy of the heart, and enhance aortic rings vasodilation (Huang et al., 2015). Furthermore, grapes have the impact of lowering plasma glucose, BG, and insulin levels in diabetics, as well as LDL cholesterol in blood lipids (Sano et al., 2007; Méndez-del Villar et al., 2014; Pascual-Serrano et al., 2017).

14.2.12 APIUM TOMBOLENS

Scientists have identified celery (*Apium tombolens*) as one of the most important sources of phytochemicals such as phenolic acids, flavonoids (flavons, flavonols), and other antioxidants in numerous research. These compounds were then utilized in diabetic studies to reduce BP. Table 14.12 shows some examples of *Apium tombolens'* effects.

TABLE 14.12 Effect of *Apium tombolens* on Disease Risk Associated with MS

Study	Part of Plant/ Constituent	Results	Effect	References
In vivo (Invasive rat model)	Seed extracts	• Blockade of calcium channels vasodilator • ↓SBP, DBP, MABP, and HR significantly decreased • Bradycardic effects	Antihypertensive	Tashakori-Sabzevar et al. (2016)
In vivo (HFD/STZ fed rats)	Seed extracts, 33 days	• ↓Glucose • ↑Insulin • ↓Concentrations of ALT and AST in serum • Less atrophy, necrosis, and inflammation in histopathology	Anti-diabetic	Tashakori-Sabzevar et al. (2016)
		• ↓TG, and cholesterol levels • ↑HDL	Lipid-lowering	
In vivo (HFD fed C57BL/6 mice)	Luteolin	• Inhibition of IL-6 that produced by mast cells, lowering BW • Inhibition effect on inflammatory response • Modulation of insulin resistance-related endothelial dysfunction	Anti-obesity Anti-diabetic	Xu et al. (2014)

According to Table 14.12, scientists mainly used celery seed extracts for different periods of time to investigate the effects of the plant. Tashakori's study in 2016 showed an impact on rat BP. Celery, in particular, aided animals in an invasive mouse model by blocking vasodilator calcium channels, resulting in substantial decreases in SBP, DBP, MABP, and HR, as well as the appearance of bradycardia (Tashakori-Sabzevar et al., 2016). Furthermore, with the HFD/STZ fed rats model, the tested mice produced positive findings. The anti-diabetic benefits of seed extracts after 33 days include increased insulin and reduced glucose, as well as serum ALT and AST values. In terms of the luteolin component in celery, there is an inhibition of IL-6 production by mast cells, a reduction in BW, an inhibition impact on the inflammatory response, and regulation of insulin resistance associated endothelial dysfunction (ED) (Xu et al., 2014; Tashakori-Sabzevar et al., 2016). In addition, the same model had anti-obesity and lipid-reducing benefits in test mice by lowering TG and cholesterol levels while increasing HDL levels (Tashakori-Sabzevar et al., 2016).

14.2.13 ABELMOSCHUS ESCULENTUS

Okra *(Abelmoschus esculentus L.)* is a member of the Malvaceae family. Bioactive components found in corn plant parts such as leaves, buds, flowers, bark, stems, and seeds include polyphenol compounds, oligomeric catechins and flavonoid derivatives, etc. *Okra* has recently been investigated for its impact on diabetes, obesity, dyslipidemia, and HTN. Table 14.13 summarizes the most recent Okra studies.

TABLE 14.13 Effect of *Abelmoschus esculentus* on Disease Risk Associated with MS

Study	Part of Plant/ Constituent	Results	Effect	References
In vivo (HFD/ STZ fed rats)	Okra powder (200 mg/kg), 30 days	• ↓FBS, TG, TC, HOMA-IR, PPAR-α, PPAR-γ	Anti-diabetic	Erfani Majd et al. (2018)
		• Reduction of β-cells mass		
		• ↓TC and TG level.	Lipid-lowering	
In vivo (HFD induced obese mouse)	Quercetin and its analogs, 30 g fresh okra pod/ kg/day, 6 weeks	• ↓BW (through regulation of PPARs expression)	Anti-obesity	Fan et al. (2014)
		• ↓BG	Anti-diabetic	
In vivo (HFD induced obese mouse)	Dietary okra fruit powder, 12 weeks	• Improvement glucose tolerance and insulin	Anti-diabetic	Esmaeilzadeh et al. (2020)
		• Mitigation the disturbance of hepatic lipid metabolism oxidative stress and hyperinsulinemia.	Anti-obesity	

TABLE 14.13 *(Continued)*

Study	Part of Plant/ Constituent	Results	Effect	References
In vivo (hypercholesterolemic rabbits)	Dry okra aqueous seed extract (0.5 g/kg), six weeks.	• ↓TC and lipoprotein. • Inhibition of HMG-COA reductase • Conversion of cholesterol to bile acids in the liver • Activation of α-hydroxylase	Lipid-lowering	Kzar et al. (2019)
In vivo (hypertensive rat models caused by fructose)	Okra seed ethanolic extract, 150 mg/kg, p.o, 42 days	• Improvement the cardiac functioning and heart rate. • ↓BP (SAP, DAP, and MAP) • ↓TC and TG levels	Antihypertensive	Mondal et al. (2019)

Erfani Majd et al. conducted an *in vivo* test on rats on a HFD and rats with STZ-induced diabetes in research on the effects of okra on metabolic syndrome. The okra powder (200 mg/kg) was used in the experiment for 30 days following the development of diabetes. The findings of the experiments reveal that okra powder has a substantial effect on lowering FBS, TC, and TG. At the same time, okra has been shown to reduce the HOMA-IR; PPAR-α, PPAR-γ, and β-cell mass (Erfani Majd et al., 2018). Moreover, Esmaeilzadeh et al. experimented on obese mice with dietary okra fruit powder. Okra-fed rats enhanced glucose and insulin tolerance while also lowering hepatic lipid metabolism disorders, oxidative stress, and hyperinsulinemia after 12 weeks (Esmaeilzadeh et al., 2020). Previously, Fan et al. shown in 2014 that the ethanolic extract of *Okra* and its major flavonoids (isoquercitrin and quercetin 3-O gentiobioside) help to decrease blood glucose levels in HDF/STZ-induced diabetic rats. Furthermore, *Okra* extract improves blood TG levels and liver shape in rats via inhibiting the expression of PPAR-γ, a key regulator of glucose and lipid homeostasis (Fan et al., 2014).

In a 2019 study, Kzar with the test utilizing okra extract on Hypercholesterolemic rabbits helped reduce TC and lipoproteins, inhibition of HMG-CoA reductase, converting cholesterol into bile acids in the liver, and activation of α-hydroxylase (Kzar et al., 2019). In the same year, Mondal et al. also gave an ethanolic extract of okra seeds orally at a dosage of 150 mg/kg in rats with fructose-induced HTN (day 22 to 42 – end of research period). The extract had an antihypertensive impact, according to the findings (Mondal et al., 2019).

14.2.14 ALOE VERA

Aloe vera is a Liliaceae plant that has long been utilized in traditional medicine. *A. vera* is a useful plant with two primary components: anthraquinone compounds in high concentrations in the leaves and a transparent gel. The gel is an anti-inflammatory, antioxidant, antiviral, antibacterial, and wound-healing food. *A. vera* has also been shown in clinical studies to have cholesterol-lowering, antihypertensive, anti-diabetic, anti-obesity, and cardioprotective activities. The results of the *Aloe vera* case study are summarized in Table 14.14.

TABLE 14.14 Effect of *Aloe vera* on Disease Risk Associated with MS

Study	Part of Plant/ Constituent	Results	Effect	References
In vivo	*Aloe*-emodin extract (dosage 50 and 100 mg/kg), 6 weeks	↓TC, LDL-C	Lipid-lowering	Ji et al. (2018)
Randomized controlled trial (136 obese prediabetes or early-diabetes patients)	147-mg *A. vera* gel per capsule, 8 weeks	↓BW, BFM, FBS, and insulin levels	Anti-obesity Anti-diabetic	Choi et al. (2013)

Ji et al. (2018) discovered that *Aloe*-emodin extract (dosage 50 and 100 mg/kg) had a moderating impact on the rate of hyperlipidemia. The study found that when compared to the control group, the extract lowered TC levels by more than 26.0–34.3% and LDL-C levels by 37.0–40.9% (Ji et al., 2018). Choi et al. (2013) found and proved that supplementation with the leaf gel preparation of *A. vera* (dose 147 mg *A. vera* gel per capsule) was beneficial in raising blood lipids and hyperglycemia by weight reduction in an 8-week randomized study of 136 obese prediabetes or early diabetic patients. Furthermore, it can increase insulin sensitivity in humans (Choi et al., 2013).

14.2.15 ANOTHER HERBAL MEDICALS

There has been a lot of studies recently on plants and their impact on metabolic disorders. However, the studies are still insufficient. E.g., *Piper sarmentosum* (PS), *Ipomoea hederacea Jacq., Silybum marianum, green tea, Phaseolus vulgaris, Hibiscus Sabdariffa*, etc. Table 14.15 offers an overview of plants having anti-metabolic effects that we may come into contact with on a daily basis.

If *Piper sarmentosum* has been proven to prevent obesity and decrease blood fat, or *Ipomoea hederacea Jacq* has been found to lower BP, *Silybum marianum, green tea, Phaseolus vulgaris*, and *Hibiscus Sabdariffa* have a variety of benefits in MS. Different dosages and research have shown that they have a wide range of possible effects. These different effects are described in Table 14.15.

TABLE 14.15 Effect of Another Herbal Medicals on Disease Risk Associated with MS

Study	Part of Plant/ Constituent	Results	Effect	References
***Piper sarmentosum* (PS)**				
In vivo (fructose-induced metabolic syndrome rats)	125 mg/kg PS methanolic extract for 12 weeks.	• ↓BW, FM, BF%, LDL, TG, TG, HMG-CoA reductase, leptin • ↑HDL, adiponectin	Anti-obesity Lipid-lowering	Kumar et al. (2021)
***Ipomoea hederacea Jacq.* (Family: Convolvulaceae)**				
In vivo (normotensive anesthetized rats)	Seeds (Aqueous-ethanolic extract) 0.01–100 mg/kg	• ↓TG, glucose level • ↑HDL • Mediated through α1 blocking, β blocking and iNOS/cGMP stimulating activity	Antihypertensive	Chaudhry et al. (2021)

TABLE 14.15 *(Continued)*

Study	Part of Plant/ Constituent	Results	Effect	References
Silybum marianum				
In vivo (eight-week-old C57BL6 male mice were fed an HFD)	Milk thistle cold press oil	• ↓BW, FBG, BP • ↓Oxygen consumption • Improvement hepatic insulin signaling in the liver	MS	Shen et al. (2020)
Green Tea				
Five studies	Green tea: 300, 400, 870, 928, 1,000, and 6,000 mg/day Catechins: 150, 300, 458, 468, 886, and 1,200 mg/day 8–12 weeks	• ↓BW, BMI, WC, HC, FM, BF% • ↑FFM • ↓SBP, DBP • ↑TG, HDL • ↓TC, LDL, FBS	MS	Payab et al. (2020)
Phaseolus vulgaris				
Five studies	445, 1,000, and 3,000 mg/day, 4–12 weeks	• ↓BW, BMI, WC, HC, FM, BF%, FFM • ↓SBP, DBP • ↓TG, TC, LDL, HDL, FBS	MS	
Hibiscus sabdariffa				
Three studies	75, 500, 1,000 mg/day, 4–12 weeks	• ↓BW, BMI, WC, WHR, BF% • ↓TG, TC, LDL, FBS	MS	

14.3 CONCLUSION

The metabolic syndrome, as indicated in this review, is characterized by a variety of symptoms, including raised glucose levels and dyslipidemia (particularly fat around the waist and belly), increased blood pressure, and more. Furthermore, this illness raises the risk of cardiovascular disease, diabetes, and stroke. Due to the negative effects of chemical pharmaceuticals nowadays, herbal medicines have fewer adverse effects and may be taken for a longer period. Through different mechanisms, many plants have been researched for their ability to reduce fasting glucose or glucose tolerance, insulin sensitivity while lowering blood sugar, bad cholesterol, and blood pressure, etc. These natural products are attractive drug candidates. Simultaneously, based on its influence in trials on diabetes and cardiovascular disease, more creative research should be done to improve reliability and be more extensively applied in practice.

KEYWORDS

- **cardiovascular**
- **diabetes**
- **fructooligosaccharides**
- **LDL receptor-related protein**
- **medicine plants**
- **metabolic syndrome (MS)**

REFERENCES

Abdel-Zaher, A. O., Farghaly, H. S. M., El-Refaiy, A. E. M., & Abd-Eldayem, A. M., (2017). Protective effect of the standardized extract of ginkgo biloba (EGb761) against hypertension with hypercholesterolemia-induced renal injury in rats: Insights in the underlying mechanisms. *Biomed. Pharmacother., 95*, 944–955.

Ajaiyeoba, E. O., Abiodun, O. O., Falade, M. O., Ogbole, N. O., Ashidi, J. S., Happi, C. T., & Akinboye, D. O., (2006). *In vitro* cytotoxicity studies of 20 plants used in Nigerian antimalarial ethnomedicine. *Phytomedicine, 13*(4), 295–298.

Alagawany, M., Farag, M., Abd El-Hack, M., & Dhama, K., (2015). The practical application of sunflower meal in poultry nutrition. *Advances in Animal and Veterinary Sciences, 3*, 634–648.

Arablou, T., Aryaeian, N., Valizadeh, M., Sharifi, F., Hosseini, A., & Djalali, M., (2014). The effect of ginger consumption on glycemic status, lipid profile and some inflammatory markers in patients with type 2 diabetes mellitus. *Int. J. Food Sci. Nutr., 65*(4), 515–520.

Azimi, P., Ghiasvand, R., Feizi, A., Hosseinzadeh, J., Bahreynian, M., Hariri, M., & Khosravi-Boroujeni, H., (2016). Effect of cinnamon, cardamom, saffron and ginger consumption on blood pressure and a marker of endothelial function in patients with type 2 diabetes mellitus: A randomized controlled clinical trial. *Blood Press., 25*(3), 133–140.

Aziz, T. A., Hussain, S. A., Mahwi, T. O., Ahmed, Z. A., Rahman, H. S., & Rasedee, A., (2018). The efficacy and safety of Ginkgo biloba extract as an adjuvant in type 2 diabetes mellitus patients ineffectively managed with metformin: A double-blind, randomized, placebo-controlled trial. *Drug Des. Devel. Ther., 12*, 735–742.

Bento, A. P. N., Cominetti, C., Simões, F. A., & Naves, M. M. V., (2014). Baru almond improves lipid profile in mildly hypercholesterolemic subjects: A randomized, controlled, crossover study. *Nutrition, Metabolism and Cardiovascular Diseases, 24*(12), 1330–1336.

Bolling, B. W., (2017). Almond polyphenols: Methods of analysis, contribution to food quality, and health promotion. *Compr. Rev. Food Sci. Food Saf., 16*(3), 346–368.

Brahma, N. P., Uddandrao, V. V., Ravindar, N. R., Suresh, P., Meriga, B., Begum, M. S., Pandiyan, R., & Saravanan, G., (2016). Ameliorative potential of gingerol: Promising modulation of inflammatory factors and lipid marker enzymes expressions in HFD induced obesity in rats. *Mol. Cell Endocrinol., 419*, 139–147.

Cd, L., Joel, E., Ul, E., Onwukike, & Tijjani, H., (2013). Hypoglycaemic properties of aqueous extracts of Anacardium Occidentale, moringa oleifera, Vernonia amygdalina and helianthus annuus: A comparative study on some biochemical parameters in diabetic rats. *International Journal of Pharmaceutical Science Invention, 2*, 2319–671816.

Chaudhry, M. A., Alamgeer, Mushtaq, M. N., Bukhari, I. A., & Assiri, A. M., (2021). Ipomoea hederacea Jacq.: A plant with promising antihypertensive and cardio-protective effects. *J. Ethnopharmacol., 268*, 113584.

Choi, C. I., Eom, H. J., & Kim, K. H., (2016). Antioxidant and α-glucosidase inhibitory phenolic constituents of *Lactuca indica* L. *Russian Journal of Bioorganic Chemistry, 42*(3): 310–315.

Choi, H. C., Kim, S. J., Son, K. Y., Oh, B. J., & Cho, B. L., (2013). Metabolic effects of aloe vera gel complex in obese prediabetes and early non-treated diabetic patients: Randomized controlled trial. *Nutrition, 29*(9), 1110–1114.

Choque, D. G. T., Da Silva, C. T. W. M., Maróstica, Jr. M. R., & Pastore, G. M., (2013). Yacon (*Smallanthus sonchifolius*): A functional food. *Plant Foods for Human Nutrition, 68*(3), 222–228.

Choudhary, P. R., Jani, R. D., & Sharma, M. S., (2018). Effect of raw crushed garlic (*Allium sativum* L.) on components of metabolic syndrome. *J. Diet. Suppl., 15*(4), 499–506.

Chunudom, L., Thongsom, M., Karim, N., Rahman, M. A., Rana, M. N., & Tangpong, J., (2020). *Tithonia diversifolia* aqueous fraction plays a protective role against alloxan-induced diabetic mice via modulating GLUT2 expression. *South African Journal of Botany, 133*, 118–123.

Cortez-Navarrete, M., Martínez-Abundis, E., Pérez-Rubio, K. G., González-Ortiz, M., & Méndez-Del, V. M., (2018). Momordica charantia administration improves insulin secretion in type 2 diabetes mellitus. *J. Med. Food, 21*(7), 672–677.

Durak, I., Kavutcu, M., Aytaç, B., Avci, A., Devrim, E., Ozbek, H., & Oztürk, H. S., (2004). Effects of garlic extract consumption on blood lipid and oxidant/antioxidant parameters in humans with high blood cholesterol. *J. Nutr. Biochem., 15*(6), 373–377.

Ebrahimzadeh, A. V., Asghari J. M., Zemestani, M., & Ostadrahimi, A., (2015). Effect of zingiber officinale supplementation on obesity management with respect to the uncoupling protein 1 -3826A>G and ß3-adrenergic receptor Trp64Arg polymorphism. *Phytother. Res., 29*(7), 1032–1039.

Erfani, M. N., Tabandeh, M. R., Shahriari, A., & Soleimani, Z., (2018). Okra (*Abelmoscus esculentus*) improved islets structure, and down-regulated PPARs gene expression in pancreas of high-fat diet and streptozotocin-induced diabetic rats. *Cell Journal, 20*(1), 31–40.

Esmaeilzadeh, D., & Razavi, B. M., (2020). *Effect of Abelmoschus Esculentus (okra) on Metabolic Syndrome: A Review, 34*(9), 2192–2202.

Fan, S., Zhang, Y., Sun, Q., Yu, L., Li, M., Zheng, B., Wu, X., et al., (2014). Extract of okra lowers blood glucose and serum lipids in high-fat diet-induced obese C57BL/6 mice. *J. Nutr. Biochem., 25*(7), 702–709.

Fan, Y., Jin, X., Man, C., & Gong, D., (2018). Does adjuvant treatment with ginkgo Biloba to statins have additional benefits in patients with dyslipidemia? Frontiers in pharmacology, *9*, 659.

Frei, B., Baltisberger, M., Sticher, O., & Heinrich, M., (1998). Medical ethnobotany of the Zapotecs of the isthmus-sierra (Oaxaca, Mexico): Documentation and assessment of indigenous uses. *Journal of Ethnopharmacology, 62*(2), 149–165.

Genta, S. B., Cabrera, W. M., Mercado, M. I., Grau, A., Catalán, C. A., & Sánchez, S. S., (2010). Hypoglycemic activity of leaf organic extracts from *Smallanthus sonchifolius*: Constituents of the most active fractions. *Chemico-Biological Interactions, 185*(2), 143–152.

Habib, N. C., Honoré, S. M., Genta, S. B., & Sánchez, S. S., (2011). Hypolipidemic effect of *Smallanthus sonchifolius* (yacon) roots on diabetic rats: Biochemical approach. *Chemico-Biological Interactions, 194*(1), 31–39.

Hirata, B. K. S., Banin, R. M., Dornellas, A. P. S., De Andrade, I. S., Zemdegs, J. C. S., Caperuto, L. C., Oyama, L. M., et al., (2015). Ginkgo biloba extract improves insulin signaling and attenuates inflammation in retroperitoneal adipose tissue depot of obese rats. *Mediators of inflammation, 2015*, 419106, 1–9.

Hoehn, A. N., & Stockert, A. L., (2012). The effects of *Cinnamomum cassia* on blood glucose values are greater than those of dietary changes alone. *Nutrition and Metabolic Insights, 5*, 77–83.

Hou, C. C., Lin, S. J., Cheng, J. T., & Hsu, F. L., (2003). Antidiabetic dimeric guianolides and a lignan glycoside from *Lactuca indica*. *Journal of Natural Products, 66*(5), 625–629.

Huang, L. L., Pan, C., Wang, L., Ding, L., Guo, K., Wang, H. Z., Xu, A. M., & Gao, S., (2015). Protective effects of grape seed proanthocyanidins on cardiovascular remodeling in DOCA-salt hypertension rats. *J. Nutr. Biochem., 26*(8), 841–849.

Ji, H., Liu, Y., He, F., An, R., & Du, Z., (2018). LC-MS based urinary metabolomics study of the intervention effect of aloe-emodin on hyperlipidemia rats. *J. Pharm. Biomed. Anal., 156*, 104–115.

Keservani, R. K., & Sharma, A. K., (2014). Flavonoids: Emerging trends and potential health benefits. *J. Chin. Pharm. Sci., 23*(12).; 815–822.

Keservani, R. K., Kesharwani, R. K., Sharma, A. K., Vyas, N., & Chadoker, A., (2010b). Nutritional supplements: An overview. *Int. J. of Curr. Pharm. Rev. and Res., 1*(1), 59–75.

Keservani, R. K., Kesharwani, R. K., Vyas, N., Jain, S., Raghuvanshi, R., & Sharma, A. K., (2010a). Nutraceuticals and functional food as future food. *Der. Pharm. Lett., 2*(1), 106–116.

Kouame, K., Peter, A. I., Akang, E. N., Moodley, R., Naidu, E. C., & Azu, O. O., (2019). Histological and biochemical effects of *Cinnamomum cassia* nanoparticles in kidneys of diabetic Sprague-Dawley rats. *Bosnian Journal of Basic Medical Sciences, 19*(2), 138–145.

Kumar, S. R., Mohd, R. E. S., & Abdul, N. N. A., (2021). *Methanolic Extract of Piper sarmentosum Attenuates Obesity and Hyperlipidemia in Fructose-Induced Metabolic Syndrome Rats, 26*(13).

Kzar, H. K., Mena, Y., A., Murad, M., & Ewadh, M., (2019). *Journal of Global Pharma Technology.* Study the ascorbic acid levels, lipoprotein ratio and hypocholesterolemia action of dry okra extract on experimental model of locally male rabbits.

Lachman, J., Fernández, E., & Orsák, M., (2003). Yacon [*Smallanthus sonchifolia* (Poepp. et Endl.) H. Robinson] chemical composition and use: A review. *Plant, Soil and Environment,* 49.

Lin, H. R., (2012). Sesquiterpene lactones from *Tithonia diversifolia* act as peroxisome proliferator-activated receptor agonists. *Bioorganic & Medicinal Chemistry Letters, 22*(8), 2954–2958.

Liu, Y., Hwang, H. J., Kim, H. S., & Park, H., (2018). Time and intervention effects of daily almond intake on the changes of lipid profile and body composition among free-living healthy adults. *J. Med. Food, 21*(4), 340–347.

Megías, C., Md. Yust, M., Pedroche, J., Lquari, H., Girón-Calle, J., Alaiz, M., Millán, F., & Vioque, J., (2004). Purification of an ACE inhibitory peptide after hydrolysis of sunflower (*Helianthus annuus* L.) protein isolates. *Journal of Agricultural and Food Chemistry, 52*(7), 1928–1932.

Méndez-Del, V. M., González-Ortiz, M., Martínez-Abundis, E., Pérez-Rubio, K. G., & Lizárraga-Valdez, R., (2014). Effect of resveratrol administration on metabolic syndrome, insulin sensitivity, and insulin secretion. *Metab. Syndr. Relat. Disord., 12*(10), 497–501.

Miura, T., Furuta, K., Yasuda, A., Iwamoto, N., Kato, M., Ishihara, E., Ishida, T., & Tanigawa, K., (2002). Antidiabetic effect of *Smallanthus sonchifolia* in KK-Ay diabetic mice. *Am. J. Chin. Med., 30*(1), 81–86.

Mondal, K., Gowda, S., & Manandhar, S., (2019). Anti-hypertensive effect of Abelmoschus esculentus (okra) seed extracts in fructose-induced hypertensive rats. *Indian Journal of Physiology and Pharmacology, 63,* 175–181.

Nerurkar, P. V., Lee, Y. K., Linden, E. H., Lim, S., Pearson, L., Frank, J., & Nerurkar, V. R., (2006). Lipid-lowering effects of *Momordica charantia* (bitter melon) in HIV-1-protease inhibitor-treated human hepatoma cells, HepG2. *British Journal of Pharmacology, 148*(8), 1156–1164.

Ojansivu, I., Ferreira, C. L., & Salminen, S., (2011). Yacon, a new source of prebiotic oligosaccharides with a history of safe use. *Trends in Food Science & Technology, 22*(1), 40–46.

Pascual-Serrano, A., & Arola-Arnal, A., (2017). *Grape Seed Proanthocyanidin Supplementation Reduces Adipocyte Size and Increases Adipocyte Number in Obese Rats, 41*(8), 1246–1255.

Payab, M., Hasani-Ranjbar, S., Shahbal, N., Qorbani, M., Aletaha, A., Haghi-Aminjan, H., Soltani, A., et al., (2020). *Effect of the Herbal Medicines in Obesity and Metabolic Syndrome: A Systematic Review and Meta-Analysis of Clinical Trials, 34*(3), 526–545.

Rahman, S., Begum, H., Rahman, Z., Ara, F., Iqbal, M., & Yousuf, A., (2013). Effect of cinnamon (*Cinnamomum cassia*) as a lipid-lowering agent on hypercholesterolemic rats. *Journal of Enam. Medical College,* 3.

Sano, A., Uchida, R., Saito, M., Shioya, N., Komori, Y., Tho, Y., & Hashizume, N., (2007). Beneficial effects of grape seed extract on malondialdehyde-modified LDL. *J. Nutr. Sci. Vitaminol. (Tokyo), 53*(2), 174–182.

Serrano, J., Casanova-Martí, À., Blay, M., Terra, X., Ardévol, A., & Pinent, M., (2016). Defining conditions for optimal inhibition of food intake in rats by a grape-seed derived proanthocyanidin extract. *Nutrients, 8*(10).

Shen, H. H., Alex, R., Bellner, L., Raffaele, M., Licari, M., Vanella, L., & Stec, D. E., (2020). *Milk Thistle Seed Cold Press Oil Attenuates Markers of the Metabolic Syndrome in a Mouse Model of Dietary-Induced Obesity, 44*(12), e13522.

Shih, C. C., Lin, C. H., & Lin, W. L., (2008). Effects of *Momordica charantia* on insulin resistance and visceral obesity in mice on high-fat diet. *Diabetes Res. Clin. Pract., 81*(2), 134–143.

Sun, Z., Chen, J., Ma, J., Jiang, Y., Wang, M., Ren, G., & Chen, F., (2012). Cynarin-rich sunflower (*Helianthus annuus*) sprouts possess both antiglycative and antioxidant activities. *Journal of Agricultural and Food Chemistry, 60*(12), 3260–3265.

Tashakori-Sabzevar, F., Ramezani, M., Hosseinzadeh, H., Parizadeh, S. M., Movassaghi, A. R., Ghorbani, A., & Mohajeri, S. A., (2016). Protective and hypoglycemic effects of celery seed on streptozotocin-induced diabetic rats: Experimental and histopathological evaluation. *Acta Diabetol., 53*(4), 609–619.

Tashakori-Sabzevar, F., Razavi, B. M., Imenshahidi, M., Daneshmandi, M., Fatehi, H., Y. Entezari, S. Y., & Mohajeri, S. A., (2016). Evaluation of mechanism for antihypertensive and vasorelaxant effects of hexanic and hydroalcoholic extracts of celery seed in normotensive and hypertensive rats. *Revista Brasileira de Farmacognosia, 26*(5), 619–626.

Thaweecharoen, R., Chaiwong, W., Boonma, C., Pasupa, K., Massa-Ard, P., & Kunanusont, C., (2019). *Diagnosis of Metabolic Syndrome Using Radar Chart, 15*, 11–18.

Valentová, K., Šeršeň, F., & Ulrichová, J., (2005). Radical scavenging and anti-lipoperoxidative activities of *Smallanthus sonchifolius* leaf extracts. *Journal of Agricultural and Food Chemistry, 53*(14), 5577–5582.

Varnham, K., (2005). Non-native species in UK Overseas Territories: a review. Annex 4. JNCC Report 372. Joint Nature Conservation Committee (JNCC), Smithsonian Institution and National Tropical Botanical Garden, www.jncc.gov.uk/pdf/jncc372_web.pdf (accessed on 25 October 2022).

Verspohl, E. J., Bauer, K., & Neddermann, E., (2005). Antidiabetic effect of *Cinnamomum cassia* and *Cinnamomum zeylanicum in vivo* and *in vitro. Phytother. Res., 19*(3), 203–206.

Xi, M., Hai, C., Tang, H., Chen, M., Fang, K., & Liang, X., (2008). Antioxidant and antiglycation properties of total saponins extracted from traditional Chinese medicine used to treat diabetes mellitus. *Phytother. Res., 22*(2), 228–237.

Xu, N., Zhang, L., Dong, J., Zhang, X., Chen, Y. G., Bao, B., & Liu, J., (2014). Low-dose diet supplement of a natural flavonoid, luteolin, ameliorates diet-induced obesity and insulin resistance in mice. *Mol. Nutr. Food Res., 58*(6), 1258–1268.

Zhang, C., Fan, L., Fan, S., Wang, J., Luo, T., Tang, Y., Chen, Z., & Yu, L., (2019). *Cinnamomum cassia* Presl: A review of its traditional uses, phytochemistry, pharmacology and toxicology. *Molecules (Basel, Switzerland) 24*(19), 3473.

Zimmet, P., Alberti, K. G., & Shaw, J., (2001). Global and societal implications of the diabetes epidemic. *Nature, 414*(6865), 782–787.

CHAPTER 15

Role of Plant-Based Bioactive Nutritional or Dietary Supplements in the Metabolic Syndrome of Women and Children During COVID-19

SRIVASTAVA YASHI and ANKIT AWASTHI

Central University of Punjab, Bathinda, Punjab, India,
E-mails: yashicupb@gmail.com (S. Yashi); awasthi0005@gmail.com (Ankit Awasthi)

ABSTRACT

Metabolic syndrome includes obesity, type 2 diabetes (T2D) mellitus, hypertension, dyslipidemia, and inflammation showing detrimental result in women and children on the immune response to COVID-19. The plants are the best sources for the production of bioactive compounds like polyphenol, flavonoids, etc. (Keservani and Sharma, 2014). These compounds have properties to improve the functionality of many receptors (insulin IR, AMPK, TNF, and TNFR, cytokine receptor, IL-1, ILR, TNF, TNFR, and PPARG) that can control internal factors (LDL, VLDL, HDL, triglycerides, glucose level, C-reactive protein (CRP), erythrocytes, heart rate and blood pressure) which eventually reduces the risk of metabolic syndrome. The processing, digestibility, and functionality of bioactive constituents from plants helps in the consumption of nutraceutical supplements through many government schemes and decreases the cases of metabolic syndrome in many countries.

The Metabolic Syndrome: Dietary Supplements and Food Ingredients. Raj K. Keservani, MPharm, Durgavati Yadav, PhD, Rajesh K. Kesharwani, PhD, Sippy Singh, DPhil & Kumar Sandeep, PhD (Eds.)

15.1 INTRODUCTION

15.1.1 DEFINITION

Nutritional supplements or dietary supplement are products utilized to boost the diet with vitamin, mineral, herb, amino acid, and enzyme (Joel, 2012). The nomenclature of nutritional supplements differs across nations, such as "natural and non-prescription health products" in Canada, "dietary supplements" in the USA, "foods for special health use" in Japan, "complementary medicine" in Australia, "biologically active supplements" in Russia and "food supplements" in the European Union.

15.1.2 CURRENT INDIAN NUTRITIONAL STATUS OF W&C

India is a developing nation with excellent food security study with exemplary achievement in food development. India also struggles to tackle children and women's chronic, pervasive hunger and poor health (Food and Nutrition Board, 1991). Women are the most important key for the nutritional status enhancement of the society. In India, female's diet is subsumed or neglected due to "family welfare" and "cultural constraints" (Jose and Navneetham, 2008). The life span of Indian women and men at birth is similar but Indian ladies have more mortality rates, especially during the childhood as well as reproductive phase (Kumar and Khan, 2010).

Stunting, wasting, and underweight are the main metabolic problems in different states of India. Among the northern Indian regions, the incidence of child stunting at the age of five is maximum in Bihar (48%) and lowest in Punjab and Himachal Pradesh, according to NFHS Survey 4, 2015–2016 (26%). There is an improvement within 10 years in child nutrition of few states from NFHS survey 3, 2005–2006 to NFHS survey 4, 2015–2016. The stunting has been decreased by 11% in Uttar Pradesh, 8% in Bihar, 8% in Madhya Pradesh, 5% in Rajasthan, 12% in Haryana, 11% in Punjab, 10% in Uttarakhand, 13% in Himachal Pradesh, 8% in Jammu and Kashmir and 11% in Delhi. Stunting in children under five years has been decreased in these North Indian states, and improvement observed with percent decrease in stunted children. In comparison to the other states, Himachal Pradesh has improved more than others during 10 years (NFHS 3, 2005–2006/NFHS 4, 2015–2016).

Among the north Indian states, the occurrence of wasting of children under the age 5 is minimum in Jammu and Kashmir (12%) and maximum in

Madhya Pradesh (26%) (NFHS survey 4, 2015–2016). As wasting decreased by 6% in Bihar, 9% in Madhya Pradesh, 5% in Himachal Pradesh, 3% in Jammu and Kashmir and 1% in Delhi but simultaneously increased in Uttar Pradesh by 3%, Rajasthan by 3%, Haryana by 2%, Punjab by 7% and Uttarakhand by 1% within 10 years (NFHS survey 3, 2005–2006 and NFHS survey 4, 2015–2016). All states except Delhi improved in terms of underweight population of children like in Uttar Pradesh (2%), Bihar (12%), Madhya Pradesh (17%), Rajasthan (3%), Haryana (11%), Punjab (3%), Uttarakhand (11%), Himachal Pradesh (16%), and Jammu and Kashmir (9%). Delhi has shown increment of 2% population in underweight children case of below five years (NFHS survey 3, 2005–2006/NFHS survey 4, 2015–2016). In northern Indian nations, the drop in the proportion of stunted, wasted, and underweight is a positive indicator of improving the nutritional health condition of children below the age of five years. The condition of women nutrition has improved over 10 years in northern Indian states. Uttar Pradesh has increased by 11%, Bihar by 15%, Madhya Pradesh by 14%, Rajasthan by 10%, Haryana by 15%, Punjab by 7%, Uttarakhand by 12%, Himachal Pradesh by 14%, and Jammu and Kashmir by 13%. On the other hand, no improvement has seen in Delhi within 10 years while Bihar and Haryana improved the women nutritional status. The overweight status of women in northern Indian states shows a percentage rise, it reflects a steady increase in obesity among women. The increase in obese women percentage by 8% in Uttar Pradesh, 7% in Bihar, 6% in Madhya Pradesh, 5% in Rajasthan, 4% in Haryana, 1% in Punjab, 7% in Uttarakhand, 16% in Himachal Pradesh, 12% in J and K and 10% in Delhi within 10 years. Young, pregnant, and lactating women are more prone to nutritional anemic, i.e., Iron or Folic acid deficiency by 40–60% in preschool children and 25–30% in pregnant (third trimester) (Food and Nutrition Board, 1991). The anemia cases increase with number anemic mothers and birth order of children. The risks of anemia arising among children aged 6–59 months are maximum in Haryana (72%) and Madhya Pradesh (69%) in comparison to the other north Indian states (NFHS Report 4, 2015–2016). Kerala displays the lowest underweight incidence among all the southern states, but compared to 2005–2006, where the rate is approximately 25, the rate has decreased below 20 in 2015–2016. Karnataka shows the highest rate of underweight both the years, with reduction in 2015–2016. Data from the years 2015–2016 is accessible for Andhra Pradesh and Telangana owing to the establishment of the new state of Telangana in the year 2014. Kerala has the lowest rate of stunting in children whereas Karnataka has the highest value. The numerous southern states, i.e.,

Andhra Pradesh, Karnataka, Kerala, Tamil Nadu, and Telangana, wasting in children in 2015–2016 were about 18, 26, 16, 20 and 17, respectively. Recently, the national family health survey-5 (NFHS-5) 2019–2020 first phase data has been released by the ministry of health and family welfare, phase 2 of survey was delayed due to COVID-19 pandemics. Comparative data shows the sustainable development that the country aims to achieve by 2030. Furthermore, the inclusion of some new topics like preschool education, disability, access to a toilet facility, death registration, bathing practices during menstruation, methods as well as reasons for abortion and first-time inclusion of women and men percentage in 2019, who have ever used the Internet shows the upgradation and improvement in the ministry of health and family welfare system. Increment of underweight children in big states like Gujarat, Maharashtra, West Bengal, Telangana, Assam, and Kerala receiving the attention for improvement in nutrition and diet of population.

15.1.3 COMPONENT AFFECTING NUTRITIONAL STATUS

There are numerous components that influence the nutritional status such as illiteracy (superstitions and food taboos), food unavailability, unhygienic environment (Infection and Infestation), so many closely spaced pregnancies and ill health or sickness.

15.1.4 NUTRITIONAL SUPPLEMENTS IN INDIA

There are different market factors for health supplements and nutraceuticals in India (Keservani et al., 2010a), including hunger, availability, evolving habits, eating trends and nutritional knowledge, dependent on India's nutritional status (World food India, 2017). In India, the health supplements and nutraceuticals regulations are governed by the Food and Safety Requirements Act, 2006 supervised by FSSAI. According to the Gazette of India (23 December 2016), the Food Safety and Standards Regulations, 2016 informed eight food groups based on substance and application, including nutritional additives, nutraceuticals, dietary foods, foods containing prebiotic ingredients, special foods containing plant or botanical ingredients, and novel foods. In addition, FSSAI, released on 19 May 2017 by the FSSAI Regulations (Fortification of Foods), recommends the fortification standards for atta, maida, milk, oil, rice, salt, and vanaspati and was rendered operational on 17 April 2017. In addition, laws in other countries, including India

(Keservani et al., 2014, 2019), break nutritional supplements and nutraceuticals into 3 fields, i.e., nutritional or enhanced diets, dietary supplements, and functional drinks. Researchers investigate the impact of dietary consumption of nutrients and have indicated that diet should be adequate to discourage therapeutic interventions such as medications and vast amounts of usage of conventional medicines.

15.1.5 TYPES OF NUTRITIONAL SUPPLEMENTS

Multivitamin/multi-mineral products and functional drinks are essentially used together with herbs as dietary supplement in India, for the treatment of health conditions and promotion. These are basically extracted from Keto acids (viz. pyruvate), sugars (viz. glucosamine, chondroitin sulfate), precursors of steroid hormone, etc. The functions of supplements such as health enhancement, weight-loss, and healthy alternatives for diet and ingredients including plant botanicals, nutrients, and medicinal herbs (Keservani et al., 2010b). The supplements are available in numerous medication dosage form, like capsules, pills, semi-solids, powders, and extracts (Bansal and Dhiman, 2019).

It includes:

- Riboflavin and coenzyme Q10 (effective for migraine);
- Supplements for weight loss (chromium, conjugated linoleic acid, L-carnitine, pyruvate, vitamin B6);
- Supplements for antioxidant activity (vitamin C, vitamin E and zinc);
- Supplementation for spinal bone loss in postmenopausal women (calcium with additional trace minerals such as Cu, Zn, manganese);
- Supplementation with probiotics for promotion of gut health (bio-yogurts containing lactobacillus, acidophilus, and bifidobacterium);
- Supplementation for diabetes and prevention of gastrointestinal cancers (refined grains, and cereals fortified with calcium).

The most commonly utilized nutritional supplements in India are Nutrilite (Amway product), Seacod capsules (Aventis Pharma), Glucose-D (Dabur), Protinex for different age (Danone), Calcium Sandoz for man and women (Novartis), Herbal powders (Herbalife International) and Pure Himalaya Herbs (Himalaya Drug Company).

15.2 COMMON METABOLIC SYNDROME OF WOMEN AND CHILD IN INDIA

Metabolic syndrome may increase the risk of obesity, T2D mellitus, high BP (hypertension), hyperglycemia, myocardial infarction, cardiovascular disease, stroke, and visceral obesity (Kylin et al., 1923; Vague et al., 1947).

15.2.1 CENTRAL (ABDOMINAL) OBESITY

Obesity is a prevalent and inflated condition that affects 13% of the global adult population and linked with adipose tissue issues such as hypertrophy, hyperplasia, and chronic systemic inflammation, that is triggering aspect of insulin resistance and metabolic syndrome. The growth of obesity is attributed to causes such as constant viewing TV and the usage of cell phones with high fat food, hereditary history, life style and climatic condition. In addition, the association between obesity and chronic low-grade inflammation can cause other chronic pathophysiological conditions, like T2D, cardiac disease, and metabolic syndrome, due to pathological changes in the metabolism of white adipose tissue and adipokine secretion, which directly influence the health status of the organism (Ellulu and Khazaai, 2017). Several therapeutic medications (phentermine, diethylpropion, sibutramine, and lorcaserin) was developed for the obesity treatment, but several simple and clinical research indicate that bioactive compounds found in food are adequate to inhibit inflammation of both obesity and adipose tissue, as well as improvement in the metabolic syndromes associated with human and animal models (Ahima et al., 2016). About 70 million people in the United States are reportedly obese, including 35 million males and 35 million females, and 99 million are overweight, including 45 million females and 54 million males (NHIS, 2015). The foods containing bioactive compounds boost body metabolism and energy balance. Plant based food such as Fruits, vegetables, nuts, seeds, herbs, and spices are rich source of bioactive phytochemicals, Phenolic compounds, flavonoids, plant sterols, carotenoids, and have some well-documented researches shown improvement in body metabolism, energy balance and protective effects against the inhibition of chronic diseases (coronary heart disease) (Cicero and Fogacci, 2017).

15.2.2 HYPERTENSION

Hypertension condition examined by high systolic and diastolic blood pressure, $\geq$140 mmHg and $\geq$90 mmHg, respectively (Tabassum and Ahmad, 2011). Rajati et al. (2019) reported that 1.56 billion hypertensive patients have risen worldwide regularly. Hypertension, combined with stroke, atherosclerosis, cardiac failure, and coronary artery disease that induces kidney disorders, lack of consciousness, or vision loss. Adipocytokines (angiotensinogen, interleukin-6 (IL-6), β-leptin, tumor necrosis factor-a) and the release of non-esterified fatty acids from visceral adipose tissue are responsible for the hypertension (Katagiri et al., 2007). In addition, Hidekatsu and Yoshida (2008) also stated that the potential factors responsible for the production of hypertension are inflammatory mediators, insulin resistance, activated renin-angiotensin, oxidative stress, disruptive sleep apnea, endothelial dysfunction (ED), and visceral obesity.

15.2.3 DYSLIPIDEMIA

Dyslipidemia defined as variety of qualitative lipid abnormalities characterized by elevated triglycerides (TGs), LDL, slightly LDL-cholesterol containing (apo-B) and low HDL-cholesterol level. The insulin resistance can cause dyslipidemia by raising the development of lipolysis and VLDL. By suppressing lipolysis in adipocytes, the main mediator of VLDL clearance stabilizes ApoB protein via PI3K-dependent pathways and controls lipoprotein lipase production and resulting in increased FFA levels and TG synthesis (Kaur et al., 2014). One of the components of dyslipidemia that can predict the presence of respiratory disease, further leading to pneumonia is high-density lipoprotein (Ginsberg et al., 2005; Lewis and Steiner, 1996). In the dyslipidemia, a high level of low-density lipoprotein interaction with macrophages in atherosclerotic plaques increases inflammatory gene expression, that increases COVID-19 infection risk (Saballs et al., 2016; Timotius and Kurniawanet, 2020; Charvet et al., 2015).

15.2.4 DIABETES

Diabetes is a long-term disorder that arises from a mixture of heredity, climate, and additional risk factors such as glucose sensitivity, insulin resistance, core obesity, hormone deficiency, and lifestyle factors (alcohol, body weight (BW), nutrition, physical activity, smoking) (Wu et al., 2014; Siegel and Wilson, 2018). In addition, metabolic metrics such as fasting plasma glucose, lipid profile, blood pressure and waist circumference may be dramatically altered by lifestyle variables (Huang et al., 2015). Additionally, Oh et al. (2005); Lee et al. (2011) stated that the two significant risk issues for the growth of metabolic syndrome components are the heavy consumption of alcohol and smoking.

The higher triglyceride elevation and HDL-cholesterol decrease in diabetic individuals were recorded by Siegel et al. (1996). Experimental studies have demonstrated that changing lifestyle and dietary supplementation will reduce the development of diabetes and other metabolic syndromes (Knowler et al., 2002). Actually, Alessandro and Zheng (2020) also stated that patients with diabetes have a two to three times higher chance of COVID-19-related serious disease and mortality.

15.2.5 INFLAMMATION

Inflammation is known as an organism's physiological reaction to damaging stimuli (physical, environmental, biological) and homeostasis is restored (Monteiro and Azevedo, 2010). The physiological reaction includes the transport of plasma components and leucocytes or the action of higher concentrations of white cells in the blood that have operated on trouble areas, which are ultimately triggered by mast cells and macrophages with the coordination of several other cell types and inflammatory mediators (cytokines, chemokines, eicosanoids, proteolytic cascade products and vasoactive amines) (Qatanani and Lazar, 2007). The high fat diets often eaten by obese individuals will induce direct inflammation of fats rich in saturated fatty acid and indirect inflammation by intestinal permeability (Monteiro and Azevedo, 2010). In addition, study between obesity and inflammation by Welsh et al. (2010) indicates that the elevated fattiness correlated with the obesity-associated gene and higher degree of C-reactive protein (CRP) is not suggested by any contrary pathway (Robertson et al., 2010). In addition, the positive correlation between insulin resistance and metabolic syndrome risk aspects with pro-inflammatory markers such as (cytokines (TNF-a, IL1, IL6), Chemokines (MCP-1) CD40L) was reported by Natali et al. (2013) and Nathalie and Sylvie (2014).

15.3 CAUSE OF METABOLIC SYNDROME

15.3.1 INSULIN RESISTANCE

Insulin hormone, that is significantly secreted by beta cells and pancreatic cells in the pancreas, maintains the level of blood glucose. Insulin resistance has been described as an impairment of insulin-mediated glucose disposal sensitivity. Hormones such as glucagon, glucocorticoids, and catecholamines influence insulin action. Excess secretion of the hormones glucagon and catecholamine can lead to resistance to insulin. Insulin release is regulated by several influences, including physical structure, carbohydrate type and glycemic index, and protein. The amount of insulin secretion was reduced when food was eaten with a low glycemic index. There are several other causes that influence action and insulin release, such as dietary fiber Zn, Cr, and Fe. Insulin-resistant people are typically diagnosed with abdominal obesity owing to subcutaneous abdominal fat (Jensen et al., 1989). Ginsberg et al., (2000) demonstrated the relationships between insulin resistance and dyslipidemia dependent on the modulation and secretion of VLDLs that further result in hypertriglyceridemia, but on the other side, reducing the mass of adipose tissue which reduced the secretion of VLDL, plasma TG and normalize the composition of LDL particles. Syndrome x is usually referred to as insulin resistance, a set of cardiovascular risk factors such as glucose sensitivity, hyper-triacylglycerolemia, HDL-cholesterol dropping and hypertension that may be associated with nutrients, insulin resistance, central obesity, and physical inactivity (Haffner et al., 1992; Daly et al., 1997; Purnell and Brunzell, 1997; Wareham et al., 1998).

15.3.2 HORMONAL IMBALANCE

Nutrients play a key role in releasing hormones (insulin and glucagon) and peptides, activating signaling mechanisms and multiple neurotransmitters that regulate neuronal function, learning, and memory, following absorption of food into the blood supply and enters a brain area (hypothalamus and hippocampus) (Pistollato and Battino, 2014). A link has been identified between metabolic syndrome and hypothalamic changes through feeding rabbits with high fat diets. The research shows a decline in gonadotropin-releasing hormone (GnRH), an opportunity for interleukin-6 (IL-6) and glucose transporter 4 (GLUT-4) expression, and a substantial decrease in KISS1 receptor immunopositivity, which is negatively associated to glucose intolerance (Morelli and Adorini, 2014). Ventimiglia and Mirone (2016) stated that metabolic syndrome had lowering effect on general male health status by minimizing the overall anti-Mullerian hormone, inhibin B, sex hormone-binding globulin and testosterone. Similarly, Saeideh and Mohseni (2013) reported that low level of sex hormone binding globulin and high testosterone in postmenopausal women having metabolic syndrome than normal. In addition, the risk of higher estrogen-androgen ratio and estrogen deficiency increases in postmenopausal women having metabolic syndrome (Ridker et al., 2006).

15.3.3 GLUCOSE INTOLERANCE

Glucose intolerance may be characterized as dysglycemia that involves both prediabetes and diabetes with risk factors such as pre-pregnancy BMI, diabetes mellitus (DM) history, gestational week, fasting blood glucose and poor insulin response during pregnancy (Kim et al., 2002; Ekelund et al., 2010). Canadian population research represented that food categories such as fruits, vegetables, cereals, legumes, and whole grains, nuts, green leafy plants, nutritious breakfast foods, and hot meals are emerging as protective against impaired tolerance to glucose and diabetes (Gittelsohn et al., 1998). The glucose level abnormalities are explained as glucose intolerance and impaired fasting glucose those are covered by the term "prediabetes." In which the glucose intolerance is deeply connected with risk of other diseases because of large vessel occlusive, stroke, myocardial infarction, and finally death (Harris et al., 1998).

15.3.4 ENVIRONMENTAL EFFECT

Many studies in human and animal models suggest that polluted environment and pollutants are the reason of promotion in metabolic syndrome and excessive exposer may leads to the higher risk of metabolic syndromes such as insulin resistance, hyperglycemia, hypertension, and obesity. Workers present in refineries and its surrounding areas are mostly targeted by metabolic syndromes (Cotrim et al., 2005; Sirit and Portillo, 2008; Mohan and Reddy, 2008), Continuous exposure to lead has relation with hypertension (Prozialeck et al., 2008), higher arsenic content in drinking water more than the permissible limit increases the chances of T2D (Wang et al., 2003; Rahman et al., 1998; Navas et al., 2008) and some studies on animal model such as in mice found that the lasting exposure to air pollutants promote the development of obesity, insulin resistance and other metabolic syndrome (Sun et al., 2009; Hutcheson et al., 2012). Vegetarian diet involvement for two years consists of whole grains, fruits, vegetables, nuts, olive oil, and tea can decrease the level of CRP and improve insulin resistance (Esposito et al., 2004). Furthermore, the fruit and vegetable study group in oxford reported that higher consumption of fruits and vegetables and increased antioxidant concentration in body helps to reduce blood pressure and shows preventive effect on metabolic syndrome (John and Neil, 2002). Moreover, supplementation with green tea reduced BW, BMI, and had an advantageous effect on peroxidation of lipid in patients suffering from obese metabolic syndrome (Basu et al., 2010). Stancliffe et al. (2011) reported the suitable dietary intake of dairy products had positive impact on oxidative stress biomarkers in metabolic syndrome. Restriction in diet filled with carbohydrate exclusively encourages the weight and fat loss (Sackner et al., 2015) and improvement in dyslipidemia and T2DM diabetes (Volek et al., 2005, 2008; Paoli et al., 2013). Furthermore, Blair and Neill (2020) reported the effect of carbohydrate diet on metabolic health, comparative study of Carbohydrate restricted diet over Excessive consumption of dietary carbohydrate shows reduction in insulin levels, inflammation markers, visceral adipose tissue, improved lipid profile, blood pressure, diabetes, and ultimately reverse the metabolic abnormalities.

15.4 PLANT ORIGINATED BIOACTIVE SUBSTANCE

Plant origin-based by-products are used to obtain bioactive peptides and utilized in the preparation of pharmaceutical components for metabolic syndrome such as diabetes, hypertension, Inflammation, dyslipidemia, etc. (Xu and Yan, 2019; Mortazavi and Hasanzadeh, 2018). Diets rich in anthocyanins, carotenoids, dietary fiber, lectins, oligosaccharides, phenolic compounds, phytic acid, phytosterols, resistant starch, saponins, sterols, tocopherols, tocotrienols, vitamins, and some other constituents have essential role in the preparation of dietary supplements.

15.4.1 FRUITS AND VEGETABLE

Many diseases and disorders can be prevented by supplementation of bioactive components like vitamins, polyphenolic, carotenoids, potassium, calcium, magnesium, and dietary fiber (Boeing et al., 2012). There are more than 1,500 phytochemicals are identified in grains, fruits, and vegetables (Liu et al., 2013). Around one third of phenolic acids are contained in dietary phenols, which can be contained in plants as bound and free forms (Robbins et al., 2003). Phenolic acids are essentially divided into 2 classes, hydroxycinnamic acids and hydroxybenzoic acids. Gallic acid, p-hydroxybenzoic acid and protocatechuic acid are the benzoic acid derivatives and present in fresh produce. Fruits and vegetables commonly contained cinnamic acid derivative such as ferulic, caffeic, coumaric, and sinapic acids (Clifford et al., 1999). In addition, studies showed that in the external parts of ripe fruits have hydroxycinnamic acids (0.5–2 g/kg). Caffeic acid is deemed the most prevalent phenolic acid and is the bulk of the overall value of fruit hydroxycinnamic acid (Manach et al., 2004). Flavonoids, lignans, phenolic acids, stilbenes, and tannins are main type of polyphenol in fruit and vegetables (hydrolysable and condensed) (Harborne and Moss, 1999; Keservani and Sharma, 2014). In addition, the class of flavonoids including flavones, anthocyanins, isoflavones, flavanones, and flavanols was described by Tsao and Yang (2003). Pelargonidin, cyanidin, peonidin, delphinidin, petunidin, and malvidin are six classes of anthocyanins present in plants. The third major phenolic community is classified into tannins that are hydrolysable and condensed (Porter et al., 1989). A significant source of phytoestrogens is lignanic secondary plant metabolites, usually present in the form of lariciresinol, matairesinol, pinoresinol, and secoisolariciresinol. Likewise, glucosinolates are categorized into 3 key groups: aromatic, aliphatic, and indolyl glucosinolate. Wine, grape juices, peanuts, and peanut butter are also a strong source of stilbenes (Balsano and Alisi, 2009). Furthermore, β-Carotene, β-cryptoxanthin, canthaxanthin, capsanthin, capsorubin, flavoxanthin, lutein, lycopene, and zeaxanthin were found in the carotenoids.

15.4.2 CEREALS, LEGUME, AND WHOLE GRAINS

Raw material is processed through many treatments in the food manufacturing process, such as baking, frying, extrusion, milling, and polishing, resulting in a reduction in the bioactivity of the product's bioactive components (Ragaee et al., 2014). Ahmed et al. (2017) stated that cereal and pulse processing (removal of bran and germ portions) and preparation strategies such as baking, frying, puffing, and extrusion processes have a substantial consequence on the bioactivity of compounds. Improving the bioactivity of bioactive compounds and other food-based components of cereals and cereals requires advanced technologies. Food derived from refined wheat flour has a lower content of bioactive compounds (phenolics and dietary fiber) relative to multigrain flours due to the presence of mineral, protein, and certain bioactive compounds like ferulic acid in the aleurone layer of bran and other whole grain constituents which shows protective effect against several diseases such as central obesity, hypertension, DM, cardiovascular diseases (CDVs) (Keservani et al., 2016), and some cancers (Fardet et al., 2010; Lillioja et al., 2013). In addition, Baiano et al. (2015) stated that different varieties of barley wort can be used in bakery industries as a water replacement and increase the overall phenolic bread content. In addition, Bhartiya et al. (2015) reveals the ability and dietary value of horse gram (Macrotyloma uniflorum) such as excellent healing properties and common applications for the care of patients with asthma, bronchitis, cardiac failure, kidney stones, leukoderma, piles, and urinary diseases. In addition, metabolic change influences the bioavailability, palatability, and digestibility of vital nutrients during sprouting and greatly boost the activities of obesity, anti-diabetic, antioxidant, anti-ulcer, and inflammatory reduction (Masood et al., 2014; Mandle et al., 2012).

The antioxidant activity, phenolic content, and inflammation reduction potential of broad bean, common bean, chickpea, grass pea, pea, soybean, and white lupin extracts were assessed by Sibul and Dukic (2015). The phenolic profile analyzed using the LC-MS/MS system indicates that the quality of isoflavones (genistein and daidzein) is higher in soybean, common bean, and white lupine extracts collected from roots. Chickpea, white lupine, and common bean as the richest source of flavonoid compounds obtained from herb extracts.

Red mold rice or red yeast rice is a traditional fermented product of Asian countries that can enhance the blood circulation and digestion. It has cholesterol lowering efficiency equivalent to pure lovastatin (20–40 mg). Red mold rice contains mainly five bioactive compounds, i.e., monacolin J, I, M, K, gamma-aminobutyric acid, citrinin, dimerumic acid, flavonoids (Tien-Ju et al., 2019).

15.4.3 GREEN LEAFY PLANTS AND MEDICINAL PLANTS, OILS, AND NUTS

The various dose of green tea (~300–6,000 mg/day), catechins (~150–1,200 mg/day), *P. vulgaris* (~445–3,000 mg/day), *G. cambogia* (~1,667–3,000 mg/day), *N. sativa* (~1,500–3,000 mg/day), puerh tea (~1,000 mg/day), *I. gabonensis* (~150–3,150 mg/day), *C. fimbriata* (~1,000 mg/day) able to BW reduction and BMI with regular consumption at certain period of period. However, the effective doses for green tea, catechins, *P. vulgaris, G. cambogia, N. sativa*, puerh tea, *I. gabonensis* and *C. fimbriata* are 6,000 mg/day, 400–800 mg/day, 1,000 mg/day, 2,400 mg/day, 3,000 mg/day, 1,000 mg/day, 300 mg/day and 1,000 mg/day, respectively (……). The consumption of green tea (up to 6,000 mg/day), catechin (up to 1,200 mg/day), *P. vulgaris* (up to 300 mg/day), *N. sativa* (up to 3,000 mg/day), *I. gabonensis* (up to 3,150 mg/day), *C. imbriata* (up to 1,000 mg/day) may reduce the waist circumference and hip circumference in addition to weight (…….). The body fat percent, fat mass, and fat-free mass was decreased by various doses intake of green tea (6,000 mg/day), catechin (458, 468, and 886 mg/day), *I. gabonensis* (300 mg/day), *H. sabdariffa* (75 mg/day), *P. vulgaris* (445 mg/day), *G. cambogia* (2,400 and 3,000 mg/day), *Ecklonia cava* (72 and 144 mg/day), cumin (3,000 mg/day), *Coleus forskolii* (250 mg/day), *Gynostemma pentaphyllum* (450 mg/day) and cinnamon (550 mg/day). The dose of cumin (100 and 3,000 mg/day), flaxseed (2,500 and 6,000 mg/day), *H. sabdariffa* (75 mg/day), rosehip (100 mg/day), *L. Barbarum* (30, 60, and 120 ml/day), cinnamon (550 and 2,000 mg/day) is workable on weight loss and BMI. While few reported that the doses of *G. ambogia*, aloe vera, cocoa, rapeseed were 3,000 mg/day, 588 mg/day, 1.4 g/day and 3,500 mg/day increased the fat free mass.

Steve et al. (2007) selected the candidates of any age and having T2D as per guidelines of American Diabetes Association (ADA). During the clinical trial participants were not permitted to consume sulfonylureas, meglitinides, metformin, thiazolidinediones, α-glucosidase inhibitors, exenatide, hydromethylglutaryl CoA reductase inhibitors, ezetimibe, niacin, or fibric acid derivatives and insulin. Candidates were segregated by sex and consumption of cinnamon or placebo. Subjects were feeded one capsules containing 500 mg product for two times (breakfast and dinner) in day for three months. Cholesterol (total, LDL, and HDL), fasting glucose, triglyceride, and insulin levels were measured after 1, 2, 3 months. It was estimated that triglycerides changed from 11.5% to 24.7% with 80% power. The results showed 17 individuals were excluded out of 77 due to health problem (03 individual) and having A1C < 6.0% (14 individual). Among 17 individual 09 belongs to cinnamon and 08 from placebo group. Totally 43 candidates completed the study because of medicine, nondietary cinnamon supplement, accident, relocation, inconvenience, dissatisfaction. The first visit completed by 53 candidates out of which 30 belongs to cinnamon and 28 from placebo group. There was 51% women representation in the complete group of candidates. It was the first US study for analyzing cinnamon effect on blood glucose and lipid levels in individuals with T2D. The study reported no significant changes in insulin levels, fasting glucose, lipid, A1C, while Khan et al. (2003) showed that three different doses of cinnamon (1, 3, and 6 g/day) was effective for lowering fasting glucose, total cholesterol, low density lipoprotein cholesterol, and triglyceride levels in T2D individual.

A study has been performed keeping International Diabetes Federation {Fasting Blood Glucose > 5.6 mmol/L, systolic blood pressure> 130 mmHg, diastolic blood pressure > 85 mmHg, Triglycerides > 1.7 mmol/L (150 mg/dl), HDL-C < 1/3 mmol/L (50 mg/dl) (Legeay et al., 2015)} criteria on females having metabolic syndrome for 08 weeks in parallel group. The 70 women with age group of 40–70 years having metabolic syndrome were selected for clinical trial. These women were randomly separated into two groups of 35 individuals of same age group and menopausal status. The study excluded the women suffering from inflammatory diseases, heart failure, cancer, use of weight loss regimens, liver, and kidney dysfunction, pregnancy, and lactation and having six months green tea consumption record. The information about smoking, medical history, age, marital status, occupation, and medication status through questionnaire at the starting of study were collected. All the participants were feeded one green tea bag (1,500 mg) 3 times/day for 8 weeks while control group subjected to only lukewarm water without change in regular diet. The polyphenol and caffeine content of green tea were 12.6% and 2.85%, respectively. Each bag of green tea contained polyphenol and caffeine content of 180 mg and 41 mg, respectively. The diet, life style, physical activity of candidates was maintained throughout the study. The important parameters, i.e., BMI, waist–hip ratio, fasting blood glucose, lipid profile (HDL, TG, LDL, TC), hip circumference blood pressure, weight, height, waist circumference was estimated initially and end of study. The 65 participants (Green Tea Group; 32 and control group; 33) have completed the study due to constipation, knee surgery, personal concern, and heart problems. The results from Anthropometric indices showed that there was significant lowering from baseline values for weight, BMI, WC, and WHR while no changes were observed in control. There was no significant change were observed in HC in both the groups. Further, the paired t-test of systolic blood pressure, fasting blood glucose, total cholesterol and LDL for Green Tea Group were significantly lower from baseline while control has no change. There were no substantial deviations in diastolic blood pressure, HDL-C, triglycerides of both the groups. In addition, the paired t test represented that the calories intake, total fats, cholesterol, saturated fatty acid, potassium, magnesium, and caffeine were significantly higher while no significant effect observed in control than the mean baseline values for Green Tea Group. The Independent t-test represented that total fats, cholesterol, SFA, K, Mg, and caffeine change knowingly more in the green tea group than in the control group. The inference of the results represented that regular green tea consumption for 8 weeks have substantial effect on weight, body mass index (BMI), WC, WHR, SBP, FBG, and low-density lipoprotein of women having metabolic syndrome in comparison to control group (Figure 15.1).

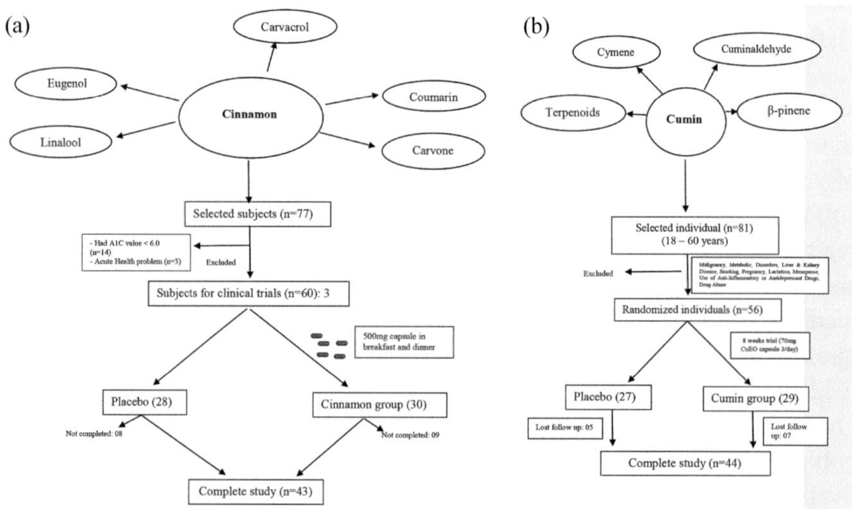

FIGURE 15.1 Representative clinical trial diagramme of spices: (a) Cinnamon; (b) Cumin.

15.4.4 MILK AND MILK PRODUCTS

It is difficult to neglect the contribution of dairy-based bioactive ingredients and their beneficial impact on metabolic syndrome. The bioactivity of the ingredients depends on their raw milk quality, manufacturing status, breeding success, and animal diet. Bioactive peptides (3 to 20 amino acids per molecule) are inactive whey protein molecules namely α-lactalbumin, β-lactoglobulin, serum albumin, immunoglobulin, lactoferrin, protein-pepton, and casein fractions (Korhonen and Pihlanto, 2007). Although, proteolytic digestion activities with naturally occurring microbial and adventitious enzymes (pepsin, trypsin, and chymotrypsin) are seen during gastrointestinal transit, the involvement of lactic acid starter bacteria in the phase of fermentation and maturation greatly enhances the milk products (Clare and Swaisgood, 2000; Gobbetti et al., 2002). The development of biologically active peptides in cheese and fermented milk is the function of the synthesis of the metabolite by bacteria and proteolytic components from milk substrates found in cell walls, bacteriocins, and extracellular environments (Fitzgerald et al., 2006; Hayes et al., 2007; Hernandez et al., 2005; Pan et al., 2006).

Dorota and Emilia (2014) examined the impact of false flax cake supplementation (120 g/d) on goat diets for the consistency and development of high bioactive components content in kefir. The higher content of bioactive peptides in goat milk is beneficial for the production of kefir. The false flax cake supplementation has a beneficial impact on the fatty acid profile due to the improved PUFA content including conjugated linoleic Acid (72% and 78%), n–3 fatty acids including ALA, EPA, and DHA (50% and 80%) and monoenic trans acids. In addition, instrumental and organoleptic evaluation indicates that the increase in bioactive components had a negligible effect on the quality of kefir, such as acidity, texture, color, flavor, aroma, basic content, and chemical composition compared to controls. Ignacio and Reyes (2020) documented the activity and ability of milk related bioactive peptides to enhance metabolic syndrome, such as blood pressure. More evidence observed when protein supplements consist of whey peptides applied to a beverage containing 50 g of glucose decreased the influence of postprandial glycemia in a dose-dependent style (Petersen et al., 2009). Furthermore, whey protein has the capacity to suppress dipeptidyl peptidase IV and raise glucagon-like peptide-1 levels, thereby decreasing incretin hormone hydrolysis or extended insulin reaction in mice (Drucker et al., 2006). The bioactivity of compounds such as genistein, daidzein, and equol obtained from commercial bovine and soybean milks on the inflammatory responses from intestinal cells (Ros et al., 2016). Hermansen et al. (2014) found that the amino acid and branched-chain amino acid structure was especially correlated with the insulinotropic impact of dairy proteins.

15.4.5 SPICES

15.4.5.1 CINNAMON

Cinnamon is able to induce glycogen synthesis by glycogen synthase production and glucose uptake by stimulating insulin-stimulation of tyrosine phosphorylation and phosphatidylinositol 3-kinase (Jarvill-Taylor et al., 2001;

Qin et al., 2003, 2004). Researchers indicated that a 40-day supplementation of 1 to 6 grams of cinnamon per day could boost the blood glucose and lipid profiles of metabolic syndrome patients. Fasting blood glucose reduced by 10.3% following filtered aqueous cinnamon extract supplementation for up to 4 months (Khan et al., 2003; Mang et al., 2006). In the suburban area of Ohio, a clinical trial was performed by recruiting volunteers aged 30–60 years with a fasting blood glucose level of 100–125 mg/dL with regular physical activity and liver and kidney function and without any infection or disease. The two Cinnulin capsules (250 mg) twice daily for 12 weeks may be adequate to boost metabolic syndromes (i.e., systolic blood pressure, fasting blood sugar, and body composition) based on age, physical exercise, fasting blood glucose, and systolic blood pressure levels (Ziegenfuss et al., 2006).

15.4.5.2 CARALLUMA FIMBRIATA

Central obesity is the trigger factor for elevated plasma glucose levels, high blood pressure, dyslipidemia, and metabolic syndrome. Research on *Caralluma fimbriata* extract with daily physical exercise and dietary consumption has been published. A clinical study was performed at Victoria University in Australia by randomly selecting 33 participants from the 29–59 years age group with BMI > 25 kg/m^2, waist circumference more than 94 cm for males, and more than 80 cm for females. They were splitted into two groups, placebo (total 16; two males, 14 females) and experimental group (total 17; 5 males, 12 females). The subjects of type 1 diabetes, heart, liver, and renal disorder, weight loss medication, smoking habit and pregnant were removed. The subjects were fed placebo and *C. fimbriata* extract capsules (500 mg) twice times/daily for 12 weeks prior to meals. The workout and hypocaloric dietary consumption were consistent. Anthropometric measures were taken in compliance with guidelines (weight, height, BMI, hip circumference, neck circumference, waist to hip ratio). The result of the clinical trials express 10 candidates (5/each group) were not completed the study. The supplementation of *C. fimbriata* extract with hypo-calorie diet was able to significantly reduced the central obesity which is one of the key components of metabolic syndrome. There was substantial decrease in waist circumference during the supplementation of *C. fimbriata* because of lipolysis of body fat or increase in lean muscle tissue with fat loss. The study showed that pregnane glycosides, i.e., essential component was played inhibition role in proliferation, differentiation, and maturation of adipocyte. The leptin in the present study has not been decreased while extract of *C. fimbriata* extract can reduced the leptin in rats. The balanced dietary intake and physical activity with *C. fimbriata* extract reduced the WHR significantly. The lowering of WHR related with the lowering of triglyceride, glucose, blood pressure and increased in HDL level (Figure 15. 2).

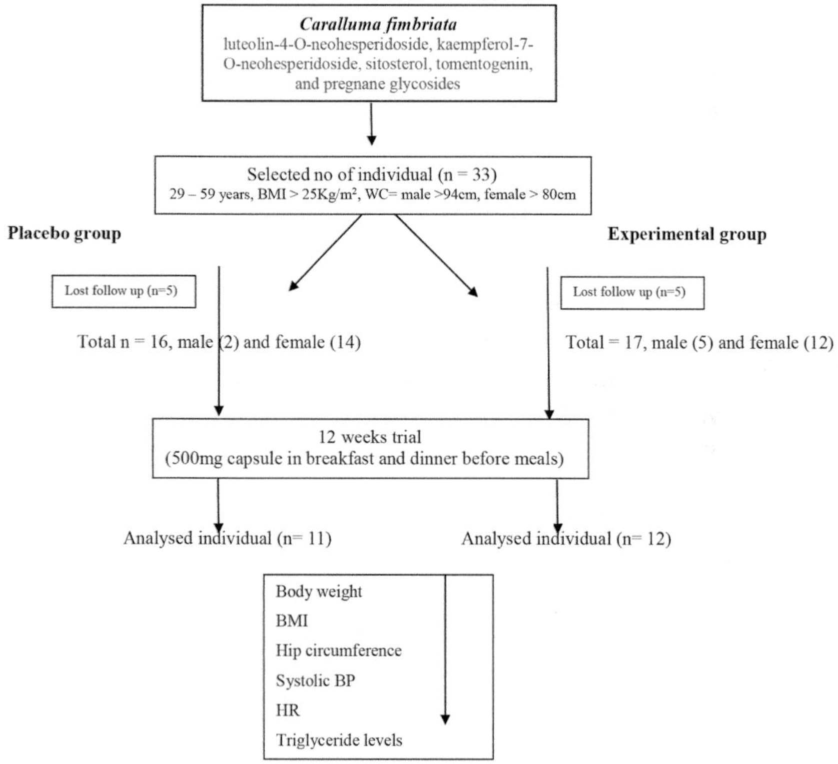

FIGURE 15.2 Representative clinical trial diagramme of *Caralluma fimbriata*.

15.4.5.3 CUMINUM CYMINUM L.

The essential oil of *Cuminum cyminum L.* supplementation has been done to analyze the inflammatory and antioxidant status of metabolic syndrome subjects. The total 81 candidates of age 18–60 years and fulfilling the metabolic syndrome criteria of blood pressure, triglycerides, BMI, sex, and waist circumference. The control and experimental group fed the soft gel (75 mg) of placebo and cumin essential oil daily for eight weeks. The subjects who consumed 90% tablets with physical activity and dietary intake were included for analysis. There was no significant difference in superoxide dismutase, glutathione peroxidase, catalase, total antioxidant capacity, malondialdehyde, and glutathione peroxidase in baseline while within the group the sharp decrease in malondialdehyde (13.3%) and increase of total antioxidant capacity (6.75%) observed after eight weeks (Figure 15. 1(b)).

15.4.6 LIPID

A concept of 'home fortification' includes micronutrient powders and lipid-based nutrient supplements (LNS) has been come into picture (De-Regil et al., 2011). In 2009 a project (International Lipid-Based Nutrient Supplements) based on supplying nutrient supplements has been started with four main objectives, i.e., LNS products development, acceptability (demand and supply) and cost evaluation for infants (6–24 months), pregnant, and lactating women in Burkina Faso, Ghana, and Malawi. The LNS supplemented products contained plant originated substances like sesame, cashew, and chickpea paste, omega-3 fatty acid, alpha-linolenic acid, dried skim milk, soybean oil, peanut. In addition to this LNS gave good result with milk in terms of preventing stunting and poor development than LNS without milk. LNS is also available in various forms like peanut flavor, low-fat grains, and legumes. The lipid-based products or Ready-to-use therapeutic foods are classified under three categories as per severity of malnutrition, i.e., acute (Ration ~180–280 g/day; 200 kcal kg^{-1}/day), moderate acute (Ration ~45–90 g/day; ~250–500 kcal/day) and undernutrition (Ration ~20 g/day; ~110 kcal/day) (Mary et al., 2015).

15.5 MARKET STATUS

India is having versatile biodiversity center for medicinal plants, biotic province, and vegetative zones. In every decade the plant-based drugs were introduced for therapeutic purposes like serpentine, vinblastine, phophyllotoxin, quinine, emetine, colchicine, etc. (Samy and Gopalakrishnakone, 2007). The data of various countries have flavonoid intake is 60–106 mg/day (Brazilian), 55 mg/day (Finland), 190 mg/day (US) and 29 mg/day (Denmark). The growth rate of herbal phytopharmaceuticals is 5–15%/

yearly that cover $60 billion US market (Naoghare and Song, 2010). In five years (2000–2006) the total of 26 plant drugs were approved (Saklani and Kutty, 2008). The several plants like Veregen, Fulyzaq, Sativex, etc., are FDA approved globally available botanicals (U.S. Food and Drug Administration, 2013). The future estimates for the pharmerging markets were 150–165 billion dollars till 2016 (Kleinrock et al., 2012). The world-wide nutraceuticals market growth was $23.8 billion in 2013 while $33.6 billion in 2018 (Markets and Markets, 2014). US is the biggest market for functional foods and nutraceuticals with 21% of growth in 2015 (Transparency Market Research, 2014).

15.6 DIFFERENT PROCESSING METHODS FOR DIETARY SUPPLEMENTS

15.6.1 ENCAPSULATION

It is the process of entrapping or covering the compounds having nutraceutical properties with matrix or wall material. These encapsulated materials released at desired place and time.

15.6.2 HIGH PRESSURE PROCESSING

It is the technique that can restore the dietary value and sensory properties of dietary supplements due to the less effect on bonding of low mass compounds. The ~600 MPas treatment may retained the phenolic contents, carotenoids, ascorbic acid, and antioxidant activity in comparison to thermal treatments (Patras et al., 2009).

15.7 BIOACTIVE SUBSTANCE CONTRIBUTION DURING COVID-19

The latest viral disease COVID-19 (cause by SARS-CoV-2) badly hit the world population 2019–2020. This virus inoculation and replication in human being happened by ACE2 (protease), TMPRSS2 (spike protein), and 3CLpro (3 chymotrypsin-like protease). The phytochemicals like tannins, flavonoid (Papyriflavonol A, Quercetin, Baicalin, Luteolin, Gallocatechin Gallate, Epigallocatechin Gallate, Scutellarein, Hesperetin, Amentoflavone,), terpenes, alkaloid, and fatty acid have common structural feature that may inhibit coronaviruses Jo et al. (2019). There are many metabolites (Ferruginol, Abietane, savinin, hinokinin) that available naturally and have antiviral action against SARS-CoV1 (Wen et al., 2007; Cui et al., 2020) (Figure 15.3; Table 15.1). Curcumin may inhibit protease as well as replication of SARS-CoV with the EC50 value of 10 μM. It has also the ability to restrict cytokine storm of Ebola virus (Wen et al., 2007). Nimbolide was reported as TNF-α inhibitor which is considered as responsible factor for pneumonia induced respiratory failure during COVID-19 (Pooladanda et al., 2019).

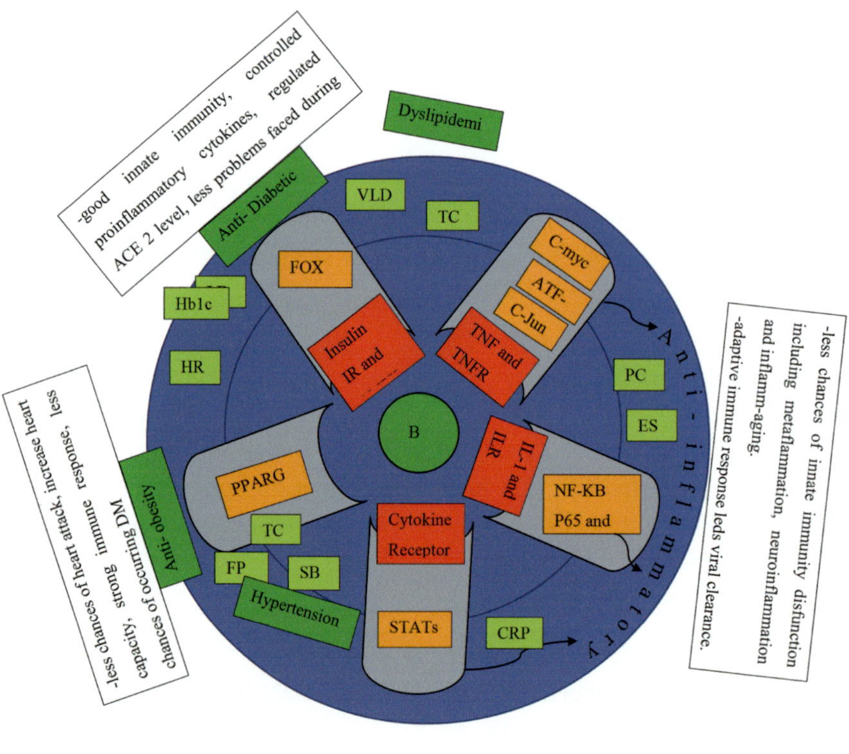

FIGURE 15.3 Diagrammatic representation of symptoms, metabolic syndrome and associate factors.

LDL – low density lipoprotein, **VLDL** – very low-density lipoprotein, **TC** – Total cholesterol, **FPG** – fasting plasma glucose, **SBP** – systolic blood pressure, **HR** – Heart rate, Hb1c – Glycated haemoglobin, **CRP** – C-Reactive Protein, **ESR** – Erythrocyte Sedimentation Rate, **PCT** – Procalcitonin

Withaferin A extracted from *Withania somnifera* has antiviral action for Herpes simplex virus 1 and 2 that may play effective role against COVID-19. Quercetin has ability to restrict the SARS-CoV entrance in the host cells with 83.4 μM concentration (EC50 value) (Yi et al., 2004). Saikosaponins B4 (triterpenoids family) has antiviral action as well as act as spike protein inhibitor that may become therapeutic molecule for corona virus 229E (Cheng et al., 2006; Goswami et al., 2020). The EC50 values of Luteolin and quercetin was 10 mM and 83 mM respectively, that might inhibit the SARS-CoV infection by restricting virus entry into Vero E6 cells. Emodin is highly target specific with less adverse effect for COVID-19 (Zhou et al., 2020).

15.8 GOVERNMENT NORMS BIOACTIVE SUBSTANCE INCORPORATION AS DIETARY SUPPLEMENTS

The specification of effective dosage of bioactive substance in patients diet is one of the major challenge, not only due to sensation perceived in the mouth and odor of the final product but also the effect of nutraceutical interactions with the environment and food matrix components, degradation in bioactivity during food processing, finished product stabilization and intended health benefit after ingestion make legal process of regulation and legislation more complicated (Augustin and Sanguansri, 2014). The government norms of bioactive supplement are changes across different countries due to the cultural facets and infrequent scientific studies of herbal supplements. So, therefore only few measures of herbal supplements tested for efficacy, quality, and safety furthermore widespread agreement exist on the requirement to explain the scientific suggestion on recommendations, regulations, and policies of supplements for safeguard the public health (Rane and Patil, 2020; Rane and Bharath, 2020). The supply of bioactive supplements has grown rapidly with the help of methodologies and skilled personnel to analyze them (Dwyer et al., 2018), for example, after announcing Dietary Supplement Health and Education Act in 1994 supplements manufacturers produced an estimated 4,000 products in USA and consequentially, it reaches up to more than 29,000 products in 2000 and 85,000 supplement products till 2017 in the US market. Furthermore, the inclusion of Natural Health Products Regulations in 2005 is the evidence of growth eruption internationally and approximately 100,000 supplements applications was approved in Canada with concerns about misidentification of ingredients, safety concerns, QA, and QC Problems in industries (Mudge et al., 2016; Orhan et al., 2016).

TABLE 15.1 Summary of Bioactive Components, Source, Function for Curing Metabolic Syndrome

S. N.	Bioactive component	Common name	Botanical Name	Function	References
1	Curcumin	Turmeric	*Curcuma longa*	Antioxidant And Anti-Inflammatory Properties, Antiviral, Antibacterial, and Anticancer Activities, Malignant Diseases, I.E., Diabetes, Allergies, Arthritis, Alzheimer's disease	Zorka Stanić (2016)
2	Gingerols, [6]-gingerol and [6]-paradol	Ginger	*Zingiber officinale*	Anti-Carcinogenic Activity, Anti-Hyperglycaemic, Anti-Diabetes, Decreasing Glucose Tolerance	*Srivastava AK et al. (2009); Son and Miura et al. (2015)*
3	Gingerone	ginger	*Zingiber officinale*	Anti-Inflammatory, Antidiabetic, Antilipolytic, Antidiarrhoeic, Antispasmodic	Venkatachalam & Subramanian et al. (2017).
4	Allicin (allyl 2-propenethiosulfinate or diallyl thiosulfinate)	garlic	*Allium sativum*	Anti-tumor and Anti-microbial	Bayan and Koulivand et al. (2014).
5	Polyphenol type-A polymers	Cinnamon	*Cinnamomum cassia*	Insulin Mimetic Activity, Antihypertensive Activity, Triglyceride-Lowering Effects	Kim et al. (2006); Preuss et al. (2006)
6	Polyphenols, mainly flavonoids and procyanidins	Hawthorn	*Crataegus species*	Reduce blood pressure, Antihypertensive Activity, Dyslipidemia	Asgary and Amiri et al. (2004); Kim and Kang et al. (2000); Shanthi and Devaraj et al. (1994)

TABLE 15.1 *(Continued)*

S. N.	Bioactive component	Common name	Botanical Name	Function	References
7	Polyphenols, citric acid, ascorbic acid, ellagitannins, ellagic acid	Pomegranate	*Punica granatum*	Antiatherogenic, antioxidant, antihypertensive, anti-inflammatory, cardioprotective, anticancer	Gumienna and Szwengiel et al. (2016); Sahebkar and Ferri et al. (2016)
8	Polyphenols (approximately 90% of which are procyanidins and 7% other polyphenol compounds	Grape	*Vitis vinifera L.*	LDL-cholesterol and triglyceride-lowering agents., dyslipidemia	Peng et al. (2005); Quesada et al. (2009)
9	Polyphenols: resveratrol, anthocyanins, ellagic acid/ ellagitannins	Jamun	*Syzygium cumini*	Antidiabetic, antioxidant, Antihyperlipidemic, antiulcer, hepatoprotective, antiallergic, antiarthritic, antimicrobial,A ntiinflammatory, antifertility, antiplaque, radioprotective, nephroprotective, antidiarrheal activities	Govindaswamy et al. (2015); Srivastava and Chandra (2013)
10	Polyphenols, myrtillin (delphinidin-3-O-glucoside), an anthocyanin, procyanidin components (colorless condensed tannins composed of flavonoid polymers)	Lowbush blueberry	*Vaccinium angustifolium Aiton*	Lowered blood glucose levels, hypoglycemic activity, antiglycemic and anti-obesogenic effects	Watson et al. (1928); Grace and Ribnicky et al. (2009); Prior and Rogers et al. (2010)
11	Flavonoids, carotenoids, polyphenolic compounds, pentacyclic triterpenoids	Guava	*Psidium guajava*	Antidiarrheal activity, antioxidant activity, antidiabetic, antimicrobial activity	Mittal and Singh et al. (2010)

TABLE 15.1 *(Continued)*

S. N.	Bioactive component	Common name	Botanical Name	Function	References
12	Flavonoids, anthocyanins, hydrolyzable tannins (gallotannins, ellagitannins), phenolic acids (hydroxycinnamic, hydroxybenzoic acids), condensed tannins, vitamin C, folate	Strawberry	*Fragaria sp.*	Antioxidant, antiinflammatory, antiatherosclerotic, anticarcinogenic, antidepressant properties	Gasparrini M et al (2017); Khan A et al (2017)
13	Flavonoids, isoflavonoids (daidzein, genistein), luteolin, quercetin, kaempferol	cluster bean	*Cyamopsis tetragonoloba*	Anticancer, antioxidant, antidiabetic, antiulcer, cytoprotective, anticholinergic, hypolipidemic	Wang et al. (2017); Sharma and Kaushik et al. (2011)
14	Flavonoid fraction	Banana	*Musa sp.*	Immunomodulatory activity	Alese and Adewole et al. (2017)
15	Sterols, terpenoids, flavonoids, saponins	bottle gourd	*Lagenaria siceraria*	Controlling diabetes, hypertension, liver diseases, weight loss, and other associated benefits	Prajapati and Sheth et al. (2010)
16	Phenolics, tannins, stilbenes, flavonoids, anthocyanins; vitamin C, alpha tocopherol; flavonol glycoside, apigenin, cyanidin-3- glucoside	blackberry, raspberry, dewberry	*Rubus sp.*	Antioxidants, caspase-mediated apoptosis in human breast adenocarcinoma cells	George et al. (2017); Abdala and Mendoza et al. (2017)

TABLE 15.1 *(Continued)*

S. N.	Bioactive component	Common name	Botanical Name	Function	References
17	Phenolic acids Ferulic acid, salicylic acid	Coffee cereal bran	*Coffea*	Anti-inflammatory	Gry et al. (2007)
18	Phenolic compounds, mangiferin, vitamins A and C,β-carotene and xanthophylls, cis-9, cis-15-octadecadienoic acid	Mango	*Mangifera sp.*	Antioxidant activity, analgesic, antidiabetic, antisclerotic, antimicrobial, antiviral; cardio-, hepato-, and neuroprotective; anti-inflammatory; antiallergic; memory improving; radioprotective against X-ray, gamma, and UV radiation	Khurana and Singh et al. (2016); López-Cobo et al. (2017); Matkowski and Woźniak et al. (2013)
19	Phenols, gallate, mucic acid 1-ethyl 6-methyl ester 2-Ogallate	Indian gooseberry	*Phyllanthus emblica*	Cardioprotective effect, Antioxidant	Kshemada K et al (2017); Zhu WF et al (2017)
20	Ascorbic acid, citric acid, carotene	Orange	*Citrus reticulata*	Antioxidant, antiallergic, improved lipid profile	Aptekmann and Cesar (2010)
21	Ascorbic acid; thiamine; riboflavinbio flavonoids; pectin; mauritine A; amphibine H; jubanine A and B; mucronine D; nummularine B; sativanine E; ziziphines A to Q; betulinic acid; jujubosides A, B, A1, and B1; saponins	Jujube	*Ziziphus jujuba*	Aphrodisiac, hypnotic-sedative, anxiolytic, anticancer (melanoma cells), antifungal, antibacterial, antiulcer, anti-inflammatory, antifertility/ contraception, hypotensive, cardiotonic, antioxidant, immunostimulant, wound healing properties	Tahergorabi and Abedini (2015); Wang et al. (2013)

TABLE 15.1 *(Continued)*

S. N.	Bioactive component	Common name	Botanical Name	Function	References
22	Ascorbic acid, folate, phylloquinone, alpha-tocopherol, and the carotenoids lutein, violaxanthin, zeaxanthin, and beta-carotene	spinach	*Spinacia oleracea*	Protective effect against metabolic syndrome, antiobesity, antiinflammatory, anticancer, hypoglycemic, and lipid-lowering effect	Lester and Hodges et al. (2010); Panda and Sudhamani et al. (2017); Roberts and Moreau et al. (2016)
23	Cucurbitane-type triterpenoids, triterpene glycosides, phenolic acids, flavonoids, essential oils, saponins, momordicosides QU, charantin	bitter gourd	*Momordica charantia*	Antidiabetic, immunomodulatory effect, antioxidant, anti-inflammatory, anticancer, antibacterial, antiobesity activity	Nekoua MP et al (2017) Anant S et al. (2016)
24	Cucurbitane-type triterpenoids, steroidal saponins called charantins, insulin-like peptides and alkaloids,	Bitter melon	*Momordica charantia L.*	Hypoglycemic activity, various diabetes complications, Insulin resistance, reduced LDL-cholesterol and triglyceride levels	Leung et al. (2009); Grover et al. (2001); Rathi et al. (2002); Ahmed et al. (2001); Chaturvedi et al. (2005); Fernandes and Lagishetty et al.(2007); Jayasooriya and Yukizaki et al. (2000); Senanayake and Shibuya et al. (2004)
25	Dietary fiber (neutral and soluble), Alkaloid trigonelline	Fenugreek	*Trigonella foenum-graecum*	hypoglycemic and hypocholesterolemic effects, reduces glycosuria, and steroidal saponins,	Srinivasan (2006)

TABLE 15.1 *(Continued)*

S. N.	Bioactive component	Common name	Botanical Name	Function	References
26	Aldose reductase inhibitors (4,5-di-O-caffeolquinic acid, davidigenin, 6-demethoxy-capillarison and 2′,4′-dihydroxy-4-methoxydihydrochalcone) protein tyrosine phosphatase-1B inhibitors(2′,4′- dihydroxy-4-methoxydihydrochalcone, 2′,4′-dihydroxy-4′-methoxydihydrochalcone and sakuranetin)	Russian tarragon	*Artemisia dracunculus L.*	Antidiabetic properties, hypoglycemic effect, insulin resistance	Ribnicky et al. (2006) Cefalu et al. (2009)
27	Steroidal pregnane glycoside, P57-AS3	Hoodia	*Hoodia gordonii*	Hypothalamic modulation, appetite-suppressing effects, Anti-obesogenic activity	MacLean and Luo (2004); Khan IA. et al. (2008)
28	Coumarins such as marmelosin, umbelliferone, and scopoletin	Bael	*Aegle marmelos*	Antihyperglycemic, chronic diarrhea, dysentery, peptic ulcers, as a laxative	Ansari et al. (2017); Laddha et al. (2014)
29	Artocarpin, moracin C	Jackfruit	*Artocarpus heterophyllus*	Chemoprevention of colorectal cancer, antioxidant activity, anti-inflammatory activity	Kang et al (2017); Pu et al (2016)

TABLE 15.1 *(Continued)*

S. N.	Bioactive component	Common name	Botanical Name	Function	References
30	Papain, carotene	Papaya	*Carica papaya*	Antioxidant activity	Krishna and Patel et al. (2008); Somanah and Bahorun et al. (2017)
31	Chlorogenic acid, epicatechin, procyanidin B2, phloretin and quercetin, vitamin C, benzoic acids (gallic acid), hydroxycinnamic acids (chlorogenic acid), flavanols (catechin), flavonols (rutin), chalcones (phloridzin), terpenes (ursolic acid)	Apple	*Malus species*	Antioxidants, antiproliferative, breast cancer treatment	González and Martínez et al. (2008); Yoon and Liu (2007)
32	Phytosterols- Campostanol, sitostanol	Wheat	*Triticum*	Cholesterol lowering	Gry et al. (2007)
33	Carotenoids Beta-carotene, lycopene, cryptaxanthine	Tomatoes carrots bell peppers	*Solanum lycopersicum* *Daucus carota* *Capsicum annuum*	Antioxidant, anti-inflammatory, anti-carcinogenic activities	Gry et al. (2007)

TABLE 15.1 *(Continued)*

S. N.	Bioactive component	Common name	Botanical Name	Function	References
34	Capsaicinoids Capsaicin, dihydrocapsaicin	Chilli peppers Sweet peppers	*Capsicum frutescens* *Capsicum annuum*	Induction of cellular antioxidant responses, inhibition of LDL-oxidation	Gry et al. (2007)
35	Raphanusamide, cyanidin, vanillic acid, kaempferol, quercetin	Radish	*Raphanus sativus*	Antiulcer, antimicrobial, antifungal, antiinflammatory, diuretic, antioxidant, anticarcinogenic	Gutierrez et al. (2004)

Furthermore, various government initiatives are being taken to increase the consumption of bioactive supplements in food and beverages globally, Key players, such as Cargill, Kerry Inc. Dupont, Archer Daniels Midland, and BASF SE, among others, have consigned a splendid amount of their revenue in research and development of fortified food and establishes novel product development in market.

In 1986 first time WHO included herbal medicines through International Conference on Drug Regulatory Authorities (ICDRA). These guiding principle offers valuable support to scientific organizations, national regulatory authorities, and producers to undertake valuation of documents and define quality evaluation of herbal medicines (Mosihuzzaman and Choudhary, 2008; Akerele et al., 1993). Furthermore, the WHO has prepared monographs and guidelines for the valuation of herbal medicines and drugs (Bulletin of the World Health Organization, 1993).

Mosihuzzaman and Choudhary (2008) reported that the herbal crops can only be promoted as food supplements in United States after the approval of FDA. Moreover, European guideline has cost and time effective assessment guidelines for approval of herbal medications as harmless and effective against metabolic syndrome and various diseases. The multifaceted legislative framework such as European Directive 91/507/EEC and 2004/24/EC of 31 March, 2004 are used for the evaluation of quality, protection, efficiency, and human use of traditional herbal Medicinal Products, like wise UK has accepted a Statutory Instrument No. 2750 entitled as "The Medicines Regulations, 2005" and the German Federal Health Agency Commission was formed in 1978 for the scientific evaluation of 380 herbal medicines. Few preparation methods were tested for the quality, safety, and efficacy of herbal products, furthermore, widespread agreement happens for the translation of scientific suggestion on supplements with suitable policies, recommendations, and regulations for benefiting the public health. The government norms of bioactive supplement changes from country to country due to the reason of cultural aspects and rare scientific studies.

15.8.1 ARGENTINA

The Argentinian National Pharmacopeia first published in 1898 manage the existence of natural extracts, composite extracts, and pure active constituents. Approximately 890 monographs, 56 crude drugs and 33 composite fractions leads towards more scientific understanding of natural medicinal extracts and fractions. The Herboristerias is a plant drug and not come under the rules and regulations (Law No. 16.463) of plant drug mixture, still improvement required in control of raw materials, wild plant, scientific criteria for plants collection, and processing methods.

15.8.2 AUSTRALIA

The Australian Parliament established the therapeutic good act in 1990 and established a committee for working on quality, safety, efficacy, and labeling of Natural and Nutritional Supplements. Furthermore, act permitted the traditional claim for herbal remedies and regulates advertising requirements.

15.8.3 BRAZIL

Brazil is hub of immense flora, cultural aspects, and herbal medicine, but the uncontrolled regulation and absence of specification for quality, safety, and efficacy deteriorate the production, processing, and marketing of valuable products. The ministry of health formed a commission in 1994 for the evaluation of phytotherapeutic agents based on regulations in Germany and France with WHO guidelines for herbal medications. In 1995; Directive No. 6 established the criteria for herbal drugs registration with the definition of phytopharmaceutical products.

15.8.4 CANADA

The regulation in Canada was established by the Canadian Health Protection Branch (HPB) in 1986, which consist a special committee with 3 pharmacists, 2 herbalists, 1 nutritionist and 1 physician and consistent with WHO guidelines for classification of herbal drugs and assessment of herbal medicines by scientific studies. Within four years committee shows its importance and listed 64 unsafe herbs in 1990, developed criteria for adulterants in 1992 and listed 64 adulterants herbs.

15.8.5 FRANCE

In France only one type of license is granted which required various criteria for granting license such as documentations, shortened application and scientific evaluation, etc., approximately 200 herbs are treated as a non-prescription medicine and 115 herbs, 31 laxatives were approved in 1990 and recently approximately 205 herbal medications are recorded.

15.8.6 GERMANY

Germany's Commission E is a scientific advisory board established in 1978 have physicians, pharmacists, pharmacologists, toxicologists, industrial representatives and laypersons, a team of 24 members collects information on herbal medicines and evaluates them for safety and efficacy. The criteria for evaluation are based on traditional use, chemical data, experimental, toxicological, clinical, field, and epidemiological studies with patient case records and additional studies. Evaluation is published in a form of monographs (mono preparations and fixed combinations) that can be approve or disapprove the herbal drug. Recently approximately 300 monographs were published including (200 positives and 100 negatives), physician prescribed 70% of the registered herbal drugs and about 700 plants are traded in Germany.

15.8.7 UNITED STATES OF AMERICA

Herbal medications were regulated under the dietary supplement health and education act, 1994 in USA. Under this act herbal medicines are not evaluated by the Food and Drug Administration (FDA) and not proposed to identify, treat, treatment or prevent diseases. The office of substitute medicine at the national institutes of health aim to explore the potential of dietary supplement, scientific study, compilation of supplements scientific data and coordinating funding for dietary supplement related to chronic disease (Keservani et al., 2015; Rane and Keservani, 2020).

The combined database of composition and biological activity of plant foods was collected and formed a record name Euro FIR-BASIS. It also includes data from processed products originated from plants like wines, bread, chocolates, soy products and provides a focus sight on compounds having beneficial health effects. In initial stage the compositional and biological information with the potential health benefit, critical valuation system and the automatic input forms were developed in BASIS included 2,000 compositional records. The EuroFIR bio actives work package comprises of 28 participants from 17 countries in the form of three teams named plant list, composition evaluators and biological effect. Furthermore, the composition inspection task group includes approximately 12 specialists for the extraction and understanding of compositional data of bioactive compounds. Presently Euro FIR-BASIS is one of the helpful resources for development, regulation, and survey of researches in area of biologically active compounds (Gry et al., 2007).

15.9 CHALLENGES FOR DIETARY SUPPLEMENTS

The scientific evidence translation, efficacy, and safety are the main challenges associated with supplementary sciences (Dwyer et al., 2018). Safety includes to sale or market hygienic products devoid of toxic compounds with detail information of dosage. Efficacy explained by study plan, significance, outcome, and clinical proof. Translation provides suitable policies, regulations, and recommendations that safeguard the public health and suitability for consumption. Monitoring is important in distribution of supplements for filling nutrition gaps in required population.

15.10 GOVERNMENT OF INDIA INITIATIVES FOR NUTRITION AND METABOLIC SYNDROME

GOI started many government interventions programs for combating malnutrition for minimizing the metabolic syndrome among the women and children, such as ICDS-1975, which is focused on child health, nutrition programs and maternal status with vaccination, additional diet, medical check-ups, nursery education, and health education involvement (UNICEF, 2016). NHM (2005) delivered the nutrition-specific interventions. The 23% supplementary food provided to children and 38% children weighed in an Aganwadi (UNICEF, 2016). Mid-day meal program for school children is very innovative for nutritional wellbeing for millions of children by supplying food of approximately 12–16 gm protein/child and 300 calories/day (Shivangi et al., 2019). State Nutrition Mission in Uttar Pradesh helped for the child and maternal nutrition improvement. Janani Suraksha Yojana is centrally sponsored intervention of the National Health Mission for motherhood safety with delivery and post-delivery care by an effective link of Accredited Social Health Activist. Indira Gandhi Matritva Sahyog Yojana is Rs. 6,000 schemes to support the delivery, healthy supplements and feeding of pregnant and lactating mothers having age of $\geq$19 for two live births. Rajiv Gandhi Scheme for Adolescent Girls empowerment, nutrition, and health status improvement, Upgrade their home-based skills (MoWCD, 2020). Govt of India launched many women empowerment schemes like Beti Bachao and Beti Padhao, scheme for Women Helpline, Ujjawala, Working women hostel, SWADHAR Greh, puruskar for Nari Shakti and Stree Shaktee, Rajya, and Zila Mahila Samman, Mahila Police Volunteer, Mahila Shakti Kendras and Nirbhaya.

15.11 VISION 2022 FOR FUTURE OF DIETARY SUPPLEMENT FOR WOMEN AND CHILD

"Kuposhan Mukt Bharat" scheme attached with Swachh Bharat and Swasth Bharat to establish a safe, healthy nourish women and children in terms of growth, productivity, learning skill for future (Nutrition Strategy, 2016).

15.12 CONCLUSION

Bioactive compounds play enormous role in the enhancement of immunity to fight against the foreign microorganism. Plant based bioactive compounds have prudential digestibility, acceptability with lesser impact on climate in comparison to animal source. These compounds extracted with appropriate techniques (high pressure processing) for better retention and yield. The compounds extracted from fruits, vegetables, cereals, legume, medicinal plants, nuts, and spices play extensive role in defending the metabolic syndrome.

KEYWORDS

- **bioactive compounds**
- **COVID-19**
- **C-reactive protein**
- **dietary supplements**
- **interleukin-6**
- **metabolic syndrome**
- **nutrition**
- **triglycerides**

REFERENCES

Ahima, R. S., (2016). Pharmacotherapy of obesity and metabolic syndrome. In: Ahima, R. S., (ed.), *Metabolic Syndrome*. Springer, Cham. https://doi.org/10.1007/978-3-319-11251-0_44.

Akerele, O., (1993). Summary of WHO guidelines for the assessment of herbal medicines. *Herbal Gram, 28*(13), 13–19.

Annual Health Survey Report Part II, (2014). *Ministry of Home Affairs, Government of India.* https://main.mohfw.gov.in/documents/Statistics update (accessed on 25 October 2022).

Annual Report (2019–2020). NITI Aayog, Govt. of India. https://niti.gov.in/sites/default/files/2020-02/Annual_Report_2019-20.pdf (accessed on 02 August 2022).

Arimond, M., Zeilani, M., Jungjohann, S., Brown, K. H., Ashorn, P., Allen, L. H., & Dewey, K. G., (2015). Considerations in developing lipid-based nutrient supplements for prevention of undernutrition: Experience from the international lipid-based nutrient supplements (ILINS) project. *Mater. Child Nutri., 11*, 31–61.

Augustin, M. A., & Sanguansri, L., (2015). Challenges and solutions to incorporation of nutraceuticals in foods. *Annual Rev. Food Sci. Tech., 6*, 463–477.

Baiano, A., Viggiani, I., Terracone, C., Romaniello, R., & Del, N. M. A., (2015). Phenolic content, physical and sensory properties of breads made with different types of barley wort. *J. Sci. Food Agri., 95*(13), 2736–2741.

Balsano, C., & Alisi, A., (2009). Antioxidant effects of natural bioactive compounds. *Curr. Pharm. Des., 15*(26), 3063–3073.

Bansal, R., & Dhiman, A., (2019). Line of progression: Indian regulatory framework for nutraceuticals and dietary supplements. *Applied Clinical Research, Clin. Trials Reg. Affa., 6*(1), 46–61.

Basu, A., Sanchez, K., Leyva, M. J., Wu, M., Betts, N. M., Aston, C. E., & Lyons, T. J., (2010). Green tea supplementation affects body weight, lipids, and lipid peroxidation in obese subjects with metabolic syndrome. *J. Am. Co. Nutri., 29*(1), 31–40.

Bhartiya, A., Aditya, J. P., & Kant, L., (2015). Nutritional and remedial potential of an underutilized food legume horsegram (*Macrotyloma uniflorum*): A review. *J. Anim. Pl. Sci., 25*(4), 908–920.

Blevins, S. M., Leyva, M. J., Brown, J., Wright, J., Scofield, R. H., & Aston, C. E., (2007). Effect of cinnamon on glucose and lipid levels in Non–insulin-dependent type 2 diabetes. *Dia. Care, 30*(9), 2236–2237.

Boeing, H., Bechthold, A., Bub, A., Ellinger, S., Haller, D., Kroke, A., & Watzl, B., (2012). Critical review: Vegetables and fruit in the prevention of chronic diseases. *Euro. J. Nutri., 51*(6), 637–663.

Borkotoky, K., Unisa, S., & Gupta, A. K., (2018). State-level dietary diversity as a contextual determinant of nutritional status of children in India: A multilevel approach. *J. Bio. Sci., 50*(1), 26.

Cais-Sokolińska, D., Pikul, J., Wójtowski, J., Danków, R., Teichert, J., Czyżak-Runowska, G., & Bagnicka, E. (2015). Evaluation of quality of kefir from milk obtained from goats supplemented with a diet rich in bioactive compounds. *Journal of the Science of Food and Agriculture*, *95*(6), 1343–1349.

Calvello, R., Aresta, A., Trapani, A., Zambonin, C., Cianciulli, A., Salvatore, R., & Panaro, M. A., (2016). Bovine and soybean milk bioactive compounds: Effects on inflammatory response of human intestinal Caco-2 cells. *Food Chem., 210*, 276–285.

Cheng, P. W., Ng, L. T., Chiang, L. C., & Lin, C. C., (2006). Antiviral effects of saikosaponins on human coronavirus 229e *in vitro. Clinical and Experimental Pharmacology and Physiology, 33*(7), 153, 612–616.

Clare, D. A., & Swaisgood, H. E., (2000). Bioactive milk peptides: A prospectus. *Journal of Dairy Science, 83*(6), 1187–1195.

Clifford, M. N., (1999). Chlorogenic acids and other cinnamates–nature, occurrence and dietary burden. *J. Sci. Food and Agri., 79*(3), 362–372.

Cotrim, H. P., Carvalho, F., Siqueira, A. C., Lordelo, M., Rocha, R., & De Freitas, L. A., (2005). Non-alcoholic fatty liver and insulin resistance among petrochemical workers. *JAMA, 294*(13), 1615–1620.

Cui, Q., Du, R., Liu, M., &. Rong, L., (2020). *Molecules, 25*, 183.

Daly, M., (2003). Sugars, insulin sensitivity, and the postprandial state. *The Amer. J. Clin. Nutri., 78*(4), 865S–872S.

De-Regil, L. M., Suchdev, P. S., Vist, G. E., Walleser, S., & Peña, R. J. P., (2013). Home fortification of foods with multiple micronutrient powders for health and nutrition in children under two years of age. *Evidence-Based Child Health: A Coch. Rev. J., 8*(1), 112–201.

Diabetes Prevention Program Research Group, (2002). Reduction in the incidence of type 2 diabetes with lifestyle intervention or metformin. *New England J. Med., 346*(6), 393–403.

Drucker, D. J., (2006). Enhancing the action of incretin hormones: A new whey forward? *Endocrinology, 147*, 3171, 3172.

Dwyer, J. T., Coates, P. M., & Smith, M. J., (2018). Dietary supplements: Regulatory challenges and research resources. *Nutrients, 10*(1), 41.

Dwyer, J. T., Holden, J., Andrews, K., Roseland, J., Zhao, C., Schweitzer, A., & Hardy, C. J., (2007). Measuring vitamins and minerals in dietary supplements for nutrition studies in the USA. *Analy. Bioanaly. Chem., 389*(1), 37–46.

Ekelund, M., Shaat, N., Almgren, P., Groop, L., & Berntorp, K., (2010). Prediction of postpartum diabetes in women with gestational diabetes mellitus. *Diabeto., 53*(3), 452–457.

Ellulu, M. S., Patimah, I., Khaza'ai, H., Rahmat, A., & Abed, Y., (2017). Obesity and inflammation: The linking mechanism and the complications. *Archives of Medical Science: AMS, 13*(4), 851–863. https://doi.org/10.5114/aoms.2016.58928.

Esposito, K., Marfella, R., Ciotola, M., Di Palo, C., Giugliano, F., Giugliano, G., & Giugliano, D., (2004). Effect of a Mediterranean-style diet on endothelial dysfunction and markers of vascular inflammation in the metabolic syndrome: A randomized trial. *JAMA, 292*(12), 1440–1446.

Fardet, A., (2010). New hypotheses for the health-protective mechanisms of whole-grain cereals: What is beyond fiber? *Nutrition Research Reviews, 23*(1), 65–134.

Fitzgerald, R. J., & Murray, B. A., (2006). Bioactive peptides and lactic fermentations. *Inter. J. Dai. Tech., 59*(2), 118–125.

Gagnier, J. J., (2011). Nutritional, herbal, and homeopathic supplements. *Evidence-Based Management of Low Back Pain-E-Book, 85*, 258.

Ginsberg, H. N., (2000). Insulin resistance and cardiovascular disease. *The Journal of Clinical Investigation, 106*(4), 453–458.

Ginsberg, H. N., Zhang, Y. L., & Hernandez-Ono, A., (2005). Regulation of plasma triglycerides in insulin resistance and diabetes. *Archives of Medical Research, 36*(3), 232–240.

Gittelsohn, J., Mookherji, S., & Pelto, G., (1998). Operationalizing household food security in rural Nepal. *Food Nutri. Bull., 19*(3), 210–222.

Gobbetti, M., Stepaniak, L., De Angelis, M., Corsetti, A., & Di Cagno, R., (2002). Latent bioactive peptides in milk proteins: Proteolytic activation and significance in dairy processing. *Cri. Rev. Food Sci. Nutri., 42*(3), 223–239.

Goswami, T., & Bagchi, B. (2020). Molecular docking study of receptor binding domain of SARS-CoV-2 Spike Glycoprotein with Saikosaponin, a triterpenoid natural product.

Haffner, C. A., & Kendall, M. J., (1992). Metabolic effects of β2-agonists. *J. Clin. Pharm. Therap., 17*(3), 155–164.

Harborne, J. B. (1989). *Methods in Plant Biochemistry. Volume 1. Plant Phenolics*. Academic Press Ltd.

Harborne, P. D., (1999). *Handbook of Bioactive Compounds from Plants* (2nd edn.). Taylor and Francis.

Hariyanto, T. I., & Kurniawan, A., (2020). Dyslipidemia is associated with severe coronavirus disease 2019 (COVID-19) infection. *Dia. Met. Synd. Clin. Res. Rev., 14*(5), 1463–1465.

Harris, R. P., Helfand, M., Woolf, S. H., Lohr, K. N., Mulrow, C. D., Teutsch, S. M., Atkins, D., & Methods Work Group, Third US Preventive Services Task Force, (2001). Current methods of the US preventive services task force: A review of the process. *American J. Preven. Med. 20*(3 Suppl), 21–35.

Hayes, M., Ross, R. P., Fitzgerald, G. F., & Stanton, C., (2007). Putting microbes to work: Dairy fermentation, cell factories and bioactive peptides. Part I: Overview. *Biotech. J. Health. Nutri. Tech., 2*(4), 426–434.

Health & Nutrition and Women & Child Development, NITI Aayog, govt. of India. [https://niti.gov.in/health-and-nutrition-and-women-and-child-development-old-2]. https://ssca.org.in/media/8._Nutrition_Situation_of_WomenCurrent_StatusImplications_on_Child_Undernutrition.pdf (accessed on 02 August 2022).

Hernandez-Ledesma, B., Dávalos, A., Bartolomé, B., & Amigo, L., (2005). Preparation of antioxidant enzymatic hydrolysates from α-lactalbumin and β-lactoglobulin. Identification of active peptides by HPLC-MS/MS. *J. Agri. Food Chem., 53*(3), 588–593.

Huang, J. H., Li, R. H., Huang, S. L., Sia, H. K., Chen, Y. L., & Tang, F. C., (2015). Lifestyle factors and metabolic syndrome among workers: The role of interactions between smoking and alcohol to nutrition and exercise. *Inter. J. Envir. Res. Pub. Health, 12*(12), 15967–15978.

Hutcheson, R., & Rocic, P., (2012). The metabolic syndrome, oxidative stress, environment, and cardiovascular disease: The great exploration. *Exper. Dia. Res. 2012*.

IIPS, I., (2017). *National Family Health Survey (NFHS-4), 2015–16*. International Institute for Population Sciences (IIPS), Mumbai, India.

Indira Gandhi Matritva Sahyog Yojna, NITI Aayog, Govt. of India. https://niti.gov.in/writereaddata/files/document_publication/IGMSY_FinalReport.pdf (accessed on 02 August 2022).

Integrated Nutrition Education Handbook (1991). Food and Nutrition Board, Ministry of Food & Civil Supplies, Govt. of India. (Combined Effort of Various Organisations). https://wcd.nic.in/sites/default/files/integrated%20nutrition%20education-%20a%20handbook.pdf (accessed on 02 August 2022).

Janani Suraksha Yojana, National Health Mission. https://nhm.gov.in/index1.php?lang=1&level=3&lid=309&sublinkid=841#:~: text=JananiSurakshaYojana%20(JSY)%20is%20a%20safe,delivery%20among%20poor%20pregnant%20women (accessed on 25 October 2022).

Jarvill-Taylor, K. J., Anderson, R. A., & Graves, D. J., (2001). A hydroxychalcone derived from cinnamon functions as a mimetic for insulin in 3T3-L1 adipocytes. *J. Amer. Coll. Nutri., 20*(4), 327–336.

Jejeebhoy, S. J., & Rao, S. R. (1995). Unsafe motherhood: a review of reproductive health. *Women's Health in India: Risk and Vulnerability, Bombay. Women of the World, 9*.

Jensen, M. D., Haymond, M. W., Rizza, R. A., Cryer, P. E., & Miles, J., (1989). Influence of body fat distribution on free fatty acid metabolism in obesity. *The J. Clin. Inves., 83*(4), 1168–1173.

John, J. H., Ziebland, S., Yudkin, P., Roe, L. S., & Neil, H. A. W., (2002). Effects of fruit and vegetable consumption on plasma antioxidant concentrations and blood pressure: A randomized controlled trial. *The Lancet, 359*(9322), 1969–1974.

Jose, S., & Navaneetham, K., (2008). A factsheet on women's malnutrition in India. *Eco. Pol. Weekly,* 61–67.

Kanjilal, B., Mazumdar, P. G., Mukherjee, M., & Rahman, M. H., (2010). Nutritional status of children in India: Household socio-economic condition as the contextual determinant. *International J. Equ. Hea., 9*(1), 19.

Katagiri, H., Yamada, T., & Oka, Y., (2007). Adiposity and cardiovascular disorders: Disturbance of the regulatory system consisting of humoral and neuronal signals. *Circulation Research, 101*(1), 27–39.

Kaur, J., (2014). A comprehensive review on metabolic syndrome. *Cardiology Research and Practice*.

Keservani, R. K., & Sharma, A. K., (2014). Flavonoids: Emerging trends and potential health benefits. *J. Chin. Pharm. Sci., 23*(12), 815–822.

Keservani, R. K., Jarouliya, U., & Sharma, A. K., (2019). Nutraceutical and functional food regulations in India. In: Bagchi, D., (ed.), *Nutraceutical and Functional Food Regulations in the United States and Around the World* (3rd edn., pp. 431–444). Academic Press, Elsevier, Waltham, USA.; Ch 28. ISBN: 9780128164679.

Keservani, R. K., Kesharwani, R. K., Sharma, A. K., & Ahmad, M. F., (2015). Stress, Parkinson's, alzemer's disease: Role of dietary supplements. In: Bagchi, D., Preuss, H. G., & Swaroop, A., (eds.), *Nutraceutical and Functional Foods in Human Life and Disease Prevention* (pp. 241–253). Boca Raton, Fl. Taylor and Francis; Ch 14. ISBN: 9781482237214.

Keservani, R. K., Kesharwani, R. K., Sharma, A. K., Vyas, N., & Chadoker, A., (2010b). Nutritional supplements: An overview. *Int. J. of Curr. Pharm. Rev. and Res., 1*(1), 59–75.

Keservani, R. K., Kesharwani, R. K., Vyas, N., Jain, S., Raghuvanshi, R., & Sharma, A. K., (2010a). Nutraceuticals and functional food as future food. *Der. Pharm. Lett, 2*(1), 106–116.

Keservani, R. K., Sharma, A. K., & Kesharwani, R. K., (2016). Nutraceutical and functional foods for cardiovascular health. In: Meghwal, M., & Goyal, M. R., (eds.), *Food Process Engineering: Emerging Technologies and Innovations in Food Engineering* (pp. 257–278). Apple Academic Press, CRC Press, Taylor and Francis; Ch 10. ISBN: 9781771884020.

Keservani, R. K., Sharma, A. K., Ahmad, F., & Baig, M. E., (2014). Nutraceutical and functional food regulations in India. In: Bagchi, D., (ed.), *Nutraceutical and Functional Food Regulations in the United States and Around the World* (2nd edn., pp. 327–342). Academic Press, Elsevier, Waltham, USA.; Ch.19. ISBN: 978-0-12-405870-5.

Khan, A., Safdar, M., Khan, M. M. A., Khattak, K. N., & Anderson, R. A., (2003). Cinnamon improves glucose and lipids of people with type 2 diabetes. *Dia. Care, 26*(12), 3215–3218.

Kleinrock, M., & Rickwood, S., (2016). *The Global Use of Medicines: Outlook Through*. IMS Institute for Healthcare Informatics, Parsippany.

Korhonen, H., & Pihlanto, A., (2007). Technological options for the production of health-promoting proteins and peptides derived from milk and colostrum. *Curr. Pharm. Des., 13*(8), 829–843.

Kumar, A., & Khan, M. E., (2010). Health status of women in India: Evidences from national family health survey-3 (2005–06) and future outlook. *Res. Prac. Soc. Sci., 6*(2), 1–21.

Kylin, E., (1923). Studien ueber das Hypertonie-Hyperglyka "mie-Hyperurika" miesyndrom. *Zentralblatt für innere Medizin, 44*, 105–127.

Lee, K. W., Park, B. J., Kang, H. T., & Lee, Y. J., (2011). Alcohol-drinking patterns and metabolic syndrome risk: The 2007 Korean national health and nutrition examination survey. *Alcohol, 45*(5), 499–505.

Legeay, S., Rodier, M., Fillon, L., Faure, S., & Clere, N., (2015). Epigallocatechin gallate: A review of its beneficial properties to prevent metabolic syndrome. *Nutrients, 7*(7), 5443–5468.

Lewis, G. F., & Steiner, G., (1996). Acute effects of insulin in the control of VLDL production in humans: Implications for the insulin-resistant state. *Dia. Care, 19*(4), 390–393.

Lillioja, S., Neal, A. L., Tapsell, L., & Jacobs, Jr. D. R., (2013). Whole grains, type 2 diabetes, coronary heart disease, and hypertension: Links to the aleurone preferred over indigestible fiber. *Biofactors, 39*(3), 242–258.

Liu, R. H., (2013). Dietary bioactive compounds and their health implications. *J. Food Sci., 78*(s1), A18–A25.

Manach, C., Williamson, G., Morand, C., Scalbert, A., & Rémésy, C., (2005). Bioavailability and bioefficacy of polyphenols in humans. I. Review of 97 bioavailability studies. *The Amer. J. Clinical Nutri., 81*(1), 230S–242S.

Mandle, V. S., Salunke, S. D., Gaikwad, S. M., Dande, K. G., & Patil, M. M., (2012). Study of nutritional value of some unique leafy vegetable grown in Latur district. *J. Anim. Sci. Adv, 2*(3.1), 296–298.

Mang, B., Wolters, M., Schmitt, B., Kelb, K., Lichtinghagen, R., Stichtenoth, D. O., & Hahn, A., (2006). Effects of a cinnamon extract on plasma glucose, HbA1c, and serum lipids in diabetes mellitus type 2. *Eur. J. Clin. Inves., 36*(5), 340–344.

Mantovani, A., Byrne, C. D., Zheng, M. H., & Targher, G., (2020). Diabetes as a risk factor for greater COVID-19 severity and in-hospital death: A meta-analysis of observational studies. *Nutri. Met. Cardio. Dis., 30*(8), 1236–1248.

Market, F. E., (2014). *Global Trends and Forecasts to 2018*. Markets And Markets.

Mason, J. B., Saldanha, L. S., Ramakrishnan, U., Lowe, A., Noznesky, E. A., Girard, A. W., & Martorell, R., (2012). Opportunities for improving maternal nutrition and birth outcomes: Synthesis of country experiences. *Food Nutri. Bull., 33*(2_suppl-1), S104–S137.

Masood, T., Shah, H. U., & Zeb, A., (2014). Effect of sprouting time on proximate composition and ascorbic acid level of mung bean (*Vigna radiate* L.) and chickpea (*Cicer Arietinum* L.) seeds. *The J. Ani. Plant Sci., 24*(3), 850–859.

Mohan, V., Deepa, M., Farooq, S., Prabhakaran, D., & Reddy, K. S., (2008). Surveillance for risk factors of cardiovascular disease among an industrial population in southern India. *National Medical Journal India, 21*(1), 8–13.

Monteiro, R., (2009). Chronic inflammation in the metabolic syndrome: Emphasis on adipose tissue. In: *Oxidative Stress, Inflammation and Angiogenesis in the Metabolic Syndrome*, (pp. 65–84). Springer, Dordrecht.

Monteiro, R., & Azevedo, I., (2010). Chronic inflammation in obesity and the metabolic syndrome. *Media. Inflam.,* 289645.

Morelli, A., Sarchielli, E., Comeglio, P., Filippi, S., Vignozzi, L., Marini, M., & Maggi, M., (2014). Metabolic syndrome induces inflammation and impairs gonadotropin-releasing hormone neurons in the preoptic area of the hypothalamus in rabbits. *Mol. Cell. Endoc., 382*(1), 107–119.

Mortazavi, F., Paknahad, Z., & Hasanzadeh, A., (2019). Effect of green tea consumption on the metabolic syndrome indices in women: A clinical trial study. *Nutrition & Food Science*,.

Mosihuzzaman, M., & Choudhary, M. I., (2008). Protocols on safety, efficacy, standardization, and documentation of herbal medicine (IUPAC Technical Report). *Pure App. Chem., 80*(10), 2195–2230.

Mudge, E. M., Betz, J. M., & Brown, P. N., (2016). The importance of method selection in determining product integrity for nutrition research. *Adv. Nutri., 7*(2), 390–398.

Naoghare, K. P., & Myong Song, J. (2010). Chip-based high throughput screening of herbal medicines. *Combinatorial Chemistry & High Throughput Screening, 13*(10), 923–931.

National and State Reports (2005–06), NFHS-3, National Family Health Survey, International Institute for Population Sciences (IIPS). http://rchiips.org/nfhs/report.shtml (accessed on 02 August 2022).

National and State Reports (2015–16), NFHS-4, National Family Health Survey, International Institute for Population Sciences (IIPS). http://rchiips.org/nfhs/NFHS-4Report.shtml (accessed on 02 August 2022).

National Sample Survey Office (NSSO). (2015). Key Indicators of Social Consumption in India Health-NSS 71st Round. Ministry of Statistics and Programme Implementation, Government of India. Available from https://www.thehinducentre.com/resources/nsso-key-indicators-of-social-consumption-in-india-health/article64936255.ece (accessed on 25 October 2022).

Navas-Acien, A., Silbergeld, E. K., Pastor-Barriuso, R., & Guallar, E., (2008). Arsenic exposure and prevalence of type 2 diabetes in US adults. *JAMA – J. Amer. Med. Ass., 300*(7), 814–822.

Nie, P., Rammohan, A., Gwozdz, W., & Sousa-Poza, A., (2019). Changes in child nutrition in India: A decomposition approach. *Inter. J. Envir. Res. Pub. Health, 16*(10), 1815.

Nourishing India, (2016). *National Nutrition Strategy*, NITI Aayog, Govt. of India. https://niti.gov.in/writereaddata/files/document_publication/Nutrition_Strategy_Booklet.pdf (accessed on 02 August 2022).

O'Neill, B. J., (2020). Effect of low-carbohydrate diets on cardiometabolic risk, insulin resistance, and metabolic syndrome. *Cur. Opin. Endoc. Dia. Obe., 27*(5), 301–307.

Oh, S. W., (2005). Korea National Health and nutrition examination survey. Association between cigarette smoking and metabolic syndrome: The Korea national health and nutrition examination survey. *Dia. Care, 28*, 2064–2066.

Orhan, I. E., Senol, F. S., Skalicka-Wozniak, K., Georgiev, M., & Sener, B., (2016). Adulteration and safety issues in nutraceuticals and dietary supplements: Innocent or risky. In: Grumezescu, A. M., (ed.), *Nutrac. Nano. Agri-Food Ind.* (pp. 153–182).

Pan, Y., Lee, A., Wan, J., Coventry, M. J., Michalski, W. P., Shiell, B., & Roginski, H., (2006). Antiviral properties of milk proteins and peptides. *Inter. Dairy J., 16*(11), 1252–1261.

Paoli, A., Rubini, A., Volek, J. S., & Grimaldi, K. A., (2014). Beyond weight loss: A review of the therapeutic uses of very-low-carbohydrate (ketogenic) diets. *Eur. J. Clin. Nutri., 68*(5), 641.

Patras, A., Brunton, N., Da Pieve, S., Butler, F., & Downey, G., (2009). Effect of thermal and high-pressure processing on antioxidant activity and instrumental color of tomato and carrot purées. *Inn. Food Sci. Emerg. Tech., 10*(1), 16–22.

Petersen, B. L., Ward, L. S., Bastian, E. D., Jenkins, A. L., Campbell, J., & Vuksan, V., (2009). A whey protein supplement decreases post-prandial glycemia. *Nutri. J., 8*(1), 47.

Pistollato, F., & Battino, M., (2014). Role of plant-based diets in the prevention and regression of metabolic syndrome and neurodegenerative diseases. *Trends Food Sci. Tech., 40*(1), 62–81.

Pooladanda, V., Thatikonda, S., Bale, S., Pattnaik, B., Sigalapalli, D. K., Bathini, N. B., ... & Godugu, C. (2019). Nimbolide protects against endotoxin-induced acute respiratory distress syndrome by inhibiting TNF-α mediated NF-κB and HDAC-3 nuclear translocation. *Cell Death & Disease, 10*(2), 1–17.

Popkin, B. M., (2003). The nutrition transition in the developing world. *Dev. Policy Rev., 21*(5, 6), 581–597.

Prozialeck, W. C., Edwards, J. R., Nebert, D. W., Woods, J. M., Barchowsky, A., & Atchison, W. D., (2008). The vascular system as a target of metal toxicity. *Toxic. Sci., 102*(2), 207–218.

Purnell, J. Q., & Brunzell, J. D., (1997). The central role of dietary fat, not carbohydrate, in the insulin resistance syndrome. *Curr. Opin. Lipid., 8*(1), 17–22.

Qatanani, M., & Lazar, M. A., (2007). Mechanisms of obesity-associated insulin resistance: Many choices on the menu. *Genes & Dev., 21*(12), 1443–1455.

Qin, B., Nagasaki, M., Ren, M., Bajotto, G., Oshida, Y., & Sato, Y., (2003). Cinnamon extract (traditional herb) potentiates *in vivo* insulin-regulated glucose utilization via enhancing insulin signaling in rats. *Diab. Res. Clin. Prac., 62*(3), 139–148.

Qin, B., Nagasaki, M., Ren, M., Bajotto, G., Oshida, Y., & Sato, Y., (2004). Cinnamon extract prevents the insulin resistance induced by a high-fructose diet. *Hor. Met. Res., 36*(02), 119–125.

Ragaee, S., Seetharaman, K., & Abdel-Aal, E. S. M., (2014). The impact of milling and thermal processing on phenolic compounds in cereal grains. *Crit. Rev. Food Sci. Nutri., 54*(7), 837–849.

Rahman, M., Tondel, M., Ahmad, S. A., & Axelson, O., (1998). Diabetes mellitus associated with arsenic exposure in Bangladesh. *Amer. J. Epid., 148*(2), 198–203.

Rajati, F., Hamzeh, B., Pasdar, Y., Safari, R., Moradinazar, M., Shakiba, E., & Najafi, F., (2019). Prevalence, awareness, treatment, and control of hypertension and their determinants: Results from the first cohort of non-communicable diseases in a Kurdish settlement. *Scien. Rep., 9*(1), 1–10.

Rajiv Gandhi Scheme for Empowerment of Adolescent Girls (RGSEAG)–'SABLA, Ministry of Women and Child development, Govt. of India. https://wcd.nic.in/sites/default/files/1-SABLAscheme_0.pdf (accessed on 02 August 2022).

Raju, K., & Shanmugam, M., (2017). Health status of the Indian women- a brief report, MOJ Proteomics *Bioinform., 5*(3), 109–111.

Rane, B. R., & Keservani, R. K., (2020). Introduction to nutraceutical & dietary supplements. In: Keservani, R. K., Sharma, A. K., & Kesharwani, R. K., (eds.), *Nutraceutical and Dietary Supplements* (pp. 35–55). Apple Academic Press, Taylor & Francis Group, Chapter 3. ISBN: 9781771888738.

Rane, B. R., Bharath, M. S., Patil, R. R., Keservani, R. K., & Jain, A. S., (2020). Novel approaches in nutraceuticals. In: Kesharwani, R. K., Keservani, R. K., & Sharma, A. K., (eds.), *Enhancing the Therapeutic Efficacy of Herbal Formulations Through Novel Drug*

Delivery Systems (pp. 241–266). IGI Global International Publisher, Pennsylvania, USA. Chapter 11. ISBN13: 9781799844532.

Rane, B. R., Patil, A. S., Keservani, R. K., & Jain, A. S., (2020). Novel approaches in herbal formulation. In: Kesharwani, R. K., Keservani, R. K., & Sharma, A. K., (eds.), *Enhancing the Therapeutic Efficacy of Herbal Formulations Through Novel Drug Delivery Systems* (pp. 43–68). IGI Global International Publisher, Pennsylvania, USA, Chapter 2. ISBN13: 9781799844532.

Rao, K. M., Balakrishna, N., Arlappa, N., Laxmaiah, A., & Brahmam, G. N. V., (2010). Diet and nutritional status of women in India. *Journal of Human Ecology, 29*(3), 165–170.

Ricci-Cabello, I., Olalla, H. M., & Artacho, R., (2012). Possible role of milk-derived bioactive peptides in the treatment and prevention of metabolic syndrome. *Nutri. Rev., 70*(4), 241–255.

Ridker, P. M., Rifai, N., Cook, N. R., Bradwin, G., & Buring, J. E., (2005). Non–HDL cholesterol, apolipoproteins AI and B100, standard lipid measures, lipid ratios, and CRP as risk factors for cardiovascular disease in women. *JAMA, 294*(3), 326–333.

Robbins, R. J., (2003). Phenolic acids in foods: An overview of analytical methodology. *J. Agri. Food Chemistry, 51*(10), 2866–2887.

Saballs, M., Parra, S., Sahun, P., Pellejà, J., Feliu, M., Vasco, C., & Castro, A., (2016). HDL-c levels predict the presence of pleural effusion and the clinical outcome of community-acquired pneumonia. *Springer Plus, 5*(1), 1–9.

Sackner-Bernstein, J., Kanter, D., & Kaul, S., (2015). Dietary intervention for overweight and obese adults: Comparison of low-carbohydrate and low-fat diets. A meta-analysis. *PloS One, 10*(10), e0139817.

Saklani, A., & Kutty, S. K., (2008). Plant-derived compounds in clinical trials. *Drug dis. Tod., 13*(3, 4), 161–171.

Saleh, A. S., Wang, P., Wang, N., Yang, S., & Xiao, Z., (2019). Technologies for enhancement of bioactive components and potential health benefits of cereal and cereal-based foods: Research advances and application challenges. *Cri. Rev. Fd. Sci. Nutri., 59*(2), 207–227.

Samy, R. P., & Gopalakrishnakone, P., (2007). Current status of herbal and their future perspectives. *Nature Prec.,* 1–1.

Sativex, (2013). http://www.sativex.co.uk/healthcare-professionals (accessed on 02 August 2022).

Shivangi, (2019). Nutrition and health policies for the children in Uttar Pradesh. *Inter. J. Res. Granth., 7*(1), 361–372.

Short Report of Nutrient Requirements for Indians Recommended Dietary Allowances and Estimated Average Requirements, (2020). ICMR-National Institute of Nutrition. https://www.nin.res.in/nutrition2020/RDA_short_report.pdf (accessed on 02 August 2022).

Sibul, F. S., Orcic, D. Z., Svircev, E., & Mimica-Dukić, N. M., (2016). Optimization of extraction conditions for secondary biomolecules from various plant species. *Hemij. Indus., 70*(4), 473–483.

Siegel, K. R., Bullard, K. M., Imperatore, G., Ali, M. K., Albright, A., Mercado, C. I., & Gregg, E. W., (2018). Prevalence of major behavioral risk factors for type 2 diabetes. *Dia. Care, 41*(5), 1032–1039.

Siegel, R. D., Cupples, A., Schaefer, E. J., & Wilson, P. W., (1996). Lipoproteins, apolipoproteins, and low-density lipoprotein size among diabetics in the Framingham offspring study. *Meta., 45*(10), 1267–1272.

Singh, S., Srivastava, S., & Upadhyay, A. K., (2019). Socio-economic inequality in malnutrition among children in India: An analysis of 640 districts from national family health survey (2015–16). *Inter. J. Equity Health, 18*(1), 203.

Sirit, Y., Acero, C., Bellorin, M., & Portillo, R., (2008). Metabolic syndrome and other factors cardiovascular risk in workers of a plant of vinyl polychloride. *Rev. de Salud Publica, 10*(2), 239–249.

Stancliffe, R. A., Thorpe, T., & Zemel, M. B., (2011). Dairy attentuates oxidative and inflammatory stress in metabolic syndrome. *The Amer. J. Clin. Nutri., 94*(2), 422–430.

Striessnig, E., & Bora, J. K., (2020). Under-five child growth and nutrition status: Spatial clustering of Indian districts. *Spat. Dem.,* 1–22.

Sun, Q., Yue, P., Deiuliis, J. A., Lumeng, C. N., Kampfrath, T., Mikolaj, M. B., & Rajagopalan, S., (2009). Ambient air pollution exaggerates adipose inflammation and insulin resistance in a mouse model of diet-induced obesity. *Circulation, 119*(4).

Tabassum, N., & Ahmad, F., (2011). Role of natural herbs in the treatment of hypertension. *Pharm. Rev., 5*(9), 30.

Tall, A. R., & Yvan-Charvet, L., (2015). Cholesterol, inflammation and innate immunity. *Nat. Rev. Immunol., 15*(2), 104e16.

Tien-Ju, W., Shin-Yu, L. A., Jiun-Liang, C., Cheng-Hui, L., Yin-Shuo, Y., & Sien-Hung, Y., (2019). A randomized clinical efficacy trial of red yeast rice (*Monascus pilosus*) against hyperlipidemia. *The American Journal of Chinese Medicine, 47*(2), 1–13.

Transparency Market Research, (2014). *Functional Foods Market: Global Industry Size, Share, Trends, Analysis and Forecast, 2012–2018*. Transparency Market Research. http://www.transparencymarketresearch.com/functional-foods-market.html (accessed on 02 August 2022).

Tsao, R., Yang, R., Young, J. C., & Zhu, H., (2003). Polyphenolic profiles in eight apple cultivars using high-performance liquid chromatography (HPLC). *J. Agri. Food Chem., 51*(21), 6347–6353.

Unicef, (2016). *State Nutrition Mission in Uttar Pradesh-Pushing the Nutrition Agenda Forward.* UNICEF. https://www.r4d.org/wp-content/uploads/Documentation-of-SNM-UP.pdf (accessed on 02 August 2022).

Vague, J., (1947). Sexual differentiation, a factor affecting the forms of obesity. *Pres. Med., 30*, 339–340.

Ventimiglia, E., Capogrosso, P., Colicchia, M., Boeri, L., Serino, A., Castagna, G., & Salonia, A., (2016). Metabolic syndrome in white European men presenting for primary couple's infertility: Investigation of the clinical and reproductive burden. *Andro., 4*(5), 944–951.

Victoria, A. V., & Arjun, A., (1998). *Women of the World, Women health in India* U.S. Department of Commerce Economics and Statistics Administration Bureau of The Census.

Vir, S. C., & Malik, R., (2015). Nutrition situation of women in India: Current status, implications on child undernutrition and challenges ahead. *Stat. Applic., 13*(1, 2), 71–84.

Volek, J. S., & Feinman, R. D., (2005). Carbohydrate restriction improves the features of metabolic syndrome. Metabolic syndrome may be defined by the response to carbohydrate restriction. *Nutri. Met., 2*(1), 31.

Volek, J. S., Fernandez, M. L., Feinman, R. D., & Phinney, S. D., (2008). Dietary carbohydrate restriction induces a unique metabolic state positively affecting atherogenic dyslipidemia, fatty acid partitioning, and metabolic syndrome. *Prog. Lipid Res., 47*(5), 307–318.

Wang, S. L., Chiou, J. M., Chen, C. J., Tseng, C. H., Chou, W. L., Wang, C. C., & Chang, L. W., (2003). Prevalence of non-insulin-dependent diabetes mellitus and related vascular diseases in southwestern arseniasis-endemic and nonendemic areas in Taiwan. *Envir. Health Perspec., 111*(2), 155–159.

Wareham, N. J., Phillips, D. I., Byrne, C. D., & Hales, C. N., (1995). The 30 minute insulin incremental response in an oral glucose tolerance test as a measure of insulin secretion. *Diab. Med. J. Bri. Dia. Asso., 12*(10), 931.

Weinberg, M. E., Manson, J. E., Buring, J. E., Cook, N. R., Seely, E. W., Ridker, P. M., & Rexrode, K. M., (2006). Low sex hormone–binding globulin is associated with the metabolic syndrome in postmenopausal women. *Metabolism, 55*(11), 1473–1480.

Welsh, P., Polisecki, E., Robertson, M., Jahn, S., Buckley, B. M., De Craen, A. J., & Sattar, N., (2010). Unraveling the directional link between adiposity and inflammation: A bidirectional mendelian randomization approach. *The J. Clin. Endo. Metab., 95*(1), 93–99.

Wen, C. C., Kuo, Y. H., Jan, J. T., Liang, P. H., Wang, S. Y., & Liu, H. G., (2007). Specific plant terpenoids and lignoids possess potent antiviral activities against severe acute respiratory syndrome coronavirus. *J. Med. Chem., 1, 50*(17), 4087–4095.

Wen, C. C., Kuo, Y. H., Jan, J. T., Liang, P. H., Wang, S. Y., Liu, H. G., Lee, C. K., et al., (2007). *J. Med. Chem., 50*, 4087–4095.

World food India, (2017). *Health Supplements and Nutraceuticals – Emerging High Growth Sector in India.* Ministry of food processing industries.

World Health Organization, (2020). *Ethics and Vector-Borne Diseases: WHO Guidance.*

Wu, Y., Ding, Y., Tanaka, Y., & Zhang, W., (2014). Risk factors contributing to type 2 diabetes and recent advances in the treatment and prevention. *Inter. J. Med. Sci., 11*(11), 1185.

Xu, Q., Hong, H., Wu, J., & Yan, X., (2019). Bioavailability of bioactive peptides derived from food proteins across the intestinal epithelial membrane: A review. *Trends Food Sci. Tech., 86*, 399–411.

Yanai, H., Tomono, Y., Ito, K., Furutani, N., Yoshida, H., & Tada, N., (2008). The underlying mechanisms for development of hypertension in the metabolic syndrome. *Nutri. J., 7*(1), 1–6.

Yaya, S., & Ghose, B., (2020). Change in nutritional status among women of childbearing age in India (1998–2016). *Obesi Sci. Prac., 6*(5), 535–543.

Yi, L., Li, Z., Yuan, K., Qu, X., Chen, J., & Wang, G., (2004). Small molecules blocking the entry of severe acute respiratory syndrome coronavirus into host cells. *J. Virol., 78*(20), 11334–11339.

Zhou, Y., Hou, Y., Shen, J., Huang, Y., Martin, W., & Cheng, F., (2020). Network based drug repurposing for novel coronavirus 2019-nCoV/SARS-CoV-2. *Cell. Discov., 6*, 14.

Ziaei, S., & Mohseni, H., (2013). Correlation between hormonal statuses and metabolic syndrome in postmenopausal women. *J. Fam. Repro. Health, 7*(2), 63.

Ziegenfuss, T. N., Hofheins, J. E., Mendel, R. W., Landis, J., & Anderson, R. A., (2006). Effects of a water-soluble cinnamon extract on body composition and features of the metabolic syndrome in pre-diabetic men and women. *J. Inter. Soc. Sports Nutri., 3*(2), 45.

Index

B

C

D

E

F

G

H

I

N

S